Springer Series in
MATERIALS SCIENCE 39

Springer
Berlin
Heidelberg
New York
Barcelona
Hong Kong
London
Milan
Paris
Singapore
Tokyo

Physics and Astronomy ONLINE LIBRARY

http://www.springer.de/phys/

Springer Series in
MATERIALS SCIENCE

Editors: R. Hull · R. M. Osgood, Jr. · H. Sakaki · A. Zunger

Springer Series in Materials Science covers the complete spectrum of materials physics, including fundamental principles, physical properties, materials theory and design. Recognizing the increasing importance of materials science in future device technologies, the book titles in this series reflect the state-of-the-art in understanding and controlling the structure and properties of all important classes of materials.

Victor E. Borisenko

Semiconducting Silicides

With 114 Figures and 70 Tables

 Springer

Professor Dr. Victor E. Borisenko

Belarusian State University
of Informatics and Radioelectronics
P. Browka 6
220027 Minsk
Belarus

Series Editors:

Prof. Robert Hull

University of Virginia
Dept. of Materials Science and Engineering
Thornton Hall
Charlottesville, VA 22903-2442, USA

Prof. R. M. Osgood, Jr.

Microelectronics Science Laboratory
Department of Electrical Engineering
Columbia University
Seeley W. Mudd Building
New York, NY 10027, USA

Prof. H. Sakaki

Institute of Industrial Science
University of Tokyo
7-22-1 Roppongi, Minato-ku
Tokyo 106, Japan

Prof. Alex Zunger

NREL
National Renewable Energy Laboratory
1617 Cole Boulevard
Golden Colorado 80401-3393, USA

ISSN 0933-033x

ISBN 3-540-66111-5 Springer-Verlag Berlin Heidelberg New York

Library of Congress Cataloging-in-Publication Data applied for.

Die Deutsche Bibliothek – CIP-Einheitsaufnahme
Borisenko, Viktor E.: Semiconducting silicides: with 69 tables / Victor E. Borisenko. - Berlin; Heidelberg; New York; Barcelona; Hong Kong; London; Milan; Paris; Singapore; Tokyo: Springer, 2000 (Springer series in materials science; 39) ISBN 3-540-66111-5

© Springer-Verlag Berlin Heidelberg 2000
Printed in Germany

The use of general descriptive names, registered names, trademarks, etc. in this publication does not imply, even in the absence of a specific statement, that such names are exempt from the relevant protective laws and regulations and therefore free for general use.

Typesetting: Camery-ready by the author
Cover concept: eStudio Calamar Steinen
Cover production: *design & production* GmbH, Heidelberg

SPIN: 10712057   57/3144/mf - 5 4 3 2 1 0 – Printed on acid-free paper

Preface

Semiconductors are well known as the main materials of modern solid-state electronics. They have held the attention of researches and engineers since the brilliant invention of the semiconductor transistor by Bardeen, Brattain and Shockley in the middle of the 20th century. Silicon, germanium, $A^{III}B^V$ and $A^{II}B^{VI}$ compounds have been widely used in discrete semiconductor devices and microelectronic and nanoelectronic integrated systems. Each of these materials has separately met specific physical and technological requirements to provide formation of solid-state structures with the best electronic or optical performance. However, attempts to combine them within integrated circuit appear to be ineffective or even technologically impossible. Thus, material and related technological compatibilities are important for further progress, particularly in microelectronics, optoelectronics and nanoelectronics. This stimulates an increasing interest in silicides and silicon-germanium alloys, which provide new prospects for silicon-based integration.

Elements from the Periodic Table form more than 180 silicides, which are chemical compounds of silicon with different metals. Most of them, except the silicides of lanthanides and actinides, are shown in Table 1. Along with appropriate compatibility with silicon and easy formation by silicidation in a metal-silicon couple, silicides are characterized by high thermal stability and resistance to oxidation. The majority of them are metallic and have low resistivity. Exactly metallic silicides were first employed for interconnections, gates in MOS structures, ohmic contacts, and Schottky barriers in silicon integrated circuits. For a comprehensive overview of their properties and general features of the formation technology the reader may address the books and reviews [1–10].

Semiconducting silicides are not as numerous as the metallic ones. There are only about a dozen silicides reported to be semiconductors. These are noted in Table 1 with the ascribed fundamental energy-gap values. It is worthwhile to stress that not all of the silicides announced in earlier research to have semiconducting properties were then confirmed to be semiconductors. An example is $LaSi_2$ mentioned in [11] as a semiconductor with an energy gap of 0.19 eV but later shown in [12] to have metallic properties. There were also reports mentioning Ca_2Si [13] and $IrSi_{-3}$ [14] as having semiconducting properties, but for the moment they have not been reproducibly observed.

There is one more tricky silicide demonstrating semiconducting behavior under certain conditions. This is cubic FeSi, which is a poor metal in the sense of conductivity above ~200 K, and a narrow-gap semiconductor (~0.06 eV) for T≤100 K [15–21]. The gap disappears at unusually low temperatures relative to its size. The anomalous charge and spin behavior suggest that a Kondo insulator description is more appropriate for FeSi. It is a class of rare-earth-derived nonmagnetic narrow-gap insulators whose insulating character is caused by hybridization of a localized f electron with one broad conduction band. More details about this material can be found in [20] and references therein.

Table 1. Semiconducting silicides

	I	II	III	IV	V	VI	VII	VIII		
1	H									He
2	Li	Be	B	C	N	O	F			Ne
3	Na	Mg Mg_2Si 0.78 eV	Al	Si	P	S	Cl			Ar
4	K	Ca	Sc	Ti	V	Cr $CrSi_2$ 0.35 eV	Mn $MnSi_{2-x}$ 0.7 eV	Fe $FeSi_2$ 0.78 eV	Co	Ni
	Cu	Zn	Ga	Ge	As	Se	Br	Kr		
5	Rb	Sr	Y	Zr	Nb	Mo $MoSi_2$ 0.07 eV	Tc	Ru Ru_2Si_3 0.8 eV	Rh	Pd
	Ag	Cd	In	Sn	Sb	Te	I	Xe		
6	Cs	Ba $BaSi_2$ 1.3 eV	La $\rightarrow$ Lu	Hf	Ta	W WSi_2 0.07 eV	Re $ReSi_{1.75}$ 0.12 eV	Os $OsSi$ 0.34 eV Os_2Si_3 2.3 eV $OsSi_2$ 1.8 eV	Ir Ir_3Si_5 1.2 eV	Pt
	Au	Hg	Tl	Pb	Bi	Po	At	Rn		
7	Fr	Ra	Ac $\rightarrow$ Lr	Ku	Ns					

$MnSi_{2-x}$: $Mn_{11}Si_{19}$ ($MnSi_{1.727}$), $MnSi_{1.730}$ ($MnSi_{1.730}$), $Mn_{15}Si_{26}$ ($MnSi_{1.733}$), $Mn_{26}Si_{45}$ ($MnSi_{1.730}$), $Mn_{27}Si_{45}$ ($MnSi_{1.741}$), Mn_4Si_7 ($MnSi_{1.750}$)

Reliable data definitely proving semiconducting properties have been obtained for the silicides formed by metals from groups II, VI, VII, and VIII of the Periodic Table, which are: Mg_2Si and $BaSi_2$ in group II; $CrSi_2$, $MoSi_2$ (hexagonal), and WSi_2 (hexagonal) in group VI; $MnSi_{2-x}$ and $ReSi_{1.75}$ in group VII; β-$FeSi_2$, Ru_2Si_3, $OsSi$, Os_2Si_3, $OsSi_2$, and Ir_3Si_5 in group VIII. This list seems to be incomplete. One can definitely note that semiconducting properties appear in the stoichiometric phases of metals with an even number of valence electrons. Representatives from groups II, VI, and VIII well illustrate this fact. Metals with an odd number of valence electrons can also form semiconducting silicides, as has been observed for group VII and iridium, but they need a certain deficit of silicon atoms in the stoichiometric composition. Thus, new semiconducting silicides are still expected to be discovered.

Energy gaps of the known semiconducting silicides range from 0.07–0.12 eV (hexagonal $MoSi_2$ and WSi_2, $ReSi_2$) to 2.3 eV (Os_2Si_3). They can obviously provide a background for energy-gap engineering identical to that achieved with $A^{III}B^V$ ternary compounds in their superlattices. The first collection of electronic properties of semiconducting silicides presented in the review [22] enhances this idea. But its practical implementation needs an overall knowledge of the fabrication peculiarities and properties of these attractive semiconductors.

This book summarizes considerable experimental and theoretical efforts undertaken within the last few decades in the investigation of semiconducting silicides. Due to the progress in the material preparation, high-purity thin films and bulk crystals with improved crystalline structure were only recently grown and analyzed. The experimental data are extended by the results of theoretical simulation. Most of the investigations performed were focused on $FeSi_2$ ($E_g = 0.78$ eV) and $CrSi_2$ ($E_g = 0.35$ eV). Meanwhile, we have also collected and presented all the available data related to other semiconducting silicides. Their critical analysis is given in this book.

There is a growing interest in ternary silicides [23], which are silicide-based compounds with metal or silicon atoms partially substituted by other metals or germanium, respectively. Because of the complexity of the formation, semiconducting ternary silicides are less well investigated than binary ones. Nevertheless, they are indeed attractive for flexible regulation of the fundamental electronic properties of silicides and for epitaxial matching of semiconducting silicide films. Systematic study of these aspects is at its very beginnings. Nevertheless, we will summarize and discuss the first results obtained.

Chapter 1 presents general material information related to the semiconducting silicides. It includes atomic structure parameters as well as the basic thermodynamic regularities and kinetics of silicidation in the metal-silicon couples. Chapter 2 summarizes methods of thin-film silicide formation with particular emphasis on their application in producing semiconducting thin-film silicides. Peculiarities of the crystal growth are discussed in Chap. 3. Chapter 4 focuses on fundamental electronic and related optical properties of the silicides as predicted by theoretical simulation and experimentall observed. Electronic band structure, density of states and their connection with optical parameters of the silicides are discussed there. Electron transport properties are presented in Chap. 5.

The book has been compiled from theoretical and experimental research performed in the Belarusian State University of Informatics and Radioelectronics (Minsk, Belarus), the Hahn-Meitner-Institute (Berlin, Germany) and the Institute of Solid State and Materials Research (Dresden, Germany). Results from other research teams are also presented and carefully analyzed.

On behalf of the contributors to the book and as its editor I express thanks to Professors F. Arnaud d'Avitaya, J. Derrien, P. Gas, F. d'Heurle, G. Krabbes, J.E. Mahan, S. Mantl, L. Miglio, and Doctors N.N. Dorozhkin, W. Henrion, W. Löser for fruitful general discussions and particular critical comments related to the material presented in the book. Technical assistance from V.F. Ledaschev and J. Werner in the figure preparation is acknowledged. We also thank authors and Publishes granted us permissions to reproduce figures from their original publications.

Minsk–Berlin–Dresden,   *Victor E. Borisenko*
January 1999

References

1 K.N. Tu, J.W. Mayer: Silicide formation, in: *Thin Films. Interdiffusion and Reactions*, edited by J.M. Poate, K.N. Tu, J.W. Mayer (Wiley, New York, 1978), pp. 359–405.

2 G.V. Samsonov, L.A. Dvorina, B.M. Rud: *Silicides* (Metallurgiya, Moscow, 1979) (in Russian); G.V. Samsonov, I.M. Vinitskii: *Handbook of Refractory Compounds* (IFI/Plenum, New York, 1980).

3 M.-A. Nicolet, S.S. Lau: Formation and characterization of transition-metal silicides, in: *VLSI Electronics Microstructure Science, Vol. 6, Materials and Process Characterization*, edited by N.G. Einspruch, G.B. Larrabee (Academic Press, New York, 1983), pp. 329–464.

4 S.P. Murarka: *Silicides for VLSI Applications* (Academic Press, New York, 1983).

5 G. Ottaviani: Metallurgical aspects of the formation of silicides, *Thin Solid Films* **140**(1), 3–21 (1986).

6 V.M. Koleshko, V.F. Belitsky, A.A. Khodin: Thin films of rare-earth metal silicides in microelectronics, *Vacuum* **136**(10), 669–676 (1986).

7 S. Mantl: *Ion Beam Synthesis of Epitaxial Silicides: Fabrication, Characterization and Application* (North-Holland, Amsterdam, 1992).

8 K. Maex: Silicides for integrated circuits: $TiSi_2$ and $CoSi_2$, *Mater. Sci. Eng.* **R11**(2-3), 53–153 (1993).

9 K. Maex, M. Van Rossum (Eds.): *Properties of Metal Silicides* (INSPEC, IEE, London, 1995).

10 V.E. Borisenko, P.J. Hesketh: *Solid State Rapid Thermal Processing of Semiconductors* (Plenum, New York, 1997).

11 M.C. Bost, J.E. Mahan: Optical properties of semiconducting iron disilicide thin films, *J. Appl. Phys.* **58**(7), 2696–2703 (1985).

12 R.G. Long, M.C. Bost, J.E. Mahan: Metallic behavior of lanthanum disilicide, *Appl. Phys. Lett.* **53**(14), 1272–1273 (1988).

13 O. Bisi,   L. Braicovich,   C. Carbone,   I. Lindau,   A. Iandelli,   G.L. Olcese,
    A. Palenzona: Chemical bond and electronic state in calcium silicides: Theory and
    comparison with synchrotron-radiation photoemission, *Phys. Rev. B* **40**(15),
    10194–10209 (1989).

14 C.E. Allevato, C.B. Vining: Phase diagram and electrical behavior of silicon-rich
    iridium silicide compounds, *J. Alloys Comp.* **200**(12), 99–105 (1993).

15 V. Jaccarino, G.K. Wertheim, J.H. Wernick, L.R. Walker, S. Arajs: Paramagnetic
    excited state of FeSi, *Phys. Rev.* **160**(3), 476–482 (1967).

16 A. Lacerda, H. Zhang, P.C. Canfield, M.F. Hundley Z. Fisk, J.D. Thompson,
    C.L. Seaman, M.B. Maple, G. Aeppli: Narrow-gap signature of $Fe_xCo_{1-x}Si$ single-
    crystals, *Physica B* **186-188**, 1043–1045 (1993).

17 Z. Schesinger, Z. Fisk, H.T. Zhang, M.B. Maple: Unconventional charge gap
    formation in FeSi, *Phys. Rev. Lett.* **71**(11), 1748–1751 (1993).

18 D. Mandrus, J.L. Sarrao, A. Migliori, J.D. Thompson, Z. Fisk: Thermodynamics of
    FeSi, *Phys. Rev. B* **51**(8), 4763–4767 (1995).

19 A. P. Samuely, P. Szabó, M. Mihalik, N. Hudáková, A.A. Menovsky: Gap
    formation in Kondo insulator FeSi-point-contact, *Physica B* **218**(1-4), 185–188
    (1996).

20 K. Breuer, S. Messerli, D. Purdie, M. Garnier, M. Hengsberger, Y. Baer,
    M. Mihalik: Observation of the gap opening in FeSi with photoelectron
    spectroscopy, *Phys. Rev. B* **56**(12), R7061–R7064 (1997).

21 S. Paschen, E. Felder, M.A. Chernicov, L. Degiogi, H. Schwer, H.R. Ott,
    D.P. Young, J.L. Sarrao, Z. Fisk: Low-temperature transport, thermodynamic, and
    optical properties of FeSi, *Phys. Rev. B* **56**(20), 12916–12930 (1997).

22 H. Lange: Electronic properties of semiconducting silicides, *Phys. Stat. Sol. (b)*
    201 (1), 3–65 (1997).

23 M. Setton: Ternary TM-TM-Si reactions, in: *Properties of Metal Silicides*, edited
    by K. Maex, M. Van Rossum (INSPEC, IEE, London, 1995), pp. 129–149.

Contributors:

Doctor Günter Behr
Institute of Solid State and Materials Research Dresden
Helmholtzstrasse 20, D-01069 Dresden, Germany

Professor Victor E. Borisenko
Belarusian State University of Informatics and Radioelectronics
P. Browka 6, 220027 Minsk, Belarus

Doctor Andrew B. Filonov
Belarusian State University of Informatics and Radioelectronics
P. Browka 6, 220027 Minsk, Belarus

Professor Armin Heinrich
Institute of Solid State and Materials Research Dresden
Helmholtzstrasse 20, D-01069 Dresden, Germany

Doctor Ludmila Ivanenko
Belarusian State University of Informatics and Radioelectronics
P. Browka 6, 220027 Minsk, Belarus

Doctor Horst Lange
Hahn-Meitner-Institute, Department of Photovoltaics
Rudower Chaussee 5, D-12489 Berlin, Germany

Doctor Victor L. Shaposhnikov
Belarusian State University of Informatics and Radioelectronics
P. Browka 6, 220027 Minsk, Belarus

Contents

Nomenclature

A atomic weight
α, β angle lattice parameters
α^* absorption coefficient
a, b, c linear lattice parameters

C_p heat capacity
c light speed in vacuum (2.997925×10^8 m/s)

D diffusivity
d thickness of a layer or film

E energy
E_F Fermi energy
E_a activation energy
E_g energy gap
E_n energy eigenvalue representing an electron in an eigenstate n
e electronic charge (1.60219×10^{-19} C)
ε dielectric function
ε_1 real part of the dielectric function
ε_2 imaginary part of the dielectric function

f volume fraction of grain boundaries

G free energy

H enthalpy
h Planck constant (6.62620×10^{-34} J s)
$\hbar = h/2\pi$ Planck constant (1.05459×10^{-34} J s)

j current density

K weight fraction
$\mathbf{k}$ wave vector
k distribution coefficient

k^* extinction index

k_B Boltzmann constant ($1.380658{\times}10^{-23}$ J/K = $8.617385{\times}10^{-5}$ eV/K)

k_{pri}, k_{cj} balance coefficients for the i-th product and the j-th initial component in chemical reactions

$\kappa_{latt}, \kappa_{el}$ lattice and electronic components of thermal conductivity

L Lorentz number

M atomic mass, mass fraction

m stoichiometric coefficient

m_0 electron rest mass ($9.10956{\times}10^{-31}$ kg)

m^* charge carrier effective mass

μ mobility of charge carriers

N atomic concentration

n stoichiometric coefficients, charge carrier concentration

n_e, n_h concentration of electrons, holes

n^* refractive index

n_A Avogadro constant ($6.0221367{\times}10^{23}$ mol^{-1})

η lattice mismatch

ω frequency of electromagnetic waves

p^* balance pressure

p_i partial pressure

Φ dose of implanted ions

R_0 universal gas constant (8.314510 J K^{-1} mol^{-1})

R reflectivity

R_H Hall coefficient

r radius, Hall factor

ρ electrical resistivity

ρ^* material density

S entropy, Seebeck coefficient

s area

σ electrical conductivity

σ^* specific surface energy

T absolute temperature

T^* transmittance

t time

τ relaxation time

τ^* time shift corresponding to the presence of the initial oxide layer during oxidation

U parabolic rate of growth

U_0 pre-exponential factor for a parabolic rate of growth

V linear rate of growth

V_0 pre-exponential factor for a linear rate of growth

v_n rate of nucleation

X mole fraction

x atomic fraction, current coordinate

y atomic fraction, current coordinate

Z figure of merit

Methods of quantum mechanic simulation of fundamental electronic and related properties of solids

ASW   augmented-spherical-waves
FLAPW   full-potential linear augmented-plane-waves
HF   Hartree–Fock model
$k \cdot p$ the $k \cdot p$ method based upon the perturbation theory
LAPW   linear augmented-plane-waves
LCAO   linear combination of atomic orbitals
LMTO   linear muffin-tin orbitals
MNDO   modified neglect of differential overlap
OPW   orthogonalized-plane-waves
PP   pseudopotential

Experimental techniques used for analysis of solids

AES   Auger electron spectroscopy
AFM   atomic force microscopy
BIS   bremsstrahlung isochromat spectroscopy
EDXS   energy dispersive x-ray spectroscopy
EELS   electron energy loss spectroscopy
EPR   electron paramagnetic resonance
FTIR   Fourier transform infrared spectroscopy
GIXD   glancing incidence x-ray diffraction
HRTEM   high resolution transmission electron microscopy
LEED   low energy electron diffraction
NAA   nuclear activation analysis
NRA   nuclear reaction analysis
RBS   Rutherford backscattering spectroscopy

RHEED	reflection high energy electron diffraction
SEM	scanning electron microscopy
STEM	scanning transmission electron microscopy
STM	scanning tunneling microscopy
SXES	soft x-ray emission spectroscopy
TEM	transmission electron microscopy
UPS	ultraviolet photoelectron spectroscopy
XRD	x-ray diffraction
XPS	x-ray photoelectron spectroscopy

1 General Material Aspects

Victor E. Borisenko and Andrew B. Filonov

Belarusian State University of Informatics and Radioelectronics
Minsk, Belarus

CONTENTS

A breakthrough in the technology and application of new materials is always based on a deep understanding of their properties. Silicides, being extensively studied within the last few decades, have provided a wealth of data regarding crystalline structure, thermodynamics and kinetics of formation, electronic and optical properties. The studies performed were mainly focused on the silicides with metallic properties. They are well presented in the previously published monographs and data reviews [1.1–1.4].

Semiconducting silicides are occasionally discussed in these books while being "dissolved" there at a very low concentration. Moreover, recent investigations as well as particular properties important for semiconductors are essential to be summarized. This chapter begins with the analysis of general material aspects providing a background for an introduction of semiconducting silicides into engineering, technology and production of novel semiconductor devices. This presentation includes the data available for crystalline structure and mechanical properties of semiconducting silicides, thermodynamics and kinetics of silicidation in the related metal–silicon systems, and thermal oxidation of the silicides. A dual approach combining information for bulk samples with that from thin films helps in determining the validity of both sets of data.

1.1   Crystalline Structure and Mechanical Properties

Semiconducting silicides demonstrate a variety of crystalline structures. The most representative general data characterizing them are listed in Table 1.1. Some of the silicides have identical in stoichiometry but different in crystalline structure nonsemiconducting phases. All of them are also presented in the table. Moreover, semiconducting germanides are added to the list accounting for prospects of ternary silicides with a partial silicon substitution by germanium. For a comparison the parameters related to crystalline silicon, which is often used as a substrate for the epitaxial growth of the silicides, are also shown.

Practical prospects of thin film application of the semiconducting silicides were taken into consideration in the summary of the mechanical parameters collected in Table 1.2. Mechanical properties of solids are known to be temperature-dependent. This is why the temperature range or the temperature at which the mentioned parameter was measures is reportred. Moreover, one should note that some of the parameters were obtained for polycrystalline samples. Different grain size in the samples used for analysis explains scatter in the data obtained, in particular for thermal conductivity.

Detailed discussion of atomic structure peculiarities of semiconducting silicides is given below in this section.

1.1.1   Silicides of Group II Metals

Mg_2Si. Magnesium silicide has a face-centered-cubic CaF_2 type lattice, with space group *Fm3m*. Its crystallographic unit cell is shown in Fig. 1.1. Primitive translation vectors of the Bravais lattice are $a_1 = a(0, ½, ½)$, $a_2 = a(½, 0, ½)$, $a_3 = a(½, ½, 0)$, where a is the lattice parameter. A range of the lattice constant may be found in the literature from 0.6338 to 0.6390 nm. The most reliable data gives a value of 0.6351 nm [1.6]. Relative to each silicon atom there are eight magnesium atoms located at the points $a(±¼, ±¼, ±¼)$ in the unit cell. The point group with respect to a silicon center of symmetry is the full cubic group *m3m*. The interatomic distance between the closest metal–silicon sites is 0.2750 nm.

At the pressure of $25×10^5$ Pa and temperatures above 900°C the face-centered-cubic lattice of Mg_2Si is transformed into the hexagonal one with parameters $a = 0.720$ nm and $c = 0.812$ nm [1.30].

$BaSi_2$. At room temperature and atmospheric pressure $BaSi_2$ can exist in three crystalline forms [1.31], one of which is stable while the other two are metastable. The stable one has an orthorhombic structure (see Table 1.1). The metastable forms are cubic and trigonal.

Table 1.1. Structural parameters of semiconducting silicides in comparison to isotypic germanides and silicon

Phase	Structure type	Space group	Number of formula units in the unit cell	Lattice	Lattice parameters (nm)			Density, ρ* (g/cm³)
					a	b	c	
Mg₂Si	CaF₂	*Fm3m*	4	cubic	0.63512 [1.1, 1.5]			1.988 [1.1]
Mg₂Ge	CaF₂	*Fm3m*	4	cubic	0.639 [1.6]			
BaSi₂	BaSi₂	$Pnma - D_{2h}^{16}$	8	orthorhombic	0.892 [1.7]	0.680	1.158	3.54 [1.7]
					0.891 [1.8]	0.672	1.153	3.68 [1.7]
					0.8942 [1.9]	0.6733	1.1555	
BaSi₂	SrSi₂	$P4_332 - O^6$		cubic	0.6715 [1.10]			
BaSi₂	EuGe₂	$P\bar{3}m1 - D_3^{3d}$		trigonal				
BaGe₂	BaSi₂	$Pnma - D_{2h}^{16}$	8	orthorhombic	0.905 [1.11]	0.683	1.124	5.14 [1.11]
BaGe₂	SrSi₂	$P4_332 - O^6$		cubic	0.688 [1.10]			
CrSi₂	CrSi₂	$P6_222 - D_{6h}^4$	3	hexagonal	0.4431 [1.1]		0.6364	4.98 [1.1]
					0.4424 [1.12]		0.6342	
					0.44281 [1.13]		0.63691	
MoSi₂	MoSi₂	$I4/mmm - D_{4h}^{17}$	2	tetragonal	0.3200 [1.13]		0.7850	6.28 [1.13]
MoSi₂	CrSi₂	$P6_222 - D_{6h}^4$ $C6_22$	3	hexagonal	0.4642 [1.1]		0.6529	6.32 [1.13]
					0.4596 [1.13]		0.6550	
WSi₂	MoSi₂	$I4/mmm - D_{4h}^{17}$	2	tetragonal	0.3211 [1.1]		0.7868	9.88 [1.13]
WSi₂	CrSi₂	$P6_222 - D_{6h}^4$	3	hexagonal	0.4614 [1.13]		0.6414	

Phase	Structure type	Space group	Number of formula units in the unit cell	Lattice	Lattice parameters (nm)			Density, ρ* (g/cm^3)
					a	b	c	
MnSi$_{2-x}$:								
Mn$_4$Si$_7$ (MnSi$_{1.750}$)	Mn$_4$Si$_7$	$P4c2$ - D_{2d}	4	tetragonal	0.5525 [1.1]		1.7463	5.186 [1.1]
Mn$_{27}$Si$_{47}$ (MnSi$_{1.741}$)	Mn$_{27}$Si$_{47}$	$P\bar{4}n2$		tetragonal	0.5530 [1.13]		11.794	5.164 [1.13]
Mn$_{15}$Si$_{26}$ (MnSi$_{1.733}$)	Mn$_{15}$Si$_{26}$	$\bar{I}42d$ - D_{2d}^{12}	4	tetragonal	0.5531 [1.1] 0.5525 [1.13]		6.5311 6.555	5.159 [1.13]
Mn$_{11}$Si$_{19}$ (MnSi$_{1.727}$)		$P\bar{4}n2$		tetragonal	0.5518 [1.13]		0.4814	
ReSi$_{1.75}$		$P1$	4	triclinic (α = 89.90°)	0.3138 [1.14]	0.3120	0.7670	
ReSi$_2$	MoPt$_2$	$Immm$	2	orthorhombic	0.3128 [1.13]	0.3144	0.7677	10.66 [1.13]
ReSi$_2$	MoSi$_2$	$I4/mmm$ - D_{4h}^{17}	2	tetragonal	0.3129 [1.1]		0.7674 [1.1]	10.71 [1.1]
α-FeSi$_2$		$P4/mmm$ - D_{4h}^{1}	1	tetragonal	0.269 [1.1] 0.2695 [1.15] 0.269392 [1.13]		0.513 0.5134 0.51361	4.99 [1.13]
β-FeSi$_2$	FeSi$_2$	$Cmca$ - D_{2h}^{18}	16	orthorhombic	0.98792 [1.1] 0.9863 [1.16]	0.77991 0.7791	0.78388 0.7833	4.93 [1.1]
γ-FeSi$_2$	CaF$_2$			cubic	0.27 [1.15]			

Phase	Structure type	Space group	Number of formula units in the unit cell	Lattice	Lattice parameters (nm)			Density, ρ^* (g/cm³)
					a	b	c	
Ru₂Si₃	Ru₂Si₃	$Pbcn - D_{2h}^{14}$	8	orthorhombic	1.1057 [1.17]	0.8934	0.5533	6.96 [1.17]
Ru₂Si₃			8	tetragonal	1.1073 [1.13]		0.8954	
Ru₂Ge₃		$Pbcn$		orthorhombic	1.1436 [1.17]	0.9238	0.5716	9.23 [1.17]
OsSi	FeSi	$P2_13$	1	cubic	0.4729 [1.13]			13.71 [1.13]
OsSi	CsCl	$Pm3m - O_h^1$	1	cubic	0.29630 [1.1]			13.98 [1.13]
Os₂Si₃	Ru₂Si₃	$Pbcn$	8	orthorhombic	1.1124 [1.13]	0.8932	0.5570	11.15 [1.13]
Os₂Ge₃	Ru₂Si₃	$Pbcn$	8	orthorhombic	1.1544 [1.17]	0.9281	0.5783	12.82 [1.17]
OsSi₂	β-FeSi₂	$Cmca$	16	orthorhombic	1.0150 [1.13]	0.8117	0.8223	9.66 [1.13]
OsSi₂	OsGe₂	$P2/m$	4	monoclinic ($\beta = 119.01°$)	0.8661 [1.18]	0.2984	0.7472	
OsSi₂	OsGe₂	$C2/m$	4	monoclinic ($\beta = 118°30'$)	0.877 [1.19]	0.300	0.738	
OsGe₂	OsGe₂	$C2/m$	4	monoclinic ($\beta = 119°10'$)	0.8995 [1.20]	0.3094	0.7685	11.9 [1.20]
Ir₃Si₅		$P2_1/c$	8	monoclinic ($\beta = 116.69°$)	0.6406 [1.21]	1.4162	1.1553	10.17 [1.21]
Si	C	$Fd\bar{3}m$	4	cubic	0.54306 [1.22]			2.33

Table 1.2. Thermal and mechanical properties of semiconducting silicides in comparison to silicon. Unless otherwise indicated the reader is referred to [1.23] for original references

Phase	Thermal expansion coefficient $(10^{-6}\ K^{-1})$	Temperature (K)	Thermal conductivity $(W\ cm^{-1}\ K^{-1})$	Temperature (K)	Microhardness (for load of 0.5 N) (GPa)
Mg$_2$Si	14.8 [1.1]		$23.47T^{-1a}$ [1.24] 0.039^c [1.26]	270–570 303	$4.0 – 4.4^b$ [1.1, 1.25]
BaSi$_2$	8.2 [1.27] 8.6 [1.27]	293 – 923 923 – 1373	0.088 [1.27]	293	9.11 [1.27]
CrSi$_2$	//a: 8.177+0.009903T //c: 8.986+7.617×10^{-4}T volumetric: 25.30+0.02055T	300 – 1400 300 – 1400 300 – 1400	0.0106 [1.1] 0.054 0.042 0.063	300 300 1000 1200	9.96 – 11.50
MoSi$_2$ (tetragonal)	//a: 5.617+0.003544T //c: 4.115+0.005565T volumetric: 15.30+0.01268T	300 – 1400 300 – 1400 300 – 1400	0.485	300	12.0 – 13.5
WSi$_2$ (tetragonal)	//a: 6.512+0.0025137T //c: 8.800+0.002404T volumetric:22.00+0.007254T	300 – 1400 300 – 1400 300 – 1400			10.74, 12.5
MnSi$_{1.73}$ (Mn$_{15}$Si$_{26}$)	//a: 5.289+0.007900T //c: 9.024+0.005006T volumetric: 19.05+0.02131T	300 – 1400 300 – 1400 300 – 1400	0.075 0.040 0.042 0.060	343 573 973 1373	
ReSi$_{1.75}$(?)	//a: 4.239+0.004396T //b: 5.604+0.003853T //c: 7.530+0.001208T volumetric: 17.39+0.009420T	300 – 1400 300 – 1400 300 – 1400 300 – 1400			15.0

Phase	Thermal expansion coefficient (10^{-6} K^{-1})	Temperature (K)	Thermal conductivity (W cm^{-1} K^{-1})	Temperature (K)	Microhardness (for load of 0.5 N) (GPa)

β-FeSi$_2$	//a: $16.697+0.001442T$
	//b: $0.026+0.001673T$
	//c: $0.415+0.001872T$
		298 – 1223
		298 – 1223
		298 – 1223
					12.6

FeSi$_2$			0.658 0.122 0.037 0.080	343 573 1273 1073 – 1173	10.74

Ru$_2$Si$_3$	//a: $7.384+0.005504T$
	//b: $4.806+0.006310T$
	//c: $6.923+0.008924T$
		298 – 1050
		298 – 1050
		298 – 1050			

OsSi					14.2[b] [1.18]
Os$_2$Si$_3$					19.6[b] [1.18]

OsSi$_2$	//a: $6.483+0.003537T$
	//b: $5.871+0.003538T$
	//c: $5.709+0.003401T$
		298 – 1552
		298 – 1552
		298 – 1552
					19.5
					13.6[b] [1.18]

Si	4.00 – 4.68 [1.4] 2.6	300 – 1400	$2.722\exp(-2.338\times10^{-3}T)$ [1.28] $0.648\exp(-7.275\times10^{-4}T)$ [1.28]	300 – 900 > 900	

[a] For a polycrystalline material, the grain size is not mentioned. [b] For load of 1 N, [c] For a polycrystalline material with the grain size of 86 nm.

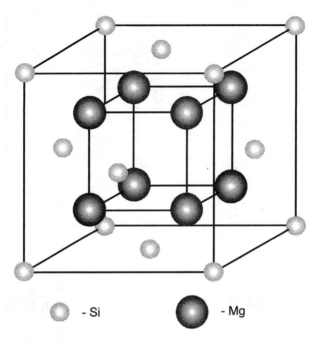

○ - Si   ● - Mg

Fig. 1.1. Crystal structure of cubic Mg_2Si [1.29]

Among the polymorphic forms only orthorhombic $BaSi_2$, space group *Pnma*, has semiconducting properties. Its crystal structure is shown in Fig. 1.2. Lattice parameters and atomic positions are listed in Tables 1.1 and 1.3, respectively.

Table 1.3. Atomic positions (in units of the primitive translation vectors) in orthorhombic $BaSi_2$ [1.7]

Atom	Site	x	y	z
Ba4	4c	0.014	0.250	0.694
Ba4	4c	0.839	0.250	0.095
Si4	4c	0.424	0.250	0.091
Si4	4c	0.205	0.250	0.969
Si8	8d	0.190	0.078	0.147

Bond angles $Si^{Si}Si$ vary approximately 1° around the ideal value of 60° and silicon–silicon bond distances lie between 0.2395 and 0.2435 nm [1.7, 1.31], which are of the same magnitude as those in elemental silicon (0.234 nm). The shortest barium–silicon distance is 0.339 nm and the barium–barium distances are 0.438 nm and 0.444 nm [1.7]. The shortest silicon–silicon distance between the tetrahedra is 0.403 nm. It is substantially longer than the distances within the tetrahedron. The isolated tetrahedra are interconnected by barium atoms and form (010) planar channels in the structure.

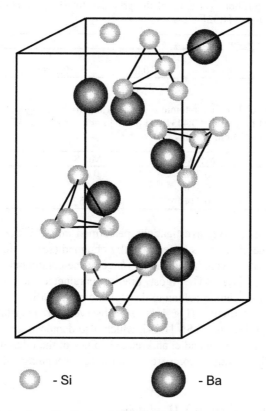

Fig. 1.2. Crystal structure of orthorhombic BaSi$_2$ [1.8]

1.1.2 Silicides of Group VI Metals

CrSi$_2$. Chromium disilicide crystallizes in the hexagonal $C40$ structure with three formula units per unit cell. The atom arrangement in it is shown in Fig. 1.3. The space group of this structure ($P6_222$) is nonsymmorphic. It contains nonprimitive translations $\tau = c/3$ and $2c/3$, which interchange three individual, hexagonal CrSi$_2$ layers with their rotation by 60° around the c axis. Thus, there are three crystallographically equivalent chromium and six silicon sites in the unit cell. The atomic positionswithin the primitive unit cell and the nearest-neighbor interatomic distances are listed in Tables 1.4 and 1.5. In addition to the hexagonal in-plane neighbors at about 0.255 nm, each chromium and silicon atom is tetrahedrally coordinated with four inter-plane neighbors at 0.247 nm. The chromium sites have four silicon inter-plane neighbors while each silicon site has a pair of chromium and silicon neighbors.

Table 1.4. Atomic positions (in units of the primitive translation vectors) in hexagonal VI group metal disilicides [1.32]

Atom	Site	x	y	z
Me	3d	0.5	0.0	0.5
Si	6j	0.0	0.2887	0.5

Table 1.5. Nearest-neighbor interatomic distances (in nm) in hexagonal VI group metal disilicides calculated for a and c: 0.4431 and 0.6364 nm in $CrSi_2$, 0.4642 and 0.6529 nm in $MoSi_2$, 0.4610 and 0.6415 nm in WSi_2

Me	Me–Me	Me–Si	Si–Si
Cr	0.306	0.247	0.247
Mo	0.318	0.256	0.256
W	0.314	0.252	0.252

In spite of the extensive investigations performed on $CrSi_2$ there still exists a scatter in the lattice parameters experimentally obtained (see Table 1.1). This can be attributed to a rather wide homogeneity range of the compound extending from $CrSi_{1.98}$ to $CrSi_{2.02}$ [1.1]. The most precise parameters $a = 0.4428$ nm and $c = 0.6369$ nm [1.13] seem to belong to the stoichiometric $CrSi_2$.

The similar $C40$ structure (Fig. 1.3) was observed for metastable hexagonal phases of **MoSi$_2$** [1.33] and **WSi$_2$** [1.34], which also demonstrate semiconducting properties. The parameters a and c, and related values of the interatomic distances (Table 1.5), are slightly higher compared with the corresponding values of $CrSi_2$.

1.1.3 Silicides of Group VII Metals

MnSi$_{2-x}$. Among semiconducting silicides this one has the most complicated structure. For $MnSi_{2-x}$ with $2-x$ in the range from 1.71 to 1.75 several tetragonal phases exist [1.1]. All of them have similar tetragonal cells elongated along the c axis. In all detected phases one may identify a cell in the manganese sublattice with four atoms and with a parameter of 0.4360 nm along the c axis. The silicon atoms form equivalent layers perpendicular to the axis. The most frequently quoted semiconducting phase is $Mn_{15}Si_{26}$ ($MnSi_{1.73}$). Its unit cell (Fig. 1.4) consists of 15 manganese subcells stacked on top of each other in the c-direction. Within the unit cell 52 silicon atom pairs are arranged as dumbells around the central axis being oriented alternately in opposite quadrants about the central axis in the c-direction. The atomic positions within the primitive unit cell and the nearest-neighbor interatomic distances are listed in Tables 1.6 and 1.7.

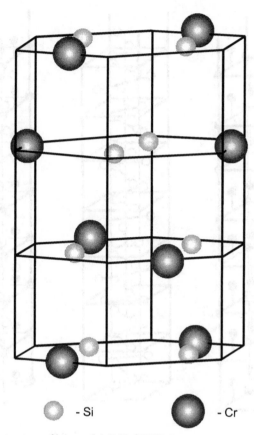

Fig. 1.3. Crystal structure of hexagonal CrSi$_2$ [1.32]

Table 1.6. Atomic positions (in units of the primitive translation vectors) in tetragonal Mn$_{15}$Si$_{26}$ [1.35]

Atom	Site	x	Y	z
Mn1	4a	0	0	0
Mn2	8c	0	0	0.0653
Mn3	8c	0	0	0.1330
Mn4	8c	0	0	0.1996
Mn5	8c	0	0	0.2669
Mn6	8c	0	0	0.3326
Mn7	8c	0	0	0.4013
Mn8	8c	0	0	0.4644
Si1	16e	0.2161	0.3445	0.0119
Si2	16e	0.8515	0.1811	0.0311
Si3	16e	0.3167	0.1642	0.0480
Si4	16e	0.6657	0.3434	0.0680
Si5	16e	0.1591	0.3122	0.0848
Si6	16e	0.7744	0.1523	0.1063
Si7	8d	0.3479	0.2500	0.1250

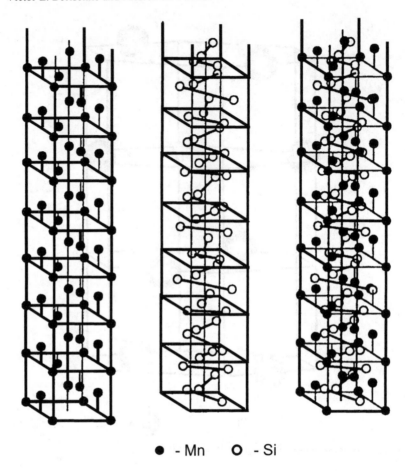

● - Mn   O - Si

Fig. 1.4. Crystal structure of tetragonal $Mn_{15}Si_{26}$ [1.35]

Table 1.7. Nearest-neighbor interatomic distances (in nm) in tetragonal $Mn_{15}Si_{26}$ calculated for $a = 0.5531$ nm and $c = 6.5311$ nm

Atom	Mn2	Mn3	Mn4	Mn5	Mn6	Mn7	Mn8
Mn1	0.4264	0.8686	1.3036	1.7431	2.1722	2.6209	3.0330
Mn2		0.4422	0.8772	1.3167	1.7458	2.1945	2.6066
Mn3			0.4350	0.8745	1.3036	1.7523	2.1644
Mn4				0.4395	0.8686	1.3173	1.8294
Mn5					0.4291	0.8778	1.2899
Mn6						0.4487	0.8608
Mn7							0.4121

Atom	Si2	Si3	Si4	Si5	Si6	Si7
Si1	0.3839	0.2618	0.4428	0.4774	0.6976	0.7440
Si2		0.3158	0.2769	0.5243	0.4932	0.6745
Si3			0.2533	0.2684	0.4573	0.5054
Si4				0.3014	0.2780	0.4148
Si5					0.3786	0.2846
Si6						0.2000

Atom	Mn1	Mn2	Mn3	Mn4	Mn5	Mn6	Mn7	Mn8
Si1	0.2379	0.4149	0.8222	1.2463	1.6805	2.1065	2.5531	2.9638
Si2	0.5225	0.5306	0.8213	1.2011	1.6134	2.0271	2.4535	2.8705
Si3	0.3703	0.2273	0.5891	1.0096	1.4432	1.8692	2.3159	2.7267
Si4	0.6073	0.4146	0.5931	0.9541	1.3634	1.7770	2.2158	2.6218
Si5	0.5867	0.2318	0.3696	0.7744	1.2049	1.6299	2.0761	2.4867
Si6	0.8200	0.5121	0.4700	0.7495	1.1361	1.5411	1.9755	2.3791
Si7	0.8499	0.4562	0.2425	0.5418	0.9565	1.3764	1.8200	2.2293

ReSi$_{1.75}$(ReSi$_2$). For a long time, semiconducting rhenium silicide was believed to have the composition of the disilicide ReSi$_2$ with a tetragonal MoSi$_2$-type structure [1.36, 1.37] (space group $I4/mmm$) or with volume-centered orthorhombic structure [1.38] (space group $Immm$, which is a subgroup of $I4/mmm$). That is why most researchers referred to this phase in their earlier studies.

The monoclinically distorted phase ReSi$_{1.75}$ was recently shown to be the only stable one within the atomic ratios close to the disilicide [1.14]. Such reduction of the symmetry from the $Immm$ to $P1$ space group due to splitting of the 4i position in the orthorhombic structure results in four formula units of ReSi$_{1.75}$ in the primitive cell and in the even number of valence electrons. This is compatible with the semiconducting behavior of the compound. Thus, semiconducting ReSi$_{1.75}$ has a triclinic phase (space group $P1$) which is only slightly distorted from the orthorhombic one. There are some silicon vacancies in the lattice in agreement with the assumption ReSi$_2$ = ReSi$_{1.75}$ + Si$_{0.25}$. The silicide unit cell contains two formula units of Re$_4$Si$_7$ (8 metal and 14 silicon sites). The reduced primitive cell is shown in Fig.1.5. The atomic positions within the primitive unit cell, the nearest-neighbor interatomic distances, and the appropriate occupancy factor are listed in Tables 1.8 and 1.9.

Table 1.8. Atomic positions (in units of the primitive translation vectors) in triclinic ReSi$_{1.75}$ [1.14]

Atom	x	y	z	Occupancy factor
Re1	0.0	0.0	0.0	1
Re2	0.508	0.501	0.518	1
Si1	−0.034	0.031	0.669	1
Si2	0.511	0.529	0.870	1
Si3	−0.044	−0.027	0.342	0.75
Si4	0.512	0.459	0.190	0.75

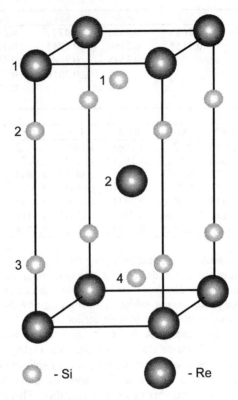

Fig. 1.5. Crystal structure of triclinic ReSi$_{1.75}$ [1.14]

Table 1.9. Nearest-neighbor interatomic distances (in nm) in triclinic ReSi$_{1.75}$ calculated for $a = 0.3138$ nm, $b = 0.3120$ nm, $c = 0.7670$ nm, and $\alpha = 89.90°$

Atom	Re2	Si1	Si2	Si3	Si4
Re1	0.4558	0.2543	0.7058	0.2628	0.2598
Re2		0.2388	0.2701	0.2746	0.2520
Si1			0.2652	0.2515	0.4187
Si2				0.4582	0.5220
Si3					0.2589

1.1.4 Silicides of Group VIII Metals

FeSi$_2$. The semiconducting β-FeSi$_2$ phase, which is stable up to about 960°C, belongs to the *Cmca* space group of the orthorhombic crystalline structure [1.16]. The conventional orthorhombic cell is rather large containing 48 atoms (16 formula units). It is generated by a cooperative Jahn–Teller distortion of the more primitive fluorite structure [1.39]. Its origin will be discussed later (see electronic property section). Such a distortion could be easily traced by considering the deformation of the tetragonal nucleus of a cubic supercell made up by 8 fcc units with a being twice the fcc edge along x and $b = c$ being the diagonals of the fcc square faces. As a result of the distortion a becomes shorter while b and c increase by slightly different values. Plane projections of the appropriate unit cell are shown in Fig. 1.6.

The crystalline structure of β-FeSi$_2$ can be also represented by a primitive cell containing 24 atoms (8 formula units) in the base-centered orthorhombic structure with D_{12h}^{18} symmetry. Both 8 iron and 16 silicon atoms are grouped into two equal sets (Fe1, Fe2, Si1, Si2) of crystallographically inequivalent sites with slightly different distances. The Fe1–Si distances vary between 0.234 and 0.239 nm whereas those of Fe2–Si vary between 0.234 and 0.244 nm. The Si–Fe1–Si and Si–Fe2–Si angles lie in the range from 62.5° to 99.5° and from 61.8° to 99.5°, respectively. The atomic positions within the primitive unit cell and the nearest-neighbor interatomic distances are listed in Tables 1.10 and 1.11.

Table 1.10. Atomic positions (in units of the primitive translation vectors) in orthorhombic FeSi$_2$ [1.16] and OsSi$_2$ [1.40]

Atom	Site	x	y	z
Fe1 (Os1)	8d	0.21465 (0.2142)	0.0	0.0
Fe2 (Os2)	8f	0.0	0.19139 (0.1881)	0.81504 (0.1812)
Si1	16g	0.37177 (0.3699)	0.27465 (0.2208)	0.44880 (0.0597)
Si2	16g	0.12729 (0.1280)	0.04499 (0.0534)	0.27392 (0.7252)

Table 1.11. Nearest-neighbor interatomic distances (in nm) in orthorhombic β-FeSi$_2$ and OsSi$_2$ calculated for a, b, and c: 0.9863, 0.7791, and 0.7833 nm in β-FeSi$_2$; 1.014, 0.811, and 0.822 nm in OsSi$_2$

Atom	Fe1	Fe2	Si1	Si2
Fe1	0.3957	0.2967	0.2338	0.2339
Fe2		0.4022	0.2333	0.2335
Si1			0.2529	0.2499
Si2				0.2449
	Os1	Os2	Si1	Si2
Os1	0.4175	0.3046	0.2439	0.2462
Os2		0.2442	0.2493	0.2474
Si1			0.2641	0.2547
Si2				0.2510

The other possible phases of iron disilicide, namely tetragonal α-, cubic γ-, and defected CsCl-type structures [1.15, 1.41], are all metallic. Their main crystallographic parameters are reported in Table 1.1.

Ru$_2$Si$_3$. Semiconducting Ru$_2$Si$_3$ has an orthorhombic structure (space group *Pbcn*) which is stable below 1000°C. It belongs to the so-called defect-type TiSi$_2$ "chimney-ladder" compounds [1.17]. The unit cell is rather large and complex. It contains 40 atoms in total (16 Ru and 24 Si atoms).

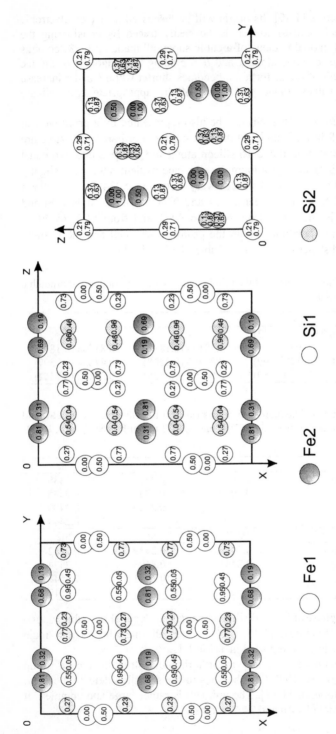

Fig. 1.6. Orthorhombic β-FeSi$_2$ unit cell projections onto *xy*, *xz*, and *yz* planes. Numbers in the circles indicate atomic site distance from the plane in the perpendicular direction [1.16]

Both ruthenium and silicon atoms are grouped into three sets (Ru1, Ru2, Ru3, Si1, Si2, Si3) of crystallographically inequivalent sites. Plane projections of atoms in the unit cell are shown in Fig. 1.7. Ruthenium atoms occupy close to ideal positions as titanium atoms in four unit cells of the $TiSi_2$ prototype, where two unit cells are stacked one on top of the other and then doubled in the horizontal direction. But the ideal $TiSi_2$ structure has four silicon atoms in a centered rectangular, which is opposite to the Ru_2Si_3 structure that has only three silicon atoms. In order to accommodate for the related silicon vacancies in the structure the remaining silicon atoms are not in the plane as in the prototype. They are shifted out of the generating plane to find ideal positions between the transition metal atoms.

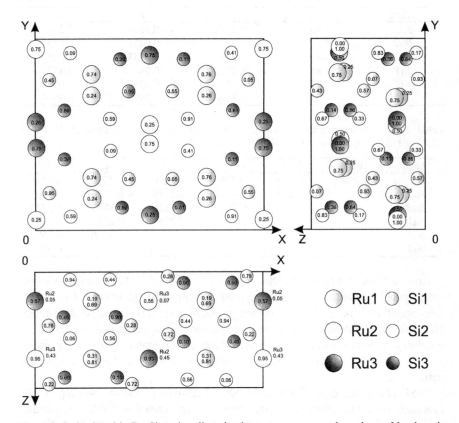

Fig. 1.7. Orthorhombic Ru_2Si_3 unit cell projections onto xy, xz, and yz planes. Numbers in the circles indicate atomic site distance from the plane in the perpendicular direction [1.17]

If one considers the structure in terms of the ruthenium–silicon coordination polyhedra, there are two $RuSi_6$ octahedra around Ru1, two deformed $RuSi_4$ tetrahedra around Ru2, and four $RuSi_7$ polyhedra around Ru3. The average interatomic ruthenium–silicon distance varies only by about 5% from

0.236–0.238 nm for tetrahedra and octahedra to 0.247 nm for the $RuSi_7$ polyhedra. Each ruthenium atom has four ruthenium neighbors. Each silicon atom has four ruthenium neighbors with an average ruthenium–silicon distance of 0.242, 0.245, and 0.242 nm for Si1, Si2, and Si3 sites, respectively. Thus, ruthenium–silicon and silicon-silicon nearest neighbor distances appear to be of the same order in the structure without any particular ruthenium-ruthenium bond length. The atomic positions within the primitive unit cell and the nearest-neighbor interatomic distances are listed in Tables 1.12 and 1.13.

Table 1.12. Atomic positions (in units of the primitive translation vectors) in orthorhombic Ru_2Si_3 [1.17]

Atom	Site	x	y	z
Ru1	8d	0.2472	0.1864	0.2401
Ru2	4c	0.0	0.0451	0.25
Ru3	4c	0.0	0.5748	0.25
Si1	8d	0.4275	0.2841	0.4537
Si2	8d	0.3253	0.4350	0.0934
Si3	8d	0.1366	0.3994	0.3946

Table 1.13. Nearest-neighbor interatomic distances (in nm) in orthorhombic Ru_2Si_3 calculated for $a = 1.1436$ nm, $b = 0.9238$ nm, and $c = 0.5716$ nm

Atom	Ru1	Ru2	Ru3	Si1	Si2	Si3
Ru1	0.2991	0.3011	0.2968	0.2476	0.2375	0.2418
Ru2		0.2881	0.4202	0.2379	0.2334	0.3597
Ru3			0.3072	0.2326	0.3852	0.2319
Si1				0.2766	0.2659	0.2956
Si2					0.3000	0.2689
Si3						0.3299

Around 1000°C the orthorhombic Ru_2Si_3 was reported [1.42] to undergo a first-order phase transition to a tetragonal Ru_2Sn_3-type structure. This phase transformation is reversible and occurs gradually over a wide temperature range. The silicon atoms displace from the positions corresponding to the *Pbcn* symmetry and pass through the intermediate orthorhombic *Pb2n* configuration into the *P4̄c2* one of the Ru_2Sn_3-type structure. The ruthenium atoms remain static.

OsSi. This phase crystallizes in the cubic FeSi-type [1.18] or CsCl-type [1.19] structure. Their main structural parameters are represented in Table 1.1. No data on precise atomic positions were published. Only the phase with the FeSi-type structure was observed to be a semiconductor [1.18].

Os_2Si_3. Semiconducting Os_2Si_3 phase has an orthorhombic structure (space group *Pbcn*) which is isostructural with the Ru_2Si_3 one [1.17]. The main structural parameters are presented in Table 1.1. No data on precise atomic positions were published.

$OsSi_2$. Semiconducting $OsSi_2$ phase has an orthorhombic structure (space group *Cmca*) which is isostructural with the β-$FeSi_2$ one [1.40]. The atomic positions within the primitive unit cell and the nearest neighbor interatomic distances are listed in Tables 1.10 and 1.11.

There is a metastable monoclinic $OsSi_2$ [1.18, 1.19]. This phase is expected to be isostructural with $OsGe_2$. It usually forms dendrites. The monoclinic structure of $OsSi_2$ seems to be stabilized by aluminum and oxygen impurity atoms [1.19].

Ir_3Si_5. An exact stoichiometry and precise crystalline parameters of the semiconducting iridium silicide remained unknown for a long time. In the first experiments [1.43–1.46] the stoichiometry of this phase was determined to be $IrSi_x$ with x varied from 1.5 to 1.75. It was believed to have a monoclinic cell with $a = 0.5542$ nm, $b = 1.4166$ nm, $c = 1.2426$ nm, $\beta = 120.61°$. Later [1.47] the compound composition was defined as $IrSi_{\sim1.6}$, which more or less corresponds to Ir_3Si_5.

The most reliable phase composition and structure data for the semiconducting iridium silicide was obtained in [1.21, 1.48]. It was stated to be monoclinic Ir_3Si_5 with the parameters listed in Table 1.1. The unit cell fragments are shown in Fig. 1.8.

The atoms in the unit cell all occupy the general 4e positions. The most characteristic feature of the structure is almost regular empty cubes formed by silicon atoms. One kind of the cubes is formed by Si1–Si4 atoms and the other one by Si5–Si8 atoms. The cubes are connected in both a and c directions. The cube centers coincide with the centers of symmetry and the distances from the centers to the eight corners are, on average, 0.233 and 0.234 nm for the two types of the cube in Ir_3Si_5. The additional Si9 and Si10 atoms complete the three-dimensional silicon network by bridging atoms in the b axis direction. The atomic positions within the primitive unit cell and the nearest-neighbor interatomic distances are listed in Tables 1.14 and 1.15.

Table 1.14. Atomic positions (in units of the primitive translation vectors) in monoclinic Ir_3Si_5 [1.21]; all atoms are in general 4e position

Atom	x	y	z
Ir1	0.4275	0.4024	0.1109
Ir2	0.5655	0.5942	0.3962
Ir3	0.9528	0.7521	0.2440
Ir4	0.0827	0.9557	0.2730
Ir5	0.2367	0.6676	0.1340
Ir6	0.2333	0.8351	0.9925
Si1	0.9263	0.8400	0.0669
Si2	0.7450	0.4903	0.5760
Si3	0.8636	0.6099	0.3272
Si4	0.3304	0.9427	0.1645
Si5	0.8596	0.6127	0.1081
Si6	0.3547	0.9354	0.5102
Si7	0.9515	0.6508	0.9084
Si8	0.2565	0.9875	0.7037
Si9	0.3849	0.7498	0.8572
Si10	0.5549	0.7797	0.1892

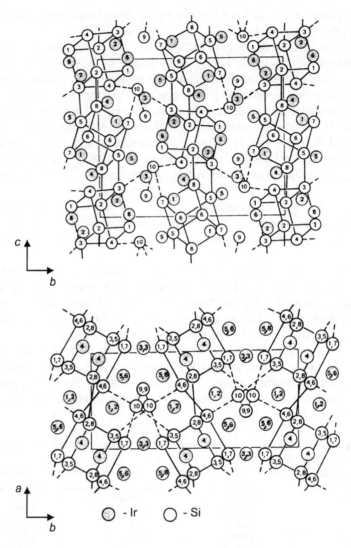

Fig. 1.8. Projections of the Ir_3Si_5 crystal structure in the [100] and [001] directions [1.21]

Table 1.15. Nearest-neighbor interatomic distances (in nm) in monoclinic Ir_3Si_5 [1.21]

Atom	Ir1	Ir2	Ir3	Ir4	Ir5	Ir6
Ir1				2.905		
Ir2					2.994	2.984
Ir3				2.979	2.905	2.891
Ir4	2.905		2.979			
Ir5		2.994	2.905			2.875
Ir6		2.984	2.891		2.873	

Atom	Si1	Si2	Si3	Si4	Si5	Si6	Si7	Si8	Si9	Si10
Si1		2.693	2.709	2.734						
Si2	2.693	2.841	2.653	2.683				2.550		
Si3	2.709	2.653		2.696	2.520					3.067
Si4	2.734	2.683	2.696					2.586		2.667
Si5			2.520			2.899	2.679	2.679		
Si6					2.894	2.696	2.612	2.683		
Si7					2.679	2.612		2.694		2.840
Si8		2.550		2.586	2.679	2.683	2.694			
Si9										2.650
Si10			3.067	2.667			2.840		2.650	

Atom	Si1	Si2	Si3	Si4	Si5	Si6	Si7	Si8	Si9	Si10
Ir1				2.419	2.377	2.425	2.456	2.416	2.415	2.852
Ir2	2.452	2.379	2.388	2.441				2.500	2.441	
Ir3	2.335		2.408		2.424		2.348		2.475	2.369
Ir4	2.686	2.455	2.565	2.431	2.552	2.519	2.572	2.444		
Ir5	3.022				2.426	2.390	2.431	2.323	2.593	2.432
Ir6	2.477	2.362	2.410	2.356			3.072		2.492	2.413

1.1.5 Ternary Silicides

There are two principal possibilities to obtain three-component silicides related to single-phase compounds. The first is a substitution of some silicon atoms by germanium ones, thus obtaining an intermediate compound between the silicide and the germanide of the particular metal. In the second case, one substitutes the host metal atoms in a silicide by the metal, which can be similar or quite different in electronic structure with respect to the hosts [1.49]. Both approaches result in a ternary silicide with modified lattice parameters and other basic properties. The crystalline structure of such compounds is focused upon here.

Ternary silicides usually preserve the type of crystalline structure of the basic binary silicide. Lattice parameters of mixed crystals are known [1.6, 1.10, 1.50, 1.51] to vary linearly with the composition showing good agreement with Vegard's law. This is valid for the substitutional solid solutions of atoms with a similar electronic structure and close size. Thus, the lattice parameter a of a ternary silicide can be calculated as

$$a(x) = (1 - x)a_1 + xa_2 , \tag{1.1}$$

where x is the atomic fraction of the substituted atoms, a_1 and a_2 are the lattice parameters of the virgin isostructural binary compounds between which, in terms of composition, the ternary is formed. The law can be applied to both metal by metal and silicon by germanium substituted silicides.

Ternary silicides produced between a semiconducting silicide and the corresponding isostructural germanide seem to be rather attractive for fine energy gap engineering. All of the related germanides are also semiconductors. Their lattice parameters are listed in Table 1.1. Energy gap parameters of the appropriate silicide–germanide couples are demonstrated in Table 1.16. While *a priori* estimated possibilities to form ternary silicides by substituting silicon atoms with

Table 1.16. Semiconducting silicides and related germanides

Silicides	Lattice structure, Miscibility		Germanides
Mg_2Si E_g=0.78•eV (indirect) [1.6, 1.29]	cubic totally miscible [1.6, 1.29]	cubic	Mg_2Ge E_g=0.70 eV (indirect) [1.6, 1.29]
$BaSi_2$ E_g=0.48•eV [1.1], 1.3 eV [1.52]	orthorhombic miscible (?)	orthorhombic	$BaGe_2$ E_g=1.0 eV [1.52]
$CrSi_2$ E_g=0.35 eV (indirect) [1.53]	hexagonal immiscible [1.54]	tetragonal	$Cr_{11}Ge_{19}$ [1.55]
$MoSi_2$ E_g=0.07 eV (indirect) [1.53]	hexagonal immiscible (?)	tetragonal	$MoGe_2$ [1.56, 1.57]
WSi_2 E_g=0.07 eV (indirect) [1.53]	hexagonal immiscible (?)	tetragonal	WGe_2 [1.57, 1.58]
$MnSi_{1.7}$ E_g=0.68 eV (direct) [1.59]	tetragonal		no candidate
$ReSi_{1.75}$ E_g=0.12 eV [1.37], 0.15 eV [1.60] (indirect)	monoclinic immiscible (?)	tetragonal	$ReGe_{1.75}$ [1.61]
•-$FeSi_2$ E_g=0.78 eV (quasidirect) [1.62]	orthorhombic immiscible [1.54] miscible [1.63]		$FeGe_2$
Ru_2Si_3 E_g=0.44 eV [1.64], 0.41 eV [1.65] (direct)	orthorhombic miscible [1.64]	orthorhombic	Ru_2Ge_3 E_g=0.34 eV [1.64], 0.31 eV [1.65] (quasidirect)
OsSi E_g=0.34 eV [1.18]	cubic		no candidate
Os_2Si_3 E_g=2.3 eV [1.18], 0.95 eV (direct) [1.66]	orthorhombic miscible (?)	orthorhombic	Os_2Ge_3 E_g=0.87 eV (quasidirect) [1.66]
$OsSi_2$ E_g=1.4 eV [1.18], 1.8 eV [1.66]	orthorhombic immiscible (?)	monoclinic	$OsGe_2$ nonsemiconductor (?)
Ir_3Si_5 E_g=1.2 eV [1.67], 1.56 eV [1.68]	monoclinic		no candidate

"?" means there are no definite data

germanium look rather simple, only a few of them were experimentally obtained and investigated. Among the silicide–germanides of interest are cubic $Mg_2Ge_xSi_{1-x}$ [1.6, 1.51] and $BaGe_xSi_{2-x}$ [1.10], but the last in the cubic structure behaves as a metal. The measured lattice parameters of the ternary silicide crystals vary linearly with the composition, showing good agreement with Vegard's law. This is demonstrated in Fig. 1.9 for $Mg_2Ge_xSi_{1-x}$.

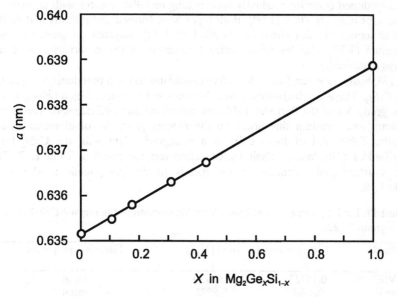

Fig. 1.9. Composition dependence of the lattice parameter of $Mg_2Ge_xSi_{1-x}$ [1.51]

In iron disilicide germanium can substitute not more than 5% of silicon in the •-$FeSi_2$ phase [1.54]. As for the •-phase, formation of thin film •-$FeSi_{1.92}Ge_{0.08}$ by reactive atomic deposition of iron onto SiGe layer was observed [1.63].

A favorable prediction can be made for osmium and ruthenium based semiconducting ternary silicides. The compounds Ru_2Si_3 and Ru_2Ge_3, Os_2Si_3 and Os_2Ge_3 are isostructural with the orthorhombic lattice (see Table 1.1). Thus, any substitution of silicon by germanium atoms into these silicides would not change the lattice type but only their lattice parameters. This could also be expected for orthorhombic $OsSi_2$ and $OsGe_2$. It is evident that the metal atoms, which form isolated squares in $OsSi_2$ [1.40], are arranged in isolated bands parallel to the b axis in $OsGe_2$. These bands are formed by repetitions of an almost square basic arrangement of metal atoms. The repeat distance along the band is the b axis (0.3094•nm) and the width of the band is 0.298•nm. In this tendency to concentrate the metal atoms, the next step is the arrangement of metal atoms in square-shaped isolated layers. The structure of $OsGe_2$ can be described as a stacking of pseudo-hexagonal nets parallel to the ab plane. The nets are alternately composed of Os–Ge structures and germanium atoms. This favors preservation of the orthorhombic lattice in $OsGe_xSi_{2-x}$.

Ternary semiconducting silicides formed by substitution of metal atoms have received more study. The parameters influencing the substitutional solid solubility are [1.49]: (i) the dimensions of the substituting atoms, (ii) its chemical potential, and (iii) the valency. The last appears to be of primary importance for semiconducting silicides distinguishing two principal groups of them. These are ternary silicides formed by isovalent metals and others. The first ones can be intrinsic semiconductors and the second group representatives demonstrate an impurity doped behavior gradually approaching metallic features with an increase of the substituted fraction [1.69]. Both types were formed and investigated on the basis of semiconducting chromium [1.50, 1.69–1.72], tungsten, rhenium, iron, and ruthenium [1.73] silicides while lattice parameters of theses ternaries were not always measured.

CrSi$_2$ basis. One can find at least five candidates to form pseudobinary systems with CrSi$_2$. These are disilicides of molybdenum and tungsten, which belong to the same group VI of the Periodic Table, as chromium and disilicides of vanadium, niobium, and tantalum that are from the nearest group V of elements in the Periodic Table. All of them exist in a hexagonal form with the CrSi$_2$ type, identified as C40, lattice. Their lattice parameters are listed in Table 1.17. The joint substitutional solubility of the metals in the compounds is shown in Table 1.18.

Table 1.17. Lattice parameters and basic electronic properties of hexagonal C40 disilicides (space group $P6_222$)

Silicide	Lattice parameters (nm) [1.13]		Electronic properties
	a	c	
VSi$_2$	0.45723	0.63730	metal
NbSi$_2$	0.4797	0.6592	metal
TaSi$_2$	0.47835	0.65698	metal
CrSi$_2$	0.44281	0.63691	E_g=0.35•eV (indirect gap)
MoSi$_2$	0.4596	0.6550	E_g≈0.07•eV (indirect gap)
WSi$_2$	0.4614	0.6414	E_g≈0.07•eV (indirect gap)

Table 1.18. Bulk substitutional solid solubility of the components in CrSi$_2$ structure type (C40) silicides [1.49]

System A-B-Si	Silicide solubility limit (mol%)	
	A in B	B in A
Cr-Mo-Si	46	32
Cr-W-Si	70	16
Cr-V-Si	100	100
Cr-Nb-Si	0	0
Cr-Ta-Si	40	12
Mo-W-Si	100	100
Mo-V-Si	65	4
Mo-Nb-Si	66	12
Mo-Ta-Si	56	16
W-V-Si	28	18
W-Nb-Si	30	15
W-Ta-Si	30	24

In these silicides, the metals may be freely substituted for each other over the mentioned compositional stability range. The only exception belongs to the Cr-Nb-Si system demonstrating complete immiscibility. For others there is no fundamental alteration of crystal structure, while the lattice parameters change smoothly with composition. Nevertheless, one should note that $MoSi_2$ and WSi_2 could have both low temperature stable hexagonal and high temperature stable tetragonal structures. In the hexagonal $C40$ structure they exist up to 550 and 650°C, respectively, possessing complete solubility in $CrSi_2$ below these temperatures and increasingly broad miscibility gap above them (for details see Sect. 1.2 and 1.3). As for the candidates from group V, vanadium was experimentally observed to have 100% substitutional solubility in $CrSi_2$ with a linear variation of the lattice parameters with the component concentration [1.50]. Both a and c lattice parameters increase as vanadium is added to $CrSi_2$ and the effect on c is much smaller [1.69]. The $Cr_{1-x}V_xSi_2$ silicide is of a pseudobinary nature. The tantalum solubility in $CrSi_2$ is limited by 12 mol %, and niobium is insoluble in it [1.49].

Ternary semiconducting n-type silicide $Cr_{1-x}Mn_xSi_2$ is found in [1.70–1.72] to be formed by substituting chromium atoms with manganese ones in $CrSi_2$. This single-phase solution with the hexagonal D_{6h}^4 symmetry of $CrSi_2$ is preserved within the composition variation in the range of $0 \leq x \leq 0.225$. In the range of $0.225 \leq x \leq 0.34$, the second phase with D_{2d}^{12} symmetry appears in addition to the D_{6h}^4 phase. At $x > 0.34$ the third phase with silicon was identified. Lattice parameters and unit cell volume as a function of composition of $Cr_{1-x}Mn_xSi_2$ are presented in Fig. 1.10. In the single phase region there is a linear decrease of the lattice parameters with increasing substituted fraction of the metal atoms. This agrees with Vegard's law accounting for those chromium atoms, which are substituted by smaller manganese ones. In the composition range $x > 0.225$ the lattice parameters of the hexagonal phase remain unchanged indicating that further substitution no longer takes place.

WSi_2 basis. There is only one example of the ternary silicide based on hexagonal WSi_2. This is $W_{1-x}Ti_xSi_2$ in which titanium is known to stabilize the hexagonal $C40$ structure [1.74–1.76]. In the single-phase thin films the lattice parameters are linearly increased with titanium content, as illustrated in Fig. 1.11. Scatter in the experimental data summarized in this plot, in particular in the low third component concentration region, seems to be attributed to accompanying impurities (oxygen, platinum).

β-$FeSi_2$ basis. The orthorhombic structure of the semiconducting β-$FeSi_2$ has been found to be preserved in $Fe_{1-x}Co_xSi_2$ for $x \leq 0.15$ [1.77, 1.78]. Meanwhile, phase separation in this ternary system was observed after long annealing, 750°C/750 h, already for $x \geq 0.1$ [1.79]. The stability range is extended to $x = 0.2$ in the layers synthesized by the component implantation into silicon with an intermediate annealing between iron and cobalt implantation [1.80]. No data related to the change of the lattice parameters were presented.

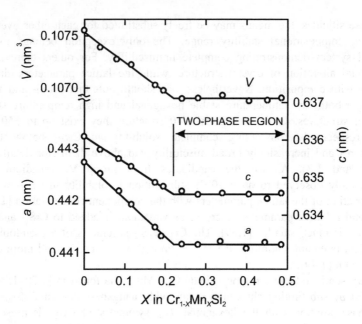

Fig. 1.10. Composition dependence of the lattice parameters and unit cell volume of hexagonal $Cr_{1-x}Mn_xSi_2$ [1.72]

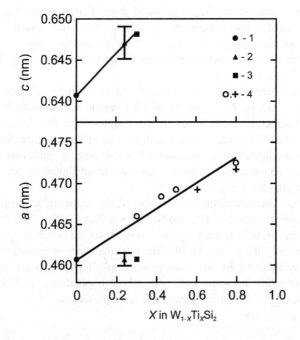

Fig. 1.11. Composition dependence of the lattice parameters of tetragonal $W_{1-x}Ti_xSi_2$: 1-[1.13], 2-[1.75], 3-[1.74], 4-[1.76]

Solid solution $Fe_{1-x}Mn_xSi_2$ with the orthorhombic structure of β-$FeSi_2$ exists at least within the composition range of $x \leq 0.1$ [1.81, 1.82]. It was observed in the bulk polycrystalline samples fabricated by powder sintering.

Ruthenium is another element tested to substitute for iron in orthorhombic β-$FeSi_2$. Single-phase β-$Fe_{1-x}Ru_xSi_2$ coexists in the composition range of $0 \leq x \leq 0.10$ [1.83]. The lattice parameters vary linearly with the ruthenium concentration. This is shown in Fig. 1.12. Ruthenium belongs to the same group of the Periodic Table as iron. Therefore, one could expect its complete substitutional solubility in β-$FeSi_2$, but in fact it is limited. The reason may be that this metal has no disilicide. Thus, it is impossible for the disilicide pseudobinary system to form. High ruthenium concentrations in β-$FeSi_2$ will inevitably bring phase separation.

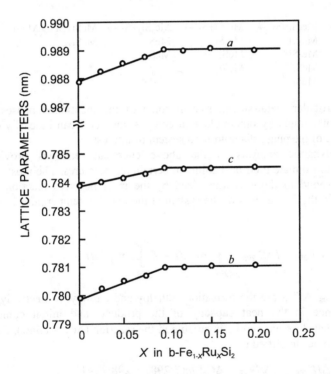

Fig. 1.12. Composition dependence of the lattice parameters of β-$Fe_{1-x}Ru_xSi_2$ [1.83]

There is one more candidate to form ternary compounds with the β-$FeSi_2$ structure. This is osmium, which belongs to the same group as iron and has a disilicide, $OsSi_2$, with identical lattice type. Ternary compounds in this couple indeed look realistic and attractive for energy gap engineering [1.84]. However, they have not been synthesized and investigated yet.

In general, ternary semiconducting silicides have been much less studied than binary ones. Nevertheless, the prospects for enhancement of the related properties are good.

1.2   Thermodynamics of Silicidation

Silicide formation can be, in general, described by the chemical reaction between metal (Me) and silicon

$$n\text{Me} + m\text{Si} = \text{Me}_n\text{Si}_m .\tag{1.2}$$

Except for direct reactions silicides may be produced by a chemical interaction of silicon or metal with already formed silicide phases. In the most practical cases of solid state silicidation in thin film structures these reactions tend to enrich the precursor phase with silicon. An extended scheme of the transformations looks like

$$
\begin{array}{llll}
\text{Me+Si} \rightarrow & \text{Me}_3\text{Si+Si} \rightarrow & \text{Me}_2\text{Si+Si} \rightarrow & \text{Me}_5\text{Si}_3\text{+Si} \rightarrow & \text{MeSi+Si} \rightarrow \text{MeSi}_2 . \\
& \text{Me}_2\text{Si} & \text{Me}_5\text{Si}_3 & \text{MeSi} & \text{MeSi}_2 \\
& \text{Me}_5\text{Si}_3 & \text{MeSi} & \text{MeSi}_2 & \\
& \text{MeSi} & \text{MeSi}_2 & & \\
& \text{MeSi}_2 & & &
\end{array}\tag{1.3}
$$

In a particular metal-silicon system some of the mentioned phases can be absent or substituted by others close in composition. This can be easily clarified with the use of the phase diagram for a system of interest.

Thermodynamic analysis of the above chemical reactions provides the possibility to estimate their probability at certain temperatures. The most probable ones correspond to those characterized by the most negative reaction induced difference in the free energy of the system at the reaction temperature T, which is [1.85]

$$\Delta G^{\circ}_T = \Delta H^{\circ}_{298} - T\Delta S^{\circ}_{298} + \int_{298}^{T} \Delta C_p dT - T \int_{298}^{T} \Delta C_p/T dT ,\tag{1.4}$$

where ΔH°_{298}, ΔS°_{298} are the formation enthalpy and entropy, respectively, ΔC_p is the difference in the heat capacity of the products and initial components. Assuming that difference to be invariable in the reaction ($\Delta C_p = const$), the above equation can be simplified to

$$\Delta G^{\circ}_T = \Delta H^{\circ}_{298} - T\Delta S^{\circ}_{298} - \Delta C_p[\ ln(T/298) + 298/T - 1]\ .\tag{1.5}$$

Handbook background is rather scanty for an appropriate choice of the thermodynamic parameters for semiconducting silicides. Therefore, theoretical approaches must be used for their calculations on the basis of the experimental data for pure elements. The most important parameters to be determined are compound heat capacities and enthalpies of formation. In silicide thermodynamics the heat capacity of a compound can be estimated on the basis of the Neumann–Kopp additivity rule [1.86]. This states that the change in the heat capacity resulting from the formation of a solid chemical compound from the solid elements is equal to zero, thus resulting in

$$C_p(Me_nSi_m) = n\,C_p(Me) + m\,C_p(Si)\,. \tag{1.6}$$

The heat capacity of silicon and most of metals are well known [1.87], so this rule is practically efficient to apply.

The analysis of theoretical efforts in intermetallic compound thermodynamics performed in [1.86] shows the way in which one can determine energies of silicide formation. One of the notable models was proposed by Miedema et al. [1.88]. It is based on the Wigner–Seitz cell approach to alloy theory. Its establishment was reported in [1.89] where the expression for estimating the enthalpy of formation of an intermetallic solution or compound was finally presented in the form

$$\Delta H = f(x)[-P(\Delta\phi^*)^2 + Q(\Delta n_{ws}{}^{1/3})^2 - Z]/(n_{ws}{}^{1/3})_{av}\,, \tag{1.7}$$

where P and Q are empirical constants applicable to specific groups of binary systems, $\Delta\phi^*$ represents the electronegativity difference, $\Delta n_{ws}{}^{1/3}$ is the difference in electron densities, Z represents the contribution due to p-d hybridization. The function $f(x)$ depicts the dependence of the contact area between dissimilar atoms in the ordered state on the component concentrations, and $(n_{ws}{}^{1/3})_{av}$ is the averaged electrostatic screening length. In the case of metal–metalloid compounds, an extra term is added to represent the enthalpy requirement for "metalizing" the latter element. For silicon this value is 33 kJ/mol. The model is semiempirical and the values of the various parameters are somewhat uncertain. Nevertheless, enthalpy of formation calculations using this approach have shown reasonable agreement with experimental data for a large number of the systems investigated.

In view of the obvious bearing of thermodynamics in proper engineering of a silicide formation technology, as well as in the simulation of their chemical and thermal stability, we will present below the experimentally observed peculiarities of the thermodynamic behavior and related available parameters for the metal-silicon systems producing semiconducting silicide phases. The most reliable data from [1.1, 1.86, 1.90] are shown and discussed.

1.2.1 Silicides of Group II Metals

Mg-Si system. The Mg-Si phase diagram presented in Fig. 1.13 shows that cubic Mg_2Si is the only stoichiometric silicide in this system. It melts congruently at 1085°C. There are two eutectics on both side of the compound. The lowest one contains 1.16 at % of silicon. It is formed at 637.6°C. The other eutectic appears at 945.6°C and includes 53 at % of silicon. Enthalpy of Mg_2Si formation is −80.02 kJ/mol [1.1] and $C_p = 73.353 + 14.989 \times 10^{-3}T - 8.834 \times 10^5 T^{-2}$ J mol^{-1} K^{-1} in the temperature range of 298–873 K [1.91].

Ba-Si system. Five stoichiometric silicides have been detected in the Ba-Si system: Ba_2Si, Ba_5Si_3, $BaSi$, Ba_3Si_4, and $BaSi_2$ [1.90]. Meanwhile, the available phase diagram of the system, shown in Fig. 1.14, definitely resolves only monosilicide BaSi and disilicide $BaSi_2$. The monosilicide has a melting point of 840°C and enthalpy of formation of −759.9 kJ/mol [1.91].

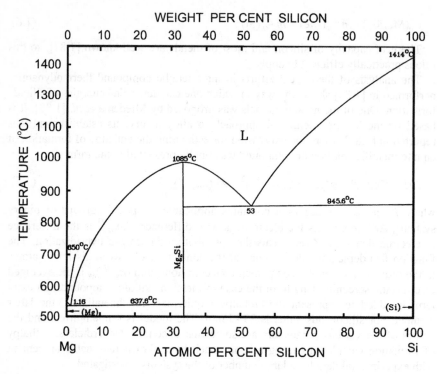

Fig. 1.13. Phase diagram of Mg-Si system [1.90]

The disilicide melts congruently at 1180°C. It can exist in three crystalline forms [1.31]. The stable one that is synthesized at normal pressure has the orthorhombic structure, and is a semiconductor. The metastable forms are cubic and trigonal. They are produced at a pressure up to 40×10^9 Pa and temperature of about 1000°C. At normal ambient pressure the metastable phases transform into the orthorhombic one [1.31, 1.92]. The trigonal–orthorhombic transition takes place in the temperature range 420–520°C and the cubic–orthorhombic transformation needs heat processing at 490–640°C. Enthalpies of the transformations are respectively −4.6 and −3.3 kJ/mol. Other thermodynamic characteristics of the system remain unknown.

1.2.2 Silicides of Group VI Metals

Cr-Si system. Four stable silicides can be found in the phase diagram of this system demonstrated in Fig. 1.15. They are: the cubic metal-rich phase Cr_3Si melting congruently at 1770°C, the tetragonal intermediate phase Cr_5Si_3 which also melts congruently at 1680°C but undergoes a polymorphic transformation at 1505°C, the cubic monosilicide CrSi decomposing peritectically at 1413°C and the disilicide $CrSi_2$ which melts congruently at 1490°C.

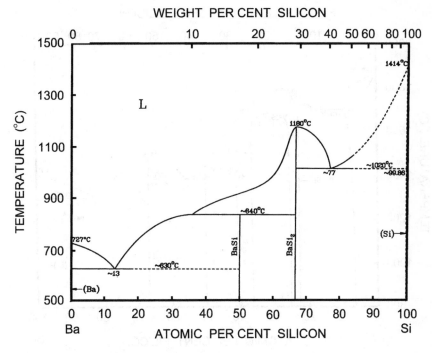

Fig. 1.14. Phase diagram of Ba-Si system [1.90]

Although the liquidus surface is not well defined, the Cr-Si phase diagram is, in general, a settled issue and is accompanied by a large body of thermodynamic research on the various related compounds. It is worthwhile to note that the phase diagram is rather flat around the $CrSi_2$ point. This compound exists over a range of compositions with the content of silicon atoms from about 66.0 to 67.5 at % [1.1]. It corresponds to $CrSi_{1.98}$–$CrSi_{2.02}$ compounds.

The relatively advanced state of knowledge on the thermodynamic properties of the system facilitates a choice of the adequate parameters [1.1, 1.86, 1.90]. They are summarized in Tables 1.19–1.21.

Table 1.19. Thermodynamic parameters of silicidation in Cr-Si system

Parameter	Cr	Cr_3Si	Cr_5Si_3	CrSi	$CrSi_2$
Melting point[a], T_m (°C)	1857	1770	1680	1413	1490
Enthalpy of melting[b], $-\Delta H_m$ (kJ/mol)			254.4		127.5
Entropy of melting[c], $-\Delta S_m$ (J mol^{-1} K^{-1})			132.8		73.8
Enthalpy of formation[c],					
$-\Delta H°_{298}$ (kJ/mol of atoms)		34.4	35.0	30.2	25.8
Entropy of formation[c], $-\Delta S°_{298}$(J mol^{-1} K^{-1})			2.14	0.09	2.06

[a] From [1.90], [b] From [1.1], [c] The values recommended in [1.86].

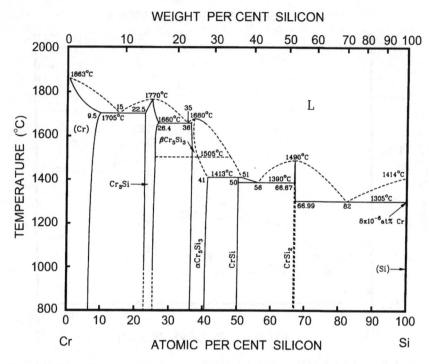

Fig. 1.15. Phase diagram of Cr-Si system [1.90]

Table 1.20. Temperature dependent thermodynamic parameters of the silicides in Cr-Si system [1.1]

Parameter	Temperature range (K)
Cr$_3$Si	
$C_p = 94.66+36.84\times10^{-3}T-2223494T^{-2}$ J mol^{-1} K^{-1}	293 – 873
Cr$_5$Si$_3$	
$C_p = 247.576+26.874\times10^{-3}T-106893T^{-2}$ J mol^{-1} K^{-1}	293 – 873
$H_T-H_{298} = -70000+198.64T+326.4\times10^{-4}T^2+25.6\times10^5T^{-1}$ J/mol	289 – 1300
$H_T-H_{298} = -236960+586.96T-2716.95\times10^{-4}T^2+75.12\times10^{-6}T^3$ J/mol	1300 – T_m
$H_T-H_{298} = -296000+508.0T$ J/mol	> T_m
CrSi	
$C_p = 50.86+14.31\times10^{-3}T-1610186T^{-2}$ J mol^{-1} K^{-1}	293 – 873
CrSi$_2$	
$C_p = 59.86+44.08\times10^{-3}T-1748199T^{-2}$ J mol^{-1} K^{-1}	293 – 873 K
$H_T-H_{298} = -23160+65.61T+11.25\times10^{-3}T^2+7.764\times10^5T^{-1}$ J/mol	< T_m
$H_T-H_{298} = -97500+90T$ J/mol	> T_m

Table 1.21. Thermodynamic values recommended in [1.86] for solid chromium silicides (per mol of atoms)

T (K)	H_T-H_{298} (J)			$S°_T$ (J/K)			$-\Delta H°_T$ (J)			$-\Delta G°_T$ (J)		
	Cr₅Si₃	CrSi	CrSi₂	Cr₅Si₃	CrSi	CrSi₂	Cr₅Si₃	CrSi	CrSi₂	Cr₅Si₃	CrSi	CrSi₂
298.15	0	0	0	19.69	21.14	22.49	34 950	30 206	25 800	34 311	30 179	26 413
331.5	311	771	287	20.71	23.59	23.42	34 942	30 198	25 802	34 282	30 176	26 446
400	2 475	2 444	2 275	26.81	28.17	29.03	34 820	30 074	25 792	34 110	30 186	26 604
500	5 075	5 029	4 670	32.61	33.94	34.37	34 663	29 902	25 774	33 950	30 234	26 782
600	7 790	7 729	7 183	37.53	38.86	39.95	34 515	29 736	25 750	33 822	30 316	26 953
700	10 597	10 512	9 496	41.88	43.15	42.98	34 390	29 593	25 721	33 716	30 424	27 118
800	13 485	13 364	12 500	45.74	46.95	46.53	34 300	29 482	25 686	33 627	30 551	27 277
900	16 447	16 276	15 288	49.22	50.39	49.87	34 342	29 407	25 645	33 546	30 690	27 290
1000	19 480	19 243	18 159	52.42	53.51	52.69	34 223	29 370	25 599	33 470	30 835	27 576
1100	22 581	22 263	21 110	55.37	56.39	55.70	34 243	29 376	25 547	33 394	30 981	27 717
1200	25 747	25 332	24 140	58.13	59.06	58.34	34 302	29 423	25 490	33 315	31 125	27 852
1300	29 007	28 450	27 248	60.72	61.56	60.83	34 403	29 514	25 428	33 229	31 264	27 981
1400	32 314	31 616	30 433	63.17	63.91	63.19	34 552	29 648	25 360	33 132	31 394	28 104
1500	35 728	34 828	33 695	65.52	66.12	65.44	34 694	29 827	25 287	33 027	31 512	28 221
1600	39 309	38 086	37 034	68.23		67.59	34 775	30 050	25 208	32 915	31 618	28 333
1687	42 603	40 958	40 000	69.84	69.97	69.40	34 751	30 280	25 135	32 817	21 697	28 426
1687	42 603	40 958	40 000	69.84	69.97	69.40	53 707	55 555	52 005	32 817	21 697	28 426
1700	43 112	41 390	40 449	70.14	70.23	69.66	53 675	55 572	51 696	32 656	31 512	28 244
1800	47 194			72.47			53 317			31 432		
1900	51 611			74.86			53 716			30 235		

Mo-Si system. There are three stoichiometric silicides in the phase diagram of the system (Fig. 1.16).

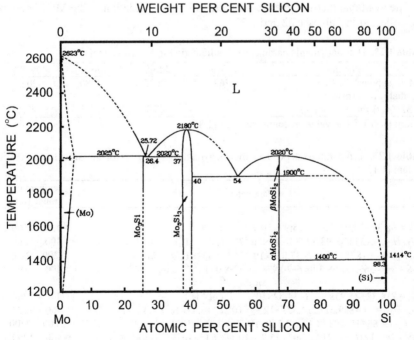

Fig. 1.16. Phase diagram of Mo-Si system [1.90]

They are: the cubic metal-rich phase Mo₃Si decomposing peritectically at 2035°C, the tetragonal intermediate phase Mo₅Si₃ which congruently melts at 2180°C and the disilicide MoSi₂ which can have tetragonal or hexagonal (C40) structure. The hexagonal $MoSi_2$ congruently melts at 2020°C. It is reported in [1.93] to undergo a polymorphic transformation into tetragonal $MoSi_2$ at 1900°C. It is a reversible process. However, experiments with alloyed silicon/molybdenum powder samples [1.94] show that the hexagonal $MoSi_2$ phase can be formed at 750°C. It is metastable and transforms into the tetragonal phase at 850°C.

In thin films the tetragonal disilicide phase with metallic properties is usually formed. The hexagonal one, which is of interest in this book as a semiconductor, is also produced in thin metal/silicon structures annealed at 500–690°C [1.95–1.97] and in ion implanted molybdenum-on-silicon ones [1.98]. Technological details will be discussed in Chap. 2, while the possibility to form hexagonal $MoSi_2$ is noted here. This phase nucleates at 300–400°C. The nucleation temperature has been observed to be the same for Si(100), Si(111), and amorphous silicon substrates [1.97]. In thin films the hexagonal phase is stable up to 675–690°C above, which the hexagonal-to-tetragonal transition takes place. The upper temperature limit is increased to 800°C when ion implantation was applied to mix the silicide forming components. Contaminants, mainly carbon, interfere with the formation of the hexagonal phase and retard the hexagonal-to-tetragonal transformation. Aluminum dopant can also be used to stabilize the hexagonal structure of $MoSi_2$ [1.94].

The available thermodynamic characteristics of silicidation in the Mo-Si system are collected in Tables 1.22 and 1.23.

Table 1.22. Thermodynamic parameters of silicidation in Mo-Si system

Parameter	Mo	Mo_3Si	Mo_5Si_3	$MoSi_2$
Melting point[a], T_m (°C)	2623	2025	2180	2020
Enthalpy of formation[b], $-\Delta H°_{298}$ (kJ/mol of atoms)		29.1	38.7	47.4

[a] From [1.90], [b] The values recommended in [1.86].

Table 1.23. Temperature dependent thermodynamic parameters of the silicides in Mo-Si system [1.1]

Parameter	Temperature range (K)
Mo_3Si	
$C_p = 98.37 + 20.97 \times 10^{-3}T - 15.99 \times 10^5 T^{-2}$ J mol^{-1} K^{-1}	400 – 1200
$H_T - H_{273} = -33450 + 98.37T + 10.46 \times 10^{-3}T^2 + 15.99 \times 10^5 T^{-1}$ J/mol	400 – 1200
$\Delta H^T_{298} = -61828 + 5.215T - 37.037 \times 10^{-3}T^2 + 18.951 \times 10^6 T^{-1}$ J/mol	1200 – 2200
$\Delta S^T_{273} = -567.33 + 98.37\ln T + 20.97 \times 10^{-3}T + 8.0 \times 10^5 T^{-2}$ J mol^{-1} K^{-1}	400 – 1200
Mo_5Si_3	
$C_p = 214.32 + 25.95 \times 10^{-3}T - 14.32 \times 10^5 T^{-2}$ J mol^{-1} K^{-1}	400 – 1200
$H_T - H_{273} = -64849.51 + 214.32T + 12.98 \times 10^{-3}T^2 + 3.42 \times 10^5 T^{-1}$ J/mol	400 – 1200
$\Delta H^T_{298} = 86445 - 203.064T - 13.558 \times 10^{-3}T^2 - 7.363 \times 10^6 T^{-1}$ J/mol	1200 – 2200
$\Delta S^T_{273} = -1216.91 + 214.32\ln T + 25.95 \times 10^{-3}T + 7.15 \times 10^5 T^{-2}$ J mol^{-1} K^{-1}	400 – 1200
$MoSi_2$	
$C_p = 67.52 + 15.53 \times 10^{-3}T - 7.41 \times 10^5 T^{-2}$ J mol^{-1} K^{-1}	400 – 1200
$H_T - H_{273} = -21725.34 + 67.52T + 7.79 \times 10^{-3}T^2 + 7.41 \times 10^5 T^{-1}$ J/mol	400 – 1200
$\Delta H^T_{298} = -15608 - 26.359T - 18.114 \times 10^{-3}T^2 + 7.471 \times 10^6 T^{-1}$ J/mol	1200 – 2200
$\Delta S^T_{273} = -387.29 + 67.52\ln T + 15.53 \times 10^{-3}T + 3.68 \times 10^5 T^{-2}$ J mol^{-1} K^{-1}	400 – 1200

Those related to the disilicide were critically evaluated and generally confirmed in [1.99].

W-Si system. Only two silicide compounds have been distinguished in this system as illustrated in Fig. 1.17. These are hexagonal W_5Si_3 melting at 2164°C and the disilicide WSi_2 with the melting point at about 2324°C. The apparent W_3Si phase, which can be expected in analogy with the Cr-Si and Mo-Si systems, has not been experimentally observed.

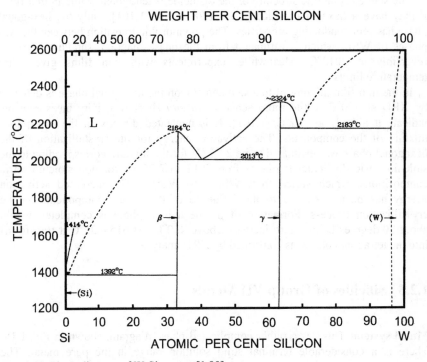

Fig. 1.17. Phase diagram of W-Si system [1.90]

The thermodynamic characteristics related to silicidation in the W-Si system are summarized in Tables 1.24 and 1.25.

Table 1.24. Thermodynamic parameters of silicidation in W-Si system

Parameter	W	W_5Si_3	WSi_2
Melting point[a], T_m (°C)	3410	2324	2164
Enthalpy of formation, $-\Delta H^o_{298}$ (kJ/mol of atoms)		23.5[b]	31.2[b], 26.5[c]

[a] From [1.90], [b] The most reasonable values from [1.86], [c] The value recommended in [1.99].

Table 1.25. Enthalpy of formation as a function of temperature for the silicides in W-Si system [1.1]

Silicide	Enthalpy of formation, ΔH^T_{298} (J/mol)	Temperature range (K)
W_5Si_3	$-11764-111.635T-37.94\times10^{-3}T^2+14.437\times10^6T^{-1}$	1200 – 2200
WSi_2	$-8074-34.294T-15.421\times10^{-3}T^2+5.864\times10^6T^{-1}$	1200 – 2200

The stable crystalline structure of the disilicide is tetragonal, while in thin films it may have a hexagonal lattice [1.34, 1.98, 1.100, 1.101]. Only the hexagonal phase has semiconducting properties. The polymorphic transition between the two phases of WSi_2, which is estimated from the phase diagram, should take place reversibly at 2013°C. Meanwhile, experiments with thin films give other temperature limits.

In the thin film mixture of tungsten and silicon the hexagonal phase is detected by XRD at 420°C, while more sensitive *in situ* sheet resistivity measurements indicate it already at 362°C [1.101]. It is nucleated directly in the amorphous mixture of the components. The activation energy for the crystallization of the hexagonal phase was estimated to be 2.05 eV for the mixture corresponding to the stoichiometric disilicide. It ranges from 2.13 to 2.25 eV for non stoichiometric compositions, which varied from $WSi_{1.79}$ to $WSi_{2.18}$. The increased activation energy has been attributed to the influence of the excess components on the crystallization process. Formation of the hexagonal phase is completed within about 20 degrees temperature increase above 420°C. At 615–635°C it transforms into the tetragonal phase, as confirmed by XRD analysis.

1.2.3 Silicides of Group VII Metals

Mn-Si system. This has a rather complicated phase diagram, shown in Fig. 1.18. There is a considerable terminal solid solution range in the pure metal. The increasing metal content in the stable silicides is apparent, as there are increasing degrees of instability and nonstoichiometry. One can find a family of silicides with a stoichiometry at about $MnSi_{1.75}$, which are: $Mn_{11}Si_{19}$ ($MnSi_{1.727}$), $Mn_{26}Si_{45}$ ($MnSi_{1.730}$), $Mn_{15}Si_{26}$ ($MnSi_{1.733}$), $Mn_{27}Si_{45}$ ($MnSi_{1.741}$), Mn_4Si_7 ($MnSi_{1.750}$). The last issue is of particular importance for us as the most frequently quoted semiconducting phase $Mn_{15}Si_{26}$ belongs to this family. Generally, three silicides, Mn_3Si, Mn_5Si_3, MnSi, and one or some representatives from the $MnSi_{1.75}$ family are considered as a result of silicidation in the Mn-Si system. The phase Mn_5Si_2 indicated in the phase diagram can exist but, in fact, is not stable in equilibrium systems [1.102]. The metal-rich silicide Mn_3Si peritectically forms at 1070°C. It has two polymorphs α-Mn_3Si and β-Mn_3Si. The transition between them occurs at 677°C. The intermediate phase Mn_5Si_3 is a compound with the highest melting point in the system. It melts congruently at 1300°C. Because the homogeneity range of the compound has not been reported, it is shown as a stoichiometric line compound. The monosilicide MnSi melts congruently at 1276°C.

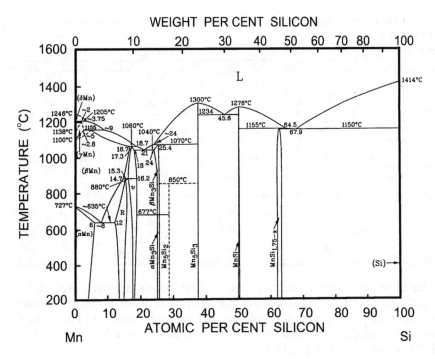

Fig. 1.18. Phase diagram of Mn-Si system [1.90]

The available thermodynamic characteristics of the system are collected in Tables 1.26–1.28.

Table 1.26. Thermodynamic parameters of silicidation in Mn-Si system

Parameter	Mn	Mn_3Si	Mn_5Si_3	MnSi	$Mn_{11}Si_{19}$
Melting point[a], T_m (°C)	1246	1040	1300	1276	1155
Enthalpy of formation[b], $-\Delta H^\circ_{298}$ (kJ/mol of atoms)		25.9	30.2	32.6[c]	35.9
Entropy of formation[c], $-\Delta S^\circ_{298}$ (J mol^{-1} K^{-1})		1.02	4.01		7.80

[a] From [1.90], [b] The values recommended in [1.86], [c] The most reasonable values from [1.86].

Table 1.27. Temperature dependence of heat capacity of manganese silicides [1.1]

Silicide	Heat capacity, C_p (J mol^{-1} K^{-1})	Temperature range (K)
Mn_3Si	$53.882 + 120.456 \times 10^{-3}T + 981701T^{-2}$	588 – 1241
Mn_5Si_3	$334.88 + 1071 \times 10^{-3}T$	743 – 1332
MnSi	$94.503 - 11.938 \times 10^{-3}T - 822130T^{-2}$	792 – 1418
$MnSi_{1.77}$	$38.736 + 35.889 \times 10^{-3}T - 4226078T^{-2}$	610 – 1402

Table 1.28. Thermodynamic values recommended in [1.86] for solid manganese silicides (per mol of atoms)

T (K)	$H_T - H_{298}$ (J)			$S°_T$ (J/K)			$-\Delta H°_T$ (J)			$-\Delta G°_T$ (J)		
	Mn₃Si	Mn₅Si₃	Mn₁₁Si₁₉	Mn₃Si	Mn₅Si₃	Mn₁₁Si₁₉	Mn₃Si	Mn₅Si₃	Mn₁₁Si₁₉	Mn₃Si	Mn₅Si₃	Mn₁₁Si₁₉
298.15	0	0	0	23.69	24.64	14.52	25 910	30 242	35 939	25 607	31 438	33 613
400	2 717	2 596	2 362	31.50	32.12	17.69	25 829	30 205	35 979	25 513	31 853	32 810
500	5 641	5 290	4 862	38.02	38.12	21.63	25 680	30 207	36 015	25 450	32 266	32 012
600	8 756	8 085	7 459	43.69	43.32	25.47	25 488	30 248	36 076	25 422	32 675	31 204
700	12 036	10 962	10 120	48.75	47.65	29.04	25 265	30 330	36 176	25 428	33 073	30 383
800	15 469	13 914	12 827	53.33	51.59	32.31	25 016	30 352	36 317	25 468	33 457	29 543
900	19 047	16 934	15 570	57.54	55.15	35.36	24 746	30 616	36 507	25 541	33 823	28 683
950	20 888			59.53			24 603			25 589		
950	22 288			61.01			23 204			25 589		
980	23 449	18 777	17 231	62.21	57.43	37.52	21 712	30 735	36 695	25 689	34 102	27 977
980	23 449	18 777	17 231	62.21	57.43	37.52	23 381	32 122	37 510	25 689	34 102	27 977
1000	24 223	20 019	18 344	63.00	58.40	38.05	23 302	32 213	37 572	25 735	34 141	27 782
1100	28 116	23 168	21 144	66.71	61.40	40.59	22 912	32 424	37 841	25 997	34 323	26 790
1200	32 087	26 279	23 969	70.16	64.49	42.95	22 484	32 613	38 124	26 296	34 487	25 773
1300	36 179	29 651	26 817	73.44	66.81	45.16	21 976	32 782	38 422	26 634	34 636	24 732
1360		31 643	28 536		68.31	46.41		32 873	38 609		34 720	24 096
1360		31 643	28 536		68.31	46.41		34 199	39 394		34 720	24 096
1400		32 984	29 686		69.08	47.23		34 356	39 581		34 729	23 643
1410			29 974			47.43			39 628			23 529
1410			29 974			47.43			40 316			23 529
1500		33 320	32 576		69.52	49.17		34 395	40 712		34 731	22 442

Re-Si system. Discrepancies in the available phase diagrams presented in [1.103] and [1.90] suggest inadequate characterization of the system. The earliest diagram from [1.103] indicates three stable compounds, which are Re₅Si₃, ReSi, and ReSi₂. On the other hand, Re₂Si, ReSi, and ReSi₁.₈ were later mentioned in [1.90], as illustrated in Fig. 1.19.

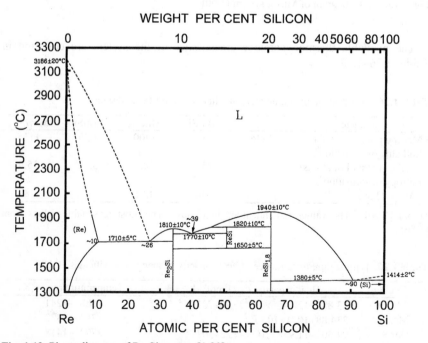

Fig. 1.19. Phase diagram of Re-Si system [1.90]

It is apparent that, while being characterized with different melting and eutectic points, both diagrams have very flat solidus–liquidus curves in the regions of the compounds under doubt. This relates to a choice between the alternatives Re_2Si or Re_5Si_3 and $ReSi_2$ or $ReSi_{1.8}$ (or $ReSi_{1.75}$). The deviation in the composition may be attributed to a possible experimental uncertainty. In the latest investigations of the rhenium silicides [1.14, 1.104] it was found that among the silicon-rich phases the $ReSi_{1.75}$ compound is the only stable one. It has well pronounced semiconducting properties. Thus, the thermodynamic characteristics previously obtained and published for the apparent disilicide $ReSi_2$ can be projected to the real phase $ReSi_{1.75}$. The data available are listed in Tables 1.29 and 1.30.

Table 1.29. Thermodynamic parameters of silicidation in Re-Si system

Parameter	Re	Re_2Si	Re_5Si_3	ReSi	$ReSi_2$
Melting point, T_m (°C)					
from [1.103]:	–	–	1810	1880	1940
from [1.90]:	3186	1810	–	1820	1940
Enthalpy of formation[a],					
$-\Delta H°_{298}$ (kJ/mol of atoms)			19.7	26.4	30.1

[a] The most reasonable values from [1.86].

Table 1.30. Formation energy characteristics for silicides in Re-Si system [1.1]

Parameter	Temperature range (K)
Re_3Si	
$\Delta G°_T = 94.20 - 23.03T$ J/mol	1700 – 2000
ReSi	
$\Delta G°_T = 110.4 - 3.94T$ J/mol	1700 – 2000
$ReSi_2$	
$\Delta G°_T = 230.3 - 3.09T$ J/mol	1700 – 2000
$\Delta H^T_{298} = 6119 + 33.391T + 16.574 \times 10^{-3}T^2 - 5.232 \times 10^6 T^{-1}$ J/mol	1200 – 1900

1.2.4 Silicides of Group VIII Metals

Fe-Si system. The latest phase diagram of the system is presented in Fig. 1.20. It has numerous specific phases in both metal-rich and silicon-rich regions, which definitely prove the complexity of the system. Some of the melting points and transformation temperatures are not entirely established. This is illustrated by the scatter of the experimental data obtained by different research groups [1.1, 1.90]. Nevertheless, four compounds are thought to form in the system. They are the metal-rich silicide Fe_3Si, the intermediate phase Fe_5Si_3, the monosilicide FeSi, and the disilicide $FeSi_2$. Tables 1.31 and 1.32 summarize their most reliable thermodynamic characteristics.

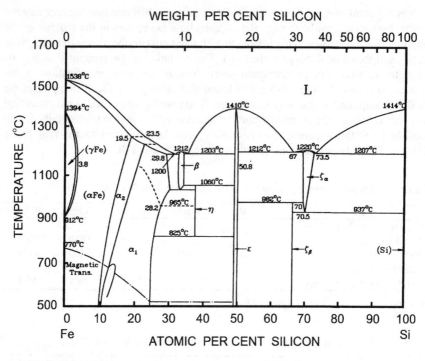

Fig. 1.20. Phase diagram of Fe-Si system [1.90]

Table 1.31. Thermodynamic parameters of silicidation in Fe-Si system

Parameter	Fe	Fe₃Si	Fe₅Si₃	FeSi	FeSi₂
Melting point[a], T_m (°C)	1538	1240	1203	1410	1212
Enthalpy of melting[b], $-\Delta H°_{298}$ (kJ/mol)		57.01		70.41	88.70
Enthalpy of formation[c], $-\Delta H°_{298}$ (kJ/mol of atoms)		25.8	30.6[d]	39.3	30.6

[a] From [1.90], [b] From [1.1], [c] The values recommended in [1.86], [d] From [1.91].

In the region of the disilicide composition at least three phases are distinguished [1.1, 1.15]. These were already mentioned in Sect. 1.1: tetragonal α-FeSi₂, orthorhombic β-FeSi₂, and cubic γ-FeSi₂. The α-phase is a high temperature form of the compound. It is stable in the temperature range from about 950°C up to the melting point. It has vacancies in the positions of metal atoms.

Table 1.32. Temperature dependent thermodynamic parameters of the silicides in Fe-Si system [1.1]

Parameter	Temperature range (K)
Fe₃Si	
$C_p = 71.33 + 87.44 \times 10^{-3}T + 0.146 \times 10^5 T^{-2}$ J mol^{-1} K^{-1}	273 – 800
$C_p = 47.18 + 103.31 \times 10^{-3}T + 38.9 \times 10^5 T^{-2}$ J mol^{-1} K^{-1}	850 – 1520
$C_p = 144.92$ J mol^{-1} K^{-1}	1540 – 1800
$\Delta H^T_{273} = 22688 + 71.31T + 43.66 \times 10^{-3}T^2 - 0.1465 \times 10^5 T^{-1}$ J/mol	273 – 800
$\Delta H^T_{273} = -24.79 + 47.18T + 51.66 \times 10^{-3}T^2 - 38.9 \times 10^5 T^{-1}$ J/mol	850 – 1520
$\Delta H^T_{273} = 34936 + 144.9T$ J/mol	1540 – 1800
Fe₅Si₃	
$C_p = 176.38 + 88.81 \times 10^{-3}T - 4.077 \times 10^5 T^{-2}$ J mol^{-1} K^{-1}	273 – 1350
$\Delta H^T_{273} = -69052 + 176.4T - 44.41 \times 10^{-3}T^2 + 4.077 \times 10^5 T^{-1}$ J/mol	273 – 1350
$S = 825.14 + 88.81 \times 10^{-3}T + 2.04 \times 10^5 T^{-2} + 405.54 \lg T$ J mol^{-1} K^{-1}	273 – 1350
FeSi	
$C_p = 44.70 + 15.03 \times 10^{-3}T - 0.816 \times 10^5 T^{-2}$ J mol^{-1} K^{-1}	273 – 1923
$\Delta H^T_{273} = 1300.46 + 44.7T + 7.07 \times 10^{-3}T^2 + 0.816 \times 10^5 T^{-1}$ J/mol	273 – 1923
$S = 210.51 + 15.03 \times 10^{-3}T + 0.406 \times 10^5 T^{-2} + 102.091 \lg T$ J mol^{-1} K^{-1}	273 – 1923
α-FeSi₂	
$C_p = 34.74 + 65.95 \times 10^{-3}T + 71.41 \times 10^{-4}T^2$ J mol^{-1} K^{-1}	> 1200
$\Delta H^T_{273} = -9899.9 + 34.74T + 32.973 \times 10^{-3}T^2 - 7.141 \times 10^5 T^{-1}$ J/mol	> 1200
β-FeSi₂	
$C_p = 60.70 + 17.2 \times 10^{-3}T$ J mol^{-1} K^{-1}	< 1233
$\Delta H^T_{273} = -142.99 + 60.7T + 8.602 \times 10^{-3}T^2$ J/mol	< 1233
$S = 296.58 + 17.2 \times 10^{-3}T + 140.23 \lg T$ J mol^{-1} K^{-1}	< 1233

Below 950°C it transforms into the β-phase according to the eutectoid reaction [1.1]

$$\alpha\text{-FeSi}_2 \rightarrow \beta\text{-FeSi}_2 + \text{Si}. \tag{1.8}$$

While the temperature of this transition determined by different researchers [1.1, 1.105] is not the same, it definitely belongs in the range 915–960°C. The β-phase is stoichiometric and transforms into the α-phase above 970°C [1.16] as follows

$$\beta\text{-FeSi}_2 \rightarrow \alpha\text{-FeSi}_2 + \text{FeSi}. \tag{1.9}$$

The cubic γ-FeSi₂ is a metastable low temperature phase appearing during polymorphic transformation of β-FeSi₂ at 650°C [1.1]. Among the above disilicide phases only β-FeSi₂ has semiconducting properties.

Formation of β-FeSi₂ phase is sensitive to the presence of impurities in the system [1.106, 1.107]. Addition of copper (0.1–1 at %) increases the rate of the eutectoid decomposition (α → β + Si) and the initial stage of the peritectoid reaction (α + ε → β). The effect of copper depends on the annealing temperature. Thus, the decomposition rate in the doped samples was observed to be fifteen times higher at 800°C with respect to an undoped one and it exceeded by more than thirty times at 600°C.

Ru-Si system. This is poorly investigated. Its phase diagram is presented in Fig.1.21 in comparison with the early published data [1.1] and definitely proves the existence of the monosilicide RuSi and the silicon-richest phase Ru_2Si_3. Both of them have polymorphs but exact transformation temperatures remain unknown. Below 1000°C Ru_2Si_3 crystallizes in an orthorhombic lattice (space group *Pbcn*) [1.42]. It is a semiconducting phase. At higher temperatures, it undergoes a phase transition into a tetragonal structure. Among metal-rich and intermediate phases Ru_2Si, Ru_5Si_3, Ru_4Si_3 were reported to form while details of the phase diagram in this region have to be analyzed more carefully. Only the enthalpy of formation for RuSi can be find in the literature. The value recommended in [1.86] is $\Delta H^\circ_{298} = 58.1$ kJ/mol of atoms.

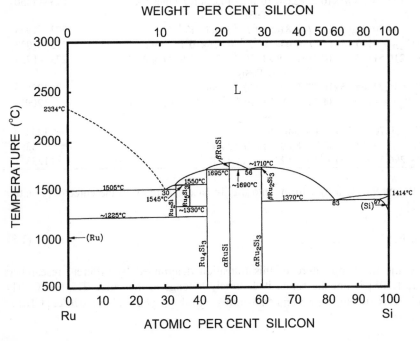

Fig. 1.21. Phase diagram of Ru-Si system [1.90]

Os-Si system. The phase diagram of the system (Fig. 1.22) is the only data collection for its thermodynamic characterization. Three stable phases, i.e. the monosilicide OsSi, the silicon-rich phase Os_2Si_3, and the disilicide $OsSi_2$ with the orthorhombic structure are observed. Moreover, metastable monoclinic $OsSi_2$ was detected on both sides of the compound concentration in the phase diagram [1.18]. It is obtained in the form of dendrites by quenching of the liquid state. Aluminum addition stabilizes the monoclinic disilicide.

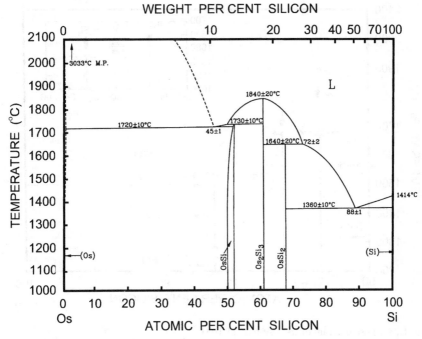

Fig. 1.22. Phase diagram of Os-Si system [1.90]

Ir-Si system. There are as many as six or even more silicides formed in the system. Unfortunately, the equilibrium phase diagram has not yet been precisely determined. Looking through the published data one can find the following phases reported: Ir_3Si, Ir_2Si, Ir_3Si_2, $IrSi$, Ir_4Si_5, Ir_4Si_7, Ir_3Si_5, and $IrSi_{-3}$.

Among the metal-rich silicides, Ir_3Si and Ir_2Si have been confidently detected [1.1, 1.2]. Contrary to the metal-rich compounds, the silicon-rich part of the iridium-silicon phase diagram is better studied, and is demonstrated in Fig. 1.23. The monosilicide IrSi seems to be the best investigated from the thermodynamic point of view. It has a melting point at 1980°C [1.48] and enthalpy of formation $\Delta H^{\circ}_{298} = -63.8$ kJ/mol of atoms [1.86].

The compound with the composition Ir_4Si_7 was suggested in [1.45, 1.47, 1.109] to be a semiconductor. Later studies performed by Engström et al. [1.21], Allevato and Vining [1.48] identified the correct composition of this compound to be Ir_3Si_5. It forms peritectically at 1402°C.

The most silicon-rich phase $IrSi_{-3}$ contains about 72.5% of silicon [1.48]. It has a peritectic decomposition temperature of 1460°C and transforms from a high temperature orthorhombic structure to a low temperature monoclinic one at about 975°C. These two forms were reported in [1.48] to exhibit semi metallic or heavily doped semiconducting behavior. Detailed investigations of the polymorphs are important to be sure whether both or at least one of them really have semiconducting properties.

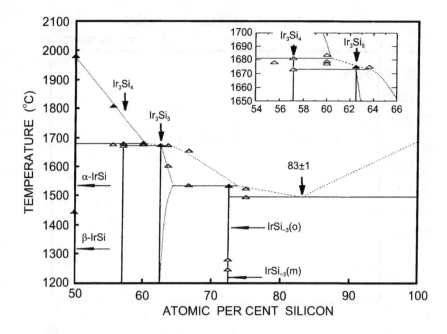

Fig. 1.23. Phase diagram of Ir-Si system [1.108]

1.2.5 Ternary Silicides

The thermodynamics of semiconducting ternary silicide formation have been poorly studied. For metal-silicon-germanium systems the available phase diagrams show total miscibility for magnesium [1.6] and immiscibility for chromium, manganese, and iron [1.54, 1.110]. Since silicon and germanium are extremely close chemically, the crystalline structure of the binary silicide and corresponding germanide is important for their solubility. The compounds should be mutually soluble if they crystallize in the same structure and if the unit cell volume difference is below the limit of 15% given by the Hume–Rothery rule for the extended solubility between elements [1.54]. Real possibilities to synthesize semiconducting metal-silicon-germanium ternaries are evident from the summary of the crystalline structure of the candidates given in Table 1.33.

Regularities of silicidation into two metals (Me1 and Me2) and silicon contained systems have received more study. Several factors are known [1.49] to influence the formation of such ternaries. First, the free energy of formation, ΔG_f, of the ternary silicon richest phases has to be lower than that of binary disilicides. Only in a few cases does a tie line exist between the compound and silicon, indicating a high thermal stability related to $\Delta G_f(\text{tern}) > \Delta G_f(\text{Me1Si}_2)$ and $\Delta G_f(\text{tern}) > \Delta G_f(\text{Me2Si}_2)$.

Table 1.33. Phase separation upon silicidation in alloyed (Me1+Me2) and bilayer (Me1/Me2) metal films on silicon

As-deposited structure	The phases detected upon annealing at 800°C for 600 min	Refs
Cr+Pt/Si	Cr+Pt/PtSi/Si[a]	1.111
Cr+Pt/Si	$CrSi_2$/PtSi/Si	1.112
Cr+Pd/Si	$CrSi_2$/Pd_2Si/Si	1.112
Cr+Ni/Si	$CrSi_2$/$NiSi_2$/Si	1.112
Cr+Ta/Si	$CrSi_2$+$TaSi_2$/ $CrSi_2$ /Si	1.113
Cr/Ta/Si	$CrSi_2$/ $TaSi_2$/Si	1.113
Ta/Cr/Si	$TaSi_2$/$CrSi_2$/Si	1.113
W+Pt/Si	WSi_2/PtSi/Si	1.112
W+Pd/Si	WSi_2/Pd_2Si/Si	1.112

[a] Annealing temperature was 400°C.

Secondly, if one of the metals, let it be Me1, starts the silicidation process, the equilibrium at the interface will be moving along the Me1-Si line in the phase diagram until the richest in silicon phase is obtained. Even if some unreacted metal Me2 is left, it will react with silicon before it can react with the silicide to form a ternary. Ternary compound growth is more likely when two metals react first. Once the intermetallic phase is formed, silicon can move into this phase (with a free energy of formation higher than the ΔG of both the first nucleated phases in the Me1-Si and Me2-Si systems) to cause its transformation to a ternary silicide either by a diffusion- or by a nucleation-controlled process. Silicon might also react with an excess metal to reach an equilibrium between ternary, binary, and silicon phases. Formation of a ternary silicide is often followed by the appearance of an intermetallic phase or/and phase separation, which are indeed pronounced in thin film structures.

Phase separation inevitably occurs in the systems composed of metals with significantly different silicidation temperatures. Their examples obtained in thin films on silicon are collected in Table 1.33. In the reaction of a metal alloy with a silicon substrate silicides with the lowest formation temperature are synthesized first at the interface. These typically are PtSi and Pd_2Si.

Restricted substitutional solubility of metals can also result in a phase separation in pseudobinary systems. For hexagonal $C40$ disilicides solubility limits are presented in Table 1.18.

Only Cr-V-Si and Mo-W-Si systems have complete miscibility of the components in their disilicides. A limited substitutional metal solubility for others should be accounted for in attempts at ternary silicide formation. Some useful details can be extracted from the data presented below.

In the W-Ti-Si system ternary compounds $W_{1-x}Ti_xSi_2$ with the hexagonal $C40$ structure can be produced in thin films with titanium substituting fraction x ranging from 0 to 0.8 [1.74–1.76]. Existence of one (hexagonal) or two (hexagonal and tetragonal) fractions in the system is a function of formation temperature and composition [1.76]. The phase diagram distinguishing the single phase and two-phase regions is presented in Fig. 1.24. It obeys all the characteristics of a true equilibrium diagram: change in crystalline parameter in the one region, coexistence of two phases in the right proportions in the two-phase region, and reversibility. The form of the diagram also implies that the hexagonal-to-tetragonal transition of WSi_2 is an allotropic transformation and that the hexagonal low temperature form of this disilicide is stable.

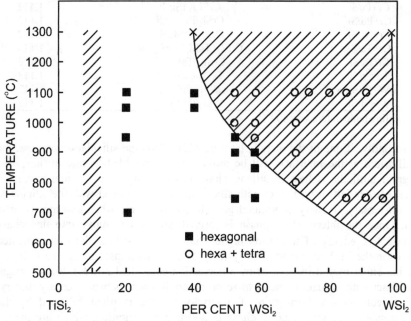

Fig. 1.24. Phase diagram for the pseudobinary system WSi_2-$TiSi_2$ [1.76]

Availing the fact that hexagonal WSi_2 is a very narrow-gap semiconductor, low titanium content $W_{1-x}Ti_xSi_2$ compounds seem to be of practical interest. Thin film single phase hexagonal $W_{0.7}Ti_{0.3}Si_2$ and $W_{0.76}Ti_{0.24}Si_2$ are shown to be formed at 700–750°C [1.74, 1.75]. They preserve only their hexagonal structure up to 900°C. The tetragonal phase appears only upon processing at higher temperatures. It considerably exceeds the stability limit of pure hexagonal WSi_2, which is not higher than 635°C. This can be attributed to the titanium addition but also to oxygen, which was registered at amounts up to some atomic per cents at the interfaces between the ternary films and substrates. An increased thermal stability of $W_{1-x}Ti_xSi_2$ due to titanium content can be explained by a very limited solubility of $TiSi_2$ in the tetragonal WSi_2.

Ternary silicides $Fe_{1-x}Co_xSi_2$ preserve the orthorhombic structure of β-$FeSi_2$ for $x \leq 0.15$ [1.77, 1.114]. It is formed in the thin film Co+Fe+Si (10 nm) structure on a $CoSi_2$ template layer on Si(111) during annealing at 650°C [1.77] or directly on a Si(111) substrate from a Fe-Co alloy film annealed at 700–900°C [1.114]. At a higher cobalt content ($x > 0.15$) the ternary phase exhibits a cubic structure with a partial CaF_2 long-range order.

The thermal stability of ternary silicides is usually increased with respect to the original ones. The stability range of semiconducting β-Fe$_{1-x}$Ru$_x$Si$_2$ is extended to the higher temperature region by increasing the ruthenium content [1.83]. This phase remains single and unchanged up to 1050°C at a composition of $x = 0.1$, while basic β-FeSi$_2$ is stable only below 960°C. In general, the valence electron concentration and atomic radius ratio of the constituent elements are important factors determining the structural stability of alloys and intermetallic compounds [1.115, 1.116]. The enhancement of atomic radius ratio r_{Me}/r_{Si} with partial substitution of host metal atoms by a similar number of valence electrons but larger ones seems to stabilize the structure, as takes place with ruthenium in β-Fe$_{1-x}$Ru$_x$Si$_2$. In this way, one can expect further increase of thermal stability of the β-phase with an osmium additive the atoms of which are much larger than the ruthenium ones. Experiments with substitution of iron in β-FeSi$_2$ by cobalt, which is similar in size to the host, show the decrease of the β-to-α transition temperature to 925°C for $x = 0.1$ and to 875°C for $x = 0.2$ [1.80]. This demonstrates that other properties of a ternary system have to be accounted for in an accurate prediction of crystalline structure stability.

1.3   Kinetics of Silicidation

Solid state reactions producing silicides are usually the result of heat processing of a metal-silicon couple. In thin film structures they start at the phase interfaces. In the structures fabricated by layer-by-layer deposition these interfaces are formed between pure metal and silicon films as well as between a metal film and silicon substrate. The latter is a more frequently realized case. In the structures produced by simultaneous metal and silicon deposition such interfaces appear between agglomerates or grains of the virgin components inside the film. From a thermodynamic point of view, any thin film deposited at room or lower temperatures is far from equilibrium. Heating of the structure allows atoms of the components to diffuse and chemically react with each other in order to lower the free energy of the system by formation of intermetallic compounds. Thus, nucleation of particular phases and their subsequent growth determine the silicidation kinetics in a metal-silicon couple.

Due to the fact that a typical metal-silicon phase diagram includes several stable silicides, the sequence of phase formation appears to be an intriguing issue. There are two principal approaches predicting the first phase to be nucleated in silicide forming systems. The first, described in [1.117–1.119] is based on classical thermodynamics. The other, developed in [1.120–1.122] is mostly empirical while corresponding equilibrium phase diagrams are referenced.

According to the classical thermodynamic nucleation theory [1.117, 1.118] formation of a nucleus is driven by a lowering of the energy of the system ΔG. It is opposed by the increase of the interfacial area with a specific surface energy σ^*. With ΔG calculated per unit volume of a silicide phase, a nucleus of average radius r will provide a difference in the free energy

$$\Delta G_n = r^2 \sigma^* - r^3 \Delta G. \tag{1.10}$$

The free energy difference of a nucleus has its maximum value for an atomic cluster of the critical radius r^*. This corresponds to

$$r^* = 2\sigma^*/3\Delta G. \tag{1.11}$$

The population of nuclei with a size smaller than r^* will exist in equilibrium distribution whereas larger nuclei will tend to grow. The rate of nucleation v_n is proportional to the concentration of critical nuclei and to the rate at which these nuclei originate. It can be described as follows

$$v_n \sim \exp(-\Delta G^*/k_B T)\exp(-E_a/k_B T), \tag{1.12}$$

where ΔG^* is the free energy of the critical nuclei, k_B is the Boltzmann constant, T the absolute temperature. The term $\exp(-E_a/k_B T)$ represents atomic interdiffusion characterized by the activation energy E_a. Since ΔG^* is inversely proportional to T^2, the rate of nucleation will vary as $\exp(-1/T^3)$. Thus, it is evident that nucleation of any particular phase takes place within narrow temperature limits, below which nothing occurs and above which the reaction is extremely fast.

Practical application of the above approach is limited by the fact that values of the parameters needed for the calculation do not have sufficient precision or are even unknown for the silicide systems of interest. Another more practically adopted thermodynamic procedure was proposed by Hyatt and Northwood [1.119]. This is based on the analysis of an entropy change associated with the formation of a compound from starting components given by the respective phase diagram.

The entropy change of a system under isothermal conditions is given by

$$\Delta S = \Delta S_{mix} + \Delta S_{trans}, \tag{1.13}$$

where ΔS_{mix} is the change in mixing entropy and ΔS_{trans} is the entropy change due to a phase transformation. In a silicide forming system, the mixing contribution to the entropy change is a complex function of the initial and final phase composition of the system as soon as the reactions schematically represented by (1.3) have to be selected and balanced. The entropy of transformation from phase 1 of the composition X_{1A} to the phase 2 of the composition X_{2A} is

$$\Delta S_{trans} = X_{2A}(S_{2A} - S_{1A}) + X_{2B}(S_{2B} - S_{1B}), \tag{1.14}$$

where X_{ki} and S_{ki} denote the mole fraction and excess entropy of the i-th component in the k-th phase, respectively. There is proportionality between the magnitude of ΔS and the sequence of phases observed in any binary system. This suggests an increasing nucleation probability of each of the phases from the bounding compositions.

The adequacy of predictions based on classical thermodynamic approaches to a great extent depends on the accuracy of the parameter values used in the calculations. Some corrections should be made to account for the high interfacial energy and large misfit strains associated with the phase transformation process. This is more or less overcome in the empirical rules predicting the first silicide phase to appear in a metal-silicon system by taking into consideration experimentally observed peculiarities.

Concerning conventional isothermal processing several attempts were made to determine general regularities predicting the first silicide phase to nucleate. However, only empirical rules have been generated. One such attempt, proposed by Walser and Bené [1.123], suggests that such a phase is a silicide with the highest melting point near the lowest eutectic temperature on the binary phase diagram. Another rule, formulated by Pretorius [1.121], employs as a criterion the compound formation heat. It postulates that the first nucleating phase is a congruently melting silicide possessing the highest negative effective formation heat when component concentrations are close to the point of the lowest eutectic temperature. The kinetic approach, developed by Bené [1.124], predicts the phase having the highest product $D\Delta G$ to be the first to nucleate and grow. Here D is the effective diffusivity of the most mobile atoms in the growing silicide and ΔG is the free energy difference related to the nucleating phase and the initial state of the system. This reflects the statement that nucleation of a new compound at an interface between compositionally dissimilar phases must be preceded by interdiffusion [1.125]. The diffusion process controls formation of critically sized clusters. This is indeed important in the transient regime, which precedes steady state nucleation. The related transient time, dependent on the slowest component diffusivity, provides a useful time scale for comparing interdiffusion and nucleation processes. It is important to account for this, particularly in the cases when rapid thermal processing is employed. However, there is no universal rule among the above mentioned, experimentally observed exceptions still exist.

A summary of predicted and experimentally observed first silicide phases for the metal-silicon systems providing formation of silicides with semiconducting properties is given in Table 1.34. Theoretical estimation of the formation temperature T_f is performed according to $T_f = 0.4T_m$ [1.126], where T_m is the melting point presented in the Tables of Section 1.2. The T_f corresponds to the parabolic rate of formation of about 10^{-14}–10^{-13} cm²/s.

One should be careful analyzing experimentally observed first nucleated phases in general and these are also listed in Table 1.34. The point is that most experiments performed ignored the presence of the thin silicon dioxide layer between the silicon and metal [1.144]. This layer can be instrumental in inhibiting metal-silicon interaction and, therefore, in modifying the intermetallic formation. Thus, the first reported nucleated phase might not be the one that would have formed in the absence of the interfacial oxide. The identical effect can be expected from the oxygen and other impurities introduced into the metal film during deposition. Obviously, only well controlled experiments in which the silicon surface was *in situ* cleaned before metal deposition, and the deposition was performed from a high purity source in high vacuum, seem to be really adequate.

As the theoretical approaches discussed are based on equilibrium thermodynamics and related phase diagrams, exceptions are inevitable.

Table 1.34. Predicted and experimentally observed first silicide phases and related formation temperatures

System	First nucleating phase		Formation temperature, T_f (°C)	
	predicted[a]	observed	estimated	observed
Mg-Si	Mg$_2$Si	Mg$_2$Si	434	200 [1.127]
Ba-Si	BaSi		336	
Cr-Si	CrSi$_2$ CrSi$_2$ [1.119, 1.121]	CrSi$_2$ [1.123, 1.128–1.132]	596	450 [1.128–1.130] 425 [1.131] 350 [1.132]
Mo-Si	MoSi$_2$	MoSi$_2$ [1.123, 1.128] h-MoSi$_2$ [1.97] Mo$_3$Si [1.133]	872 810	~400 [1.97] 400 [1.133]
W-Si	WSi$_2$	WSi$_2$ [1.123, 1.128]	930	650 [1.128] ~360 [1.101]
Mn-Si	Mn$_5$Si$_3$ MnSi [1.119, 1.121]	MnSi [1.128, 1.134]	520 510	400 [1.128, 1.134]
Re-Si	ReSi$_2$	ReSi$_2$ [1.36]	776	600 [1.36]
Fe-Si	FeSi$_2$ FeSi [1.121]	FeSi [1.123, 1.128, 1.135–1.139]	485 564	450 [1.128] 400 [1.135, 1.136, 1.138] 370 [1.137] 325 [1.139]
Ru-Si	Ru$_2$Si$_3$	Ru$_2$Si$_3$ [1.128, 1.140]	684	375 [1.140]
Os-Si	OsSi$_2$	Os$_2$Si$_3$ [1.128, 1.140]	656 736	600 [1.140] 500[b] [1.140]
Ir-Si		IrSi [1.128, 1.141, 1.142]	683	~ 500 [1.141] 400 [1.128, 1.143] 300 [1.142]

[a] Predicted according to the Walser and Bené rule [1.123] unless otherwise stated, [b] In the thin film Os/Ru/Si structure.

The list seems to be extended when thermodynamic conditions are far from equilibrium how, for example, it takes place during rapid thermal processing [1.4] or when the first formed phase is determined by concentration of atoms at the growth interface available to participate in the reaction how it takes place during reactive atomic deposition [1.145]. In any case, the initial phase sequence is important at the beginning of silicidation when its kinetics are controlled by the rate of chemical reactions. The reaction-controlled silicidation will always change with time to diffusion-controlled kinetics as soon as a silicide layer has grown sufficiently . The "sufficient" thickness is a function of the processing temperature and chemical peculiarities of a particular metal-silicon system. It varies from a monolayer thickness to some hundred of nanometers. In the first case it is certain that diffusion-controlled silicidation is the dominant mechanism. This corresponds to a parabolic law of growth in the form

$$d^2 = U t, \tag{1.15}$$

where d is the thickness of a silicide layer, U the parabolic growth rate, t the time of isothermal processing. For the reaction-controlled silicidation a linear law of the growth is valid

$$d = V t, \tag{1.16}$$

where V is the linear growth rate. Both growth rates are temperature-dependent. They are usually represented by an Arrhenius-type approximation with the pre-exponential factor U_0 or V_0 and the activation energy E_a

$$U = U_0 \exp(-E_a/k_B T) \quad \text{and} \quad V = V_0 \exp(-E_a/k_B T). \tag{1.17}$$

Pure diffusion-controlled kinetics always lead to the simultaneous growth of all phases, while with initial rates determined by reaction constants, the growth becomes sequential when the appearance of new phases is delayed [1.146]. In the case of reactions, which are not too fast, the growing phase may be mediated by thermodynamics, whereas the rate is determined by diffusion of the components. It is related to a certain mutual solubility in many silicides. Component activities also have to be taken into account. The diffusion-controlled silicide phase growth seems to be a dominant mode for most silicide-forming systems [1.4, 1.126]. However, reaction-controlled situations are also of practical importance because they are still typical for some silicides. Besides, the presence of chemically active impurities in a system, such as oxygen and nitrogen, can change silicidation kinetics towards the reaction-controlled mode. In some silicide forming systems the two controlling processes can interfere, leading to complex kinetics.

There are some theoretical approaches developed to simulate silicidation when both reaction-controlled and diffusion-controlled regularities take place [1.4, 1.118, 1.146–1.148]. However, most of these are limited to demonstrate only the general qualitative picture, because values of the parameters used for calculations are mainly unknown. These, for example, are: the component weight coefficients accounting for a change in composition at each interface and the coefficients characterizing barriers against the growth of one silicide phase against the other in [1.118]; the free energy change per moving atom and the coefficients representing reaction rate constants in [1.146]; the interdiffusion coefficients or atomic component diffusivities in silicides in [1.118, 1.146, 1.147] and so on. The above parameters are difficult to determine experimentally and evaluate numerically.

Parabolic or linear rates of silicide phase growth are usually determined in experiments. Their values for certain temperature ranges have been obtained for many silicides. This makes it reasonable to use them for numerical simulation of silicidation. One of the related approaches has been proposed by Borisenko et al. [1.4, 1.148]. This implies that an as-deposited thin film metal/silicon structure already contains silicide phases at the interface and a dominantly growing phase rather than any of the first nucleating ones being of practical interest. Independent of the first nucleated phase, a metal-rich silicide is thought to start growing first as soon as a penetration of metal atoms into silicon is limited only by their diffusivity. On the other hand, an injection of silicon atoms into metal requires the breaking of silicon–silicon bonds at the interface, which is much more energy-consuming [1.149]. Thus, the silicon layer at the interface becomes saturated with metal atoms and if even a silicon-rich phase has nucleated first the excess of the metal will reduce it.

Initial solid state growth of metal-rich silicides was experimentally observed for most thin film metal-silicon systems, in particular when rapid thermal processing is employed [1.4]. It is not surprising because both circumstances, i.e. thin film related effects and rapid heating, introduce dramatic thermodynamic imbalance in a system and equilibrium thermodynamics is generally not applicable. In such conditions the criterion for the first nucleated phase has to be a kinetic one determined by a competition between the values of the component diffusivities and the reaction rate constants. After some short transient period, related to the nucleation of new phases, their simultaneous growth occurs with corresponding thickness increments. Thus, a general presentation of silicidation kinetics should consider the evolution of a few silicide phases, which are: a metal-rich compound at the metal/silicide interface, an intermediate phase, and a silicon-rich compound at the silicide/silicon interface. This is illustrated in Fig. 1.25 where Me_2Si represents a metal-rich silicide, $MeSi$ corresponds to an intermediate one, and $MeSi_2$ is a representative of a silicon-rich phase. In the beginning, they could be phases (1), (2), and (3), or even more, with compositions corresponding to the phase diagram of the particular metal-silicon system.

During the silicide layer growth some of them will disappear as a result of the competition regulated by growth kinetics of the developing phases. Typical thin film metal/silicon structures used for silicide formation contain a limited amount of metal with respect to silicon. This is the case for either metal or metal+silicon films on silicon substrates. Heat processing of such structures results in movement of the metal/silicide interface towards the surface with a processing time or/and temperature increase. The growing silicide layer consumes metal first. The silicidation is completed by formation of the silicon-richest phase, which is usually disilicide. Disilicide can also be the only growing phase if its growth rate is much higher than that of the other phases.

For the numerical simulation of silicidation kinetics we place the zero point of the coordinate axis at the silicide/silicon interface and mark current positions of the corresponding phase interfaces d_1, d_2, and d_3 as shown in Fig. 1.25.

Assuming diffusion-limited kinetics for the growth of all phases included in the analysis, for an arbitrary time-temperature profile $T(t)$ of heat processing one can determine current coordinates of the interfaces from the equations

$$(d_1)^2 = U_{01} \int_0^\infty \exp\{-E_{a1}/[k_B T(t)]\}dt , \qquad (1.18)$$

$$(d_2)^2 = U_{02} \int_0^\infty \exp\{-E_{a2}/[k_B T(t)]\}dt , \qquad (1.19)$$

$$\left(d_3\right)^2 = U_{03} \int_0^\infty \exp\!\left\{-E_{a3}\,/\left[k_B T(t)\right]\right\}dt \;, \tag{1.20}$$

where the U_0s are the pre-exponential factors and E_as are the activation energies for parabolic rates related to the phases (1), (2), and (3), respectively.

If the growth of the i-th silicide phase is known to follow reaction-limited kinetics, the corresponding equation in the above set will have the linear form

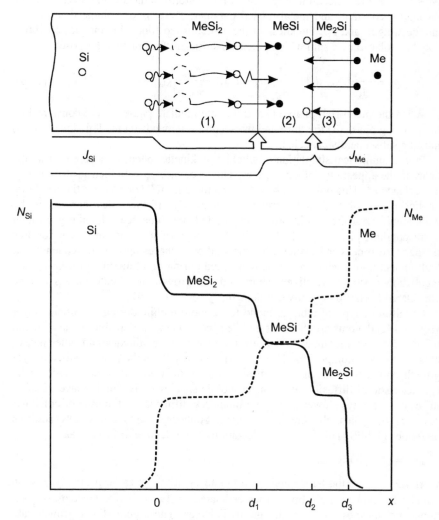

Fig. 1.25. Silicidation in a metal/silicon structure [1.4]

$$d_i = V_{0i} \int\limits_0^\infty \exp\{-E_{ai}/[k_B T(t)]\} dt . \tag{1.21}$$

The limited amount of metal in the structure implies the following boundary conditions

$$\int\limits_0^{d_3} N_{Me} dx \le \frac{\rho^*_{Me} \, n_A}{A_{Me}} d_{Me} , \tag{1.22}$$

where N_{Me} is the concentration profile of metal atoms in the silicide layer, n_A is the Avogadro constant, and metal-related parameters are: ρ^*_{Me} volume density, A_{Me} atomic weight, and d_{Me} thickness of the as-deposited film. This integral condition may be replaced by a sum with a precision sufficient for practical purposes

$$\frac{\rho^*_1}{A_1} d_1 + \frac{\rho^*_2}{A_3}(d_2 - d_1) + \frac{\rho^*_3}{A_3}(d_3 - d_2) \le \frac{\rho^*_{Me}}{A_{Me}} d_{Me} . \tag{1.23}$$

When the metal-rich phase (1) and the intermediate phase (2) are consumed by the growing phase (3) the third and the second summands on the left side of (1.23) must be subsequently canceled.

Precise numerical simulation of silicidation kinetics needs to account for a time delay in the appearance of new phases. This is connected with their growth up to a critical size r^*. This delay can be estimated to be $(r^*)2/U$ or r^*/V for the parabolic or linear law of the phase growth, respectively. If the delay appears to be comparable with the whole processing period, the time scale for the growth of corresponding phases should be shifted by this value. It is important for low temperature processing cycles when new silicide phases appear subsequently. At high temperatures, which are usually induced during rapid thermal processing, the incubation period is relatively short and the growth of silicide phases can be considered to start practically simultaneously.

Usually one type of atom is found to be more mobile during the silicide layer growth. Metal atoms are the most mobile species in metal-rich silicides and silicon in silicon-rich compounds, which are monosilicides, disilicides and other phases close to them in composition [1.4, 1.150]. This results in the rates of silicide layer growth and activation energies measured in a diffusion-controlled mode to be characteristic of diffusion of the most mobile species in the silicide layer. Lattice diffusion is always slower than grain boundary diffusion. In the case of columnar structure of grains, the lattice and grain boundary fluxes are simply additive quantities [1.147]. This provides calculation of an effective diffusivity as

$$D_{eff} = (1 - f)D_l + f D_{gb} , \tag{1.24}$$

where D_l, D_{gb} are the lattice and grain boundary diffusivities, respectively; f is the volume fraction of grain boundaries estimated as $2\delta/d^*$, if δ is the typical grain boundary thickness and d^* the grain diameter. Thus, both direct growth rate parameters and diffusivities of the most mobile species can be used for simulation of the silicidation kinetics.

The mathematics demonstrated permits us to predict phase composition and silicide layer thickness produced by thermal processing of a thin film metal-silicon couple as a function of the temperature regimes employed. Nevertheless, one should be very careful in the choice of the growth rate parameters among the experimental data available. Peculiarities of silicidation kinetics in particular metal-silicon systems, and some comments on their numerical characteristics are presented in the sections below. They were mostly investigated in metal-on-silicon (metal/silicon) thin film structures.

1.3.1 Silicides of Group II Metals

Mg-Si system. In thin film structures magnesium reacts with a silicon substrate producing Mg_2Si at a temperature as low as 200°C [1.127]. The silicidation proceeds rather rapidly at higher temperatures. Thus, at 275–300°C magnesium films of 300–400 nm thickness were observed to be completely converted into the silicide within 30–70 min [1.127, 1.151]. The metal atoms are the predominant mobile species during the solid state diffusion synthesis, as deduced from the Xe-marker experiment [1.127]. Other details of the silicidation kinetics in this system were not investigated.

Ba-Si system. Data characterizing silicidation kinetics in this system are not available.

1.3.2 Silicides of Group VI Metals

Cr-Si system. The first nucleated and growing phase in a thin film chromium/silicon structure subjected to conventional furnace annealing is disilicide $CrSi_2$ [1.128, 1.152–1.157]. It appears after processing at 350–425°C (see Table 1.34). At the beginning of the reaction metal atoms are the main mobile species. The limiting process for the reaction is diffusion of chromium into silicon [1.158]. At the early stages of phase formation, both unreacted chromium and the disilicide are present in the film. Moreover, multiphase composition of the silicide layer can be produced in the interfacial reactions as detected in ultrahigh vacuum (1.3×10^{-8} Pa) deposited chromium thin films on Si(111) [1.159]. An amorphous interlayer composed of Cr_5Si_3, $CrSi$, and $CrSi_2$ was observed to form after annealing at 375°C for 30 min. Nevertheless, the $CrSi_2$ phase grows at about 450°C consuming the metal and other silicide phases. The silicidation reaction with the metal takes place at the $Cr/CrSi_2$ interface. Experiments with radioactive ^{31}Si [1.156, 1.160] and Xe marker [1.132] have shown that silicon is the dominant diffusing species during $CrSi_2$ formation.

The data on its growth kinetics are contradictory. The available data are collected in Table 1.35 and shown graphically in Figs. 1.26 and 1.27.

A linear growth rate was observed in [1.132, 1.152, 1.154, 1.155, 1.157, 1.161, 1.162], however, a parabolic one was found in [1.153, 1.161, 1.163–1.166].

Impurities, primarily oxygen, seem to have a regulating role in the type of silicidation kinetic. This is well illustrated by the experiments performed with Kr^+ implanted and unimplanted thin film chromium/silicon structures [1.161, 1.166]. A dramatic difference in the kinetics was observed between them.

Table 1.35. Growth parameters of chromium silicides

Phase	Substrate	Temperature range (°C)	Kinetics	Pre-exponential factors, U_0 (cm^2/s), V_0 (cm/s)	Activation energy, E_a (eV)	Refs
Cr$_3$Si		1400 – 1550	$t^{1/2}$	1.06[a]	3.58	1.167
CrSi$_2$	Si(111)	425 – 525	t	200[a]	1.5	1.132
	Si(100)	425 – 475	t	4000[a]	1.7	1.161
	Cr/Si$_{am}$ multilayers	70 – 600	t	7.5×10^9	2.6	1.162
	Si(100)	450 – 525	t	5 200[a]	1.7	1.154
	Si(100)	500 – 700	t		1.7	1.155
	Si(100), Si$_{am}$	400 – 505	t	8.0×10^{5a}	2.0	1.152
	Si(111)	400 – 585	t	180[a]	1.5	1.157
	Si(100)	400 – 475	$t^{1/2}$	5000	2.3	1.163
	Si(100)	450 – 500	$t^{1/2}$	21000[a,c]	2.6[c]	1.161, 1.166
	Si(100)	800 – 1050[b]	$t^{1/2}$	0.45[a]	1.5[a]	1.164, 1.165
	Lateral growth on Al$_2$O$_3$	500 – 650	$t^{1/2}$	2.2[a]	1.4	1.153, 1.166

[a] Calculated from the experimental data presented in the reference, [b] Rapid thermal processing, [c] Silicidation took place between silicon substrate and chromium film implanted with Kr^+ ions.

The silicide growth in unimplanted samples is linear with time, indicating an interfacial-reaction-limited process. It has an activation energy of 1.7 eV. Meanwhile, parabolic growth with an activation energy of 2.6 eV was found for the implanted samples. This changeover to a diffusion-controlled process was interpreted as being due to the fact that krypton atoms segregated on grain boundaries and impeded silicon diffusion along the boundaries. Thus, reaction-controlled formation of CrSi$_2$ with activation energies of 1.7–2.0 eV can be considered as the most probable for conventional thin film structures, usually containing a certain amount of impurities in the materials and at the interface.

Accounting for scatter in the experimental data the linear rate of the CrSi$_2$ growth is best fitted by $V = 3.71×10^5 \exp(-1.98 \text{ eV}/k_B T)$ cm/s. It is shown with a dashed line in Fig. 1.26. On the other hand, impurity free structures mostly demonstrate diffusion-controlled silicidation kinetics. In this case, fitting of the experimentally obtained parabolic growth rates for the disilicide gives $U = 1.65×10^3 \exp(-2.23 \text{ eV}/k_B T)$ cm^2/s. This is plotted in Fig. 1.27 for comparison with the experimental data.

While CrSi$_2$ is predicted and experimentally confirmed to be the first nucleating and growing phase, multiphase composition of the silicide layer is observed when rapid thermal processing of the film structure has been performed or the film structure has a limited amount of silicon atoms. Rapid-heating-induced silicide formation in this system was investigated in [1.164, 1.165, 1.169]. Processing at 450–1100°C for seconds produces a multiphase silicide layer, which consists of Cr_3Si, Cr_5Si_3 and $CrSi_2$. The balance between the phases varies with the processing temperature and time.

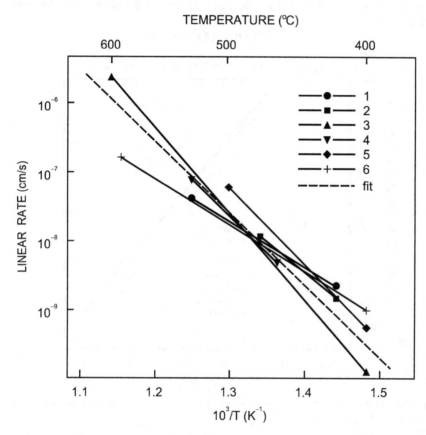

Fig. 1.26. Linear rate of $CrSi_2$ growth as a function of temperature: 1-[1.168], 2-[1.161], 3-[1.162], 4-[1.154], 5-[1.152], 6-[1.157]

Chromium disilicide and Cr_3Si are registered in the layers synthesized at 1100°C for 10 s. The increase of the processing time to 30 s results in the reduction of Cr_3Si phase. Cr_5Si_3 inevitably appears and the disilicide is monotonously increased. Phase transitions in the silicide layer are completed by uniform disilicide formation after 90 s processing at that temperature.

This takes longer at lower temperatures. Formation of metal-rich phases provides evidence of the limited silicon concentration at the metal/silicide interface. Meanwhile, an excess of chromium atoms appears on reduction of the native chromium oxide at the interface by silicon [1.170, 1.171].

Silicon atoms become bonded there with oxygen when they reduce the chromium oxide by the reaction

$$2Cr_2O_3 + 3Si \rightarrow 3SiO_2 + 4Cr. \tag{1.25}$$

Thus, the disilicide is "diluted" by free chromium atoms. Rapid-heating-induced chromium silicide layer growth is governed by the linear law at temperatures up to 650°C and by the parabolic one at temperatures higher than 800°C.

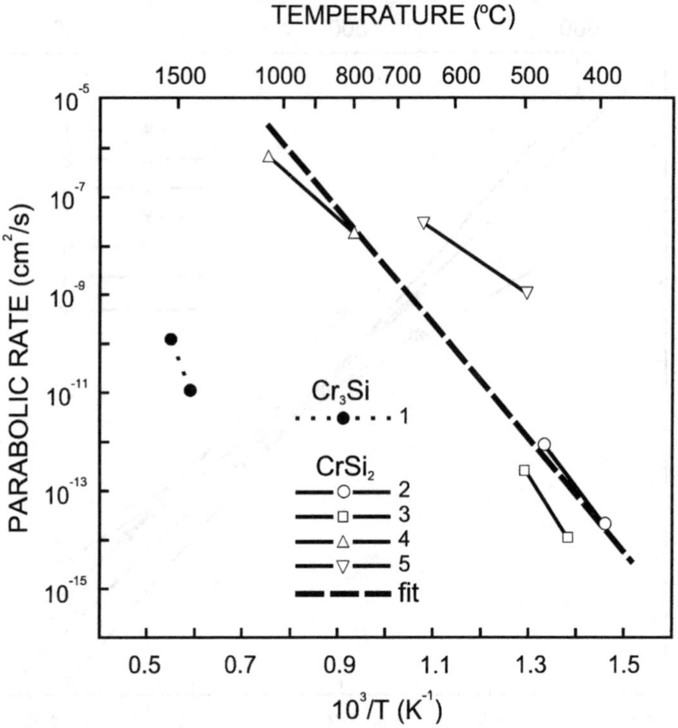

Fig. 1.27. Parabolic rate of Cr$_3$Si and CrSi$_2$ growth as a function of temperature: 1-[1.167], 2-[1.163], 3-[1.161], 4-[1.165], 5-[1.153]

Comparison of the kinetic parameters summarized in Table 1.35 shows that the activation energy of the parabolic growth under rapid thermal processing (1.5 eV) is practically identical to the value (1.4 eV) from [1.153] where the same kinetics was observed for lateral growth of the disilicide.

The absolute rate of the silicide layer growth during rapid thermal processing at 1000°C is 1–2 orders lower than could be expected from the results of conventional furnace annealing. This seems to be a consequence of the above discussed silicon oxidation. Chromium recovered from its oxide causes formation of nonequilibrium metal-rich phases, which retard the disilicide growth. However, the high activation energy of the Cr$_3$Si phase growth (3.58 eV [1.167]) and the absolute growth rates being considerably lower than for CrSi$_2$ (see Fig. 1.27) shifts the balance in the silicide layer to the dominant disilicide formation.

Simultaneous appearance of some chromium silicide phases is also observed when a thin film structure has a limited source of silicon atoms. This could be realized by codeposition of the components onto oxidized silicon substrates or in free-standing films. In such structures $CrSi_2$ is still one of the first phases to appear at about 450°C. Cr_5Si_3 is the next to be formed after processing at 550°C and then Cr_3Si at 650°C [1.129, 1.130]. This confirms that among the chromium silicide phases, $CrSi_2$ has the highest growth rate providing formation of single-phase silicide films in solid state reactions of thin metal films with silicon substrates.

Mo-Si system. In conventional furnace experiments Mo_3Si was found to nucleate first [1.133]. This occurs in the amorphous molybdenum/silicon interlayer at 400°C. The next silicide phase, which is Mo_5Si_3, appears in the system at 440°C. Molybdenum disilicide $MoSi_2$ is known to start growing at a temperature of 500°C [1.128, 1.172]. It has a hexagonal lattice. In very thin film structures, composing of 5–10 nm of the metal on silicon, the hexagonal disilicide is detected at about 400°C [1.197].

The metal-rich phase Mo_3Si and the intermediate Mo_5Si_3 can also be present in the silicide layer formed at relatively low temperatures (475–550°C). Silicidation is completed by the uniform tetragonal disilicide formation only after annealing in the temperature range 1050–1100°C when conventional furnace annealing [1.173] or rapid thermal processing within seconds [1.174–1.177] is employed. There are no phase changes observed at higher processing temperatures.

Structural analysis of the disilicide formed at 600°C shows that both hexagonal and tetragonal phases exist in the silicide films [1.95, 1.96]. Hexagonal grains of 13 nm size were observed upon processing of molybdenum film on Si(111) substrate at 600°C for 60 min [1.96]. Their size was almost doubled when the annealing temperature was chosen to be 700°C. After heat processing at 800°C, both hexagonal and tetragonal $MoSi_2$ were detected. At higher temperatures epitaxial growth of these phases was reported. Due to the fact that the hexagonal phases are metastable at temperatures below 1900°C, one can expect their formation to be related to the stresses existing in thin film structures.

Growth kinetics of all the molybdenum silicides are diffusion-controlled. Their parameters are summarized in Table 1.36 and the temperature dependence of the parabolic growth rate is presented in Fig. 1.28. Activation energies for the growth of Mo_3Si and Mo_5Si_3 phases are 1–1.5 eV higher than that for the disilicide which has a tetragonal lattice. The growth rate of $MoSi_2$ is the highest in the system at temperatures above 800°C. Silicon atoms are dominant diffusing species during the disilicide growth [1.178]. Germanium additives retard the silicidation process [1.179]. The particular kinetics of hexagonal $MoSi_2$ formation was not investigated.

In molybdenum-on-silicon film structures mixed by ion implantation, a silicide layer begins to grow at 600°C, as can be concluded from the sharp decrease of sheet resistivity [1.181]. The process is completed by uniform disilicide formation within seconds when the sample is heated to 1100°C. Data on the phase content of the silicide layer in the interval between these two temperatures are not available. Nevertheless, metal-rich phases might be expected as occurs upon conventional long-time furnace annealing.

Table 1.36. Growth parameters of molybdenum silicides

Phase	Substrate	Temperature range (°C)	Pre-exponential factor[a], U_0 (cm²/s)	Activation energy, E_a (eV)	Refs
Mo₃Si	MoSi₂		0.26	3.25	1.180
Mo₃(Si,Ge)	Mo	950 – 1150	1.61×10^{-5}	1.93	1.179
Mo₅Si₃	MoSi₂	1190 – 1715	38.4	3.58	1.180
Mo₅(Si,Ge)₃	Mo	950 – 1150	5.66×10^{-4}	2.26	1.179
MoSi₂ + Mo₃Si	Si(111)	475 – 550		2.4[b]	1.172
MoSi₂	Si-poly	850 – 1100	0.88	2.2	1.173
Mo(Si,Ge)₂	Mo	950 – 1150	8.65×10^{-3}	2.05	1.179

[a] Calculated from the experimental data presented in the reference, [b] Unusual growth kinetic t^2 is mentioned in the paper.

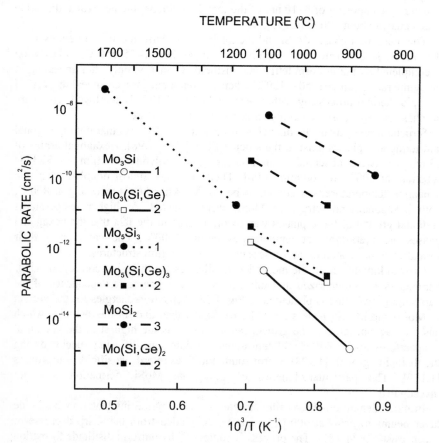

Fig. 1.28. Parabolic rate of Mo₃Si, Mo₅Si₃ and MoSi₂ growth as a function of temperature: 1-[1.180], 2-[1.179], 3-[1.173]

W-Si system. The analysis performed in [1.4] shows that silicide formation data for this system are contradictory. In conventional furnace experiments hexagonal WSi_2 was found to nucleate and grow first in the temperature range 300–420°C [1.101, 1.182]. At about 650°C it converts into the tetragonal phase which was often observed to grow rapidly starting from 650°C [1.128].

In the samples annealed at and above 800°C the metal film completely transforms into WSi_2 [1.34]. According to [1.183] its growth rate follows a linear law with an activation energy of 3.0 eV in the temperature range 700–850°C. This is an indication of a reaction-controlled silicidation. Meanwhile, tungsten disilicide formation was observed in [1.173, 1.184, 1.185] to be diffusion-controlled with a parabolic growth law.

A summary of the available growth parameters is given in Table 1.37. Experimental data for both linear and parabolic growth rates are plotted in Figs. 1.29 and 1.30 as a function of temperature. The linear growth rate of WSi_2 correlates well with that for the ternary compounds $W_{1-x}Ti_xSi_2$. This supports the idea that the presence of impurities makes the silicidation process reaction-controlled with the corresponding linear kinetics.

Table 1.37. Growth parameters of tungsten silicides

Phase	Substrate	Temperature range (°C)	Kinetics	Pre-exponential factors[a], U_0 (cm²/s), V_0 (cm/s)	Activation energy, E_a (eV)	Refs
W_5Si_3	WSi_2	1350 – 1870	$t^{1/2}$	13.2	3.58	1.180
WSi_2	Si(100)	650 – 800	$t^{1/2}$	6.6×10^3	3.4	1.184
	Si(100)	700 – 800	$t^{1/2}$	88.3	2.6	1.185
	Si-poly	850 – 1100	$t^{1/2}$	0.6^b	2.2	1.173
	Si(100)	1150 – 1350[c]	$t^{1/2}$	1.46	2.2	1.184
	Si(100)	700 – 850	t	5.8×10^6	3.0	1.183
$W_{0.7}Ti_{0.3}Si_2^d$	Si(111)	700 – 750	t	1.7×10^{14}	4.5	1.74
$W_{0.76}Ti_{0.24}Si_2^e$	PtSi on Si(100)	625 – 735	t	2.3×10^8	3.3	1.75
$W_{0.76}Ti_{0.24}Si_2^f$	PtSi on Si(100)	675 – 775	t	1.3×10^8	3.3	1.75

[a] Calculated from the experimental data presented in the reference, [b] Presented in the reference, [c] Rapid thermal processing, [d] Containing up to 15% of oxygen, [e] Containing less than 1% of oxygen, [f] Containing 5% of oxygen.

On the other hand, the pure tungsten-silicon system demonstrates diffusion-controlled regularities. The most realistic parabolic rate of the WSi_2 growth, as evident from the data plotted in Fig. 1.30, is $U = 0.6\exp(-2.2\ eV/k_BT)$ cm²/s [1.173]. Silicon atoms are dominant diffusing species during the disilicide growth [1.178]. Metal-rich silicide W_5Si_3 in quantities sufficient for analysis was found only at the tungsten-saturated interface in W/WSi_2 structures [1.180]. This is not surprising taking into consideration the dramatic difference in the W_5Si_3 and WSi_2 growth rates illustrated in Fig. 1.30.

Investigations of rapid-thermal-processing-induced synthesis of tungsten silicides [1.181, 1.184, 1.186–1.192] partly resolve the contradiction in the phase sequence specifying the picture of the silicide formation. Fig. 1.31 shows the results of x-ray phase analysis of the thin film tungsten-on-silicon structure subjected to thermal processing for 30 s at different temperatures. There are no phase changes up to 900°C. Silicides W_5Si_3 and WSi_2 (tetragonal) are reported upon processing at 950°C. Unreacted metal decreases with the processing temperature rise and it completely disappears above 1050°C.

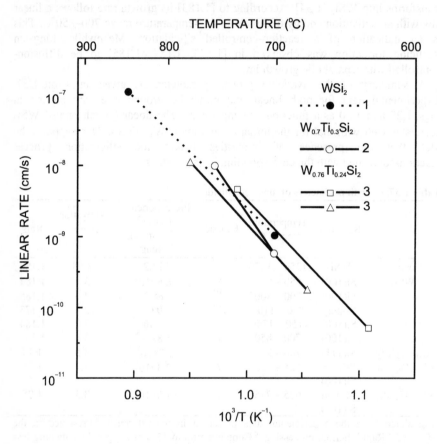

Fig. 1.29. Linear rate of WSi_2 and $W_{1-x}Ti_xSi_2$ growth as a function of temperature: 1-[1.183], 2-[1.74], 3-[1.75]

The disilicide grows in the whole temperature range while the phase W_5Si_3 reaches its maximum at 1000°C and then rapidly reduces at higher temperatures. It disappears above 1100°C. Stoichiometric tungsten disilicide is formed within seconds by processing at 1100°C and above in both metal-on-silicon [1.184, 1.186, 1.192] and codeposited mixed metal+silicon [1.187, 1.188, 1.190, 1.191, 1.193] thin film structures. The best uniformity is achieved by processing at 1200°C with the duration up to 10 s. High energy ion implantation in the film yields destruction of interface oxide layers and mixing of the atomic components, causing 50–100°C temperature decrease of uniform disilicide formation [1.181, 1.188, 1.189].

Tungsten disilicide growth kinetics under rapid thermal processing are typical diffusion-controlled [1.184, 1.186, 1.193]. The activation energy is 2.2 eV. This agrees well with the data received in conventional long-term experiments (see Table 1.37). Silicon atoms are the most mobile in the growing layer. Rapid thermal processing promotes faster silicide layer growth.

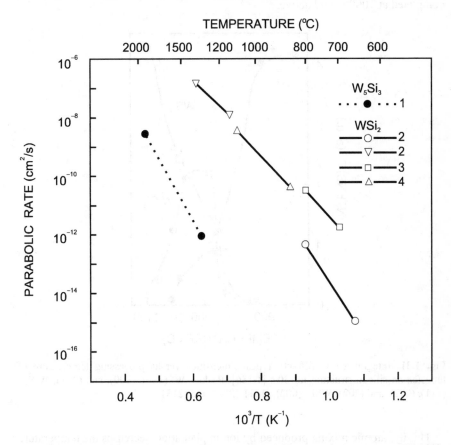

Fig. 1.30. Parabolic rate of W_5Si_3 and WSi_2 growth as a function of temperature: 1-[1.180], 2-[1.184], 3-[1.185], 4-[1.173]

The enhancement factor varies from 2.5 to 10–15 in comparison with the data from high temperature (850–1100°C [1.173]) and low temperature (650–800°C [1.184]) conventional furnace experiments extrapolated to 1100–1200°C. Elastic stresses generated in the film structure are proposed in [1.184] to be responsible for the enhancement.

Silicide formation in molybdenum-silicon and tungsten-silicon systems has much in common. The most important features are as follows. First, nucleation and growth of both semiconducting hexagonal disilicides occurs at temperatures as low as 400–600°C. They are metastable and tend to transform into tetragonal lattices at 650–800°C. Mechanical stresses in the film, impurities, or favorable epitaxial conditions at the substrate are needed to stabilize the hexagonal disilicides and preserve their growth at higher temperatures. The latter aspects will be presented and discussed in Chap. 2. Secondly, in spite of the fact that the disilicides are nucleated at relatively low temperatures, the silicidation is often completed at 1000°C and above.

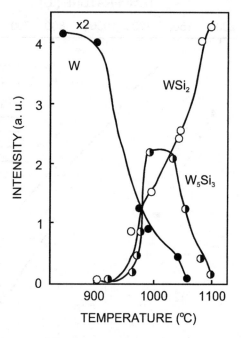

Fig. 1.31. Integral x-ray diffraction peak intensities versus processing temperature of tungsten-on-silicon structure for 30 s [1.186]. Peaks - W: (110), (200), and (211); W_5Si_3: (211), (002), and (202); WSi_2: (002), (101), (202), and (213)

Thirdly, atomic mixing produced by ion implantation decreases the temperature of uniform disilicide layer formation. The last two features, along with the linear growth rate of tungsten disilicide during conventional furnace annealing, display the influence of barrier oxide at the metal/silicon interface on the silicidation process. The oxide layer restricts the rate of the silicide produced reactions there. Ion beam mixing of the components or processing temperature increase is necessary. Metal-rich silicides of both metals grow slower than their disilicides. This is why observation of molybdenum and tungsten-rich silicides upon rapid thermal processing is difficult.

1.3.3   Silicides of Group VII Metals

Mn-Si system. The first phase observed to nucleate and grow in this system is monosilicide MnSi [1.134]. It appears after annealing of a thin film metal-on-silicon structure at 400°C. A uniform MnSi layer is formed after processing at 500°C. Its growth is diffusion-controlled. The parabolic growth rate is characterized in the Arrhenius expression by the activation energy of 1.9 eV and pre-exponential factor of 0.52 cm^2/s.

Other silicide phases appear and grow in the beginning of silicidation at 380–430°C when there is a thin silicon dioxide film between the metal film and silicon substrate [1.194]. Mn_3Si was observed to form first at the oxide/silicon interface. It extends toward the silicon substrate. Next, MnSi is produced at the Mn_3Si/Si interface and grows toward the surface. At the same time, the thickness of the Mn_3Si layer stops changing. Finally, Mn_5Si_3 begins to form at the $Mn_3Si/MnSi$ interface. Mn_5Si_3 grows by consuming Mn_3Si, while MnSi grows by consuming Mn_5Si_3, until all the Mn_3Si as well as Mn_5Si_3 are exhausted and only MnSi is left.

The next phase synthesized is $MnSi_{1.7}$. It nucleates at 485–525°C and grows in three dimensions in patches inside the MnSi layer. This type of growth process is nucleation-controlled. A linear kinetics is observed. The transformation is completed after annealing at 570–600°C [1.134, 1.195–1.197]. There is indirect evidence that silicon atoms are the dominant diffusing species during manganese silicidation.

Re-Si system. Thin film rhenium-on-silicon demonstrates no changes upon annealing up to 400°C [1.60]. Above this temperature, silicide synthesis takes place rapidly and the stable silicide composition around $ReSi_2$ is attained at 650°C. In the intermediate temperature range, a mixture of pure rhenium and $ReSi_2$ grains is observed [1.36].

1.3.4   Silicides of Group VIII Metals

Fe-Si system. Silicidation in this system has been most carefully investigated with respect to other semiconducting silicide forming metal-silicon couples. Meanwhile, some details related to the first steps of the process remain to be further clarified.

Most of the theoretical predictions and experiments (see data collected in Table 1.34) demonstrate monosilicide FeSi nucleates and grows first in thin film structures. The lowest process temperature varies from one experiment to another, but it is surely in the range of 370–400°C. However, well controled *in situ* experiments with metal films a few monolayer (ML) thick on silicon substrates demonstrate not only an amorphous mixture of the components but metal-rich phases to be precursors in the formation of the monosilicide. Note that 1 ML of iron in this case corresponds to 1.7×10^{15} atoms/cm^2, which is about 0.2 nm. Thus, the first steps of silicidation in extremely thin films are distinctly different from the processes in the film structures with a typical thickness of some tens of nanometers.

The phase sequence in silicide formation as a function of iron coverage of a Si(111) substrate and annealing temperature is illustrated in Fig. 1.32. The interface between the amorphous components is reactive even at room temperature. The stoichiometrically defected Fe_3Si_{1-x} phase is observed by Alvarez *et al.* [1.198, 1.199] to be spontaneously formed at this temperature when the deposited iron film is less than 2 ML thick. This correlates to the solubility limit of silicon into iron, which is about 25 at % [1.90]. For an initial iron coverage of about 3.5 ML and annealing temperatures below 300°C stoichiometric Fe_3Si was found by ultraviolet photoelectron spectroscopy (UPS) technique. On Si(001) substrates Hasegawa *et al.* [1.200] observed formation of polycrystalline Fe_5Si_3 after room temperature deposition of 1.5 ML of the metal. Annealing at temperatures above 300°C, as well as the use of thicker films, always leads to FeSi nucleation and growth.

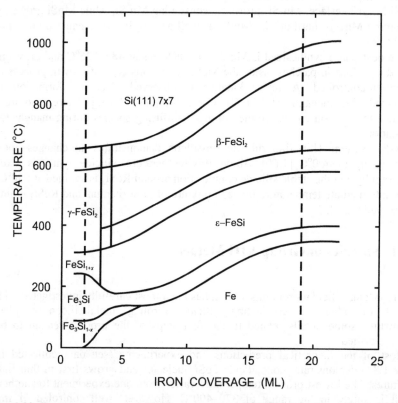

Fig. 1.32. Temperature versus initial iron coverage of Si(111) 7×7 phase diagram [1.198, 1.199]. The solid lines limit the regions between UPS spectra characteristic of different phases as well as coexistence regions of adjacent phases. Solid phase epitaxy allows us to probe the diagram vertically, while reactive deposition epitaxy allows us to probe it horizontally

Disordered or highly strained FeSi appears first [1.137]. It transforms into the bulk unstable FeSi with the CsCl-type lattice and into the stable ϵ-FeSi after annealing at 330 and 370°C, respectively. The latter grows intensively at temperatures above 400°C consuming the metal. Its formation is diffusion-controlled [1.137, 1.201, 1.202] with an activation energy in the range 1.36–1.67 eV, as is seen from the data collected in Table 1.38. The related variation of the parabolic growth rate with the annealing temperature is presented in Fig. 1.33. In spite of the notable scatter in the activation energy, the experimentally observed growth rates of FeSi are quite accurately represented by the fit $U = 3.25 \times 10^{-3} \exp(-1.43 \text{ eV}/k_B T)$ cm^2/s. A Xe-marker experiment [1.201] showed silicon to be dominantly moving species during the monosilicide growth.

Table 1.38. Growth parameters of iron silicides

Phase	Substrate	Temperature range (°C)	Pre-exponential factor[a], U_0 (cm^2/s)	Activation energy, E_a (eV)	Refs
FeSi	Si(100)	450 – 525	0.19	1.67	1.201
	Si(111)	290 – 410	0.0025	1.5	1.137
	Si(111)	500 – 625[b]	0.0014	1.36	1.202
β-FeSi$_2$	Si(111)	450 – 530	0.024	1.8	1.137
	Si-am	525 – 625	5.6×10^{-5}	1.5	1.199
	Si(111)	675 – 750[b]	16	2.3[a]	1.202

[a] Calculated from the experimental data presented in the reference, [b] Rapid thermal processing.

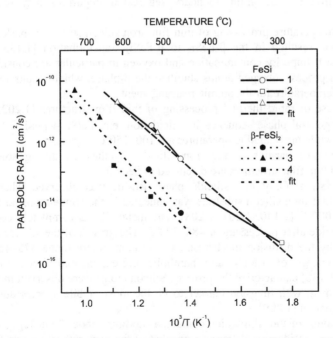

Fig. 1.33. Parabolic rate of FeSi and β-FeSi$_2$ growth as a function of temperature: 1-[1.201], 2-[1.137], 3-[1.202], 4-[1.203]

The kinetically unstable Fe_3Si phase may appear and grow simultaneously with the monosilicide when the amount of silicon is limited. It forms after most of the iron film is transformed into FeSi at 450 to 500°C [1.135, 1.138]. But with an increase of annealing time, this metal-rich phase is consumed by growing $\in$-FeSi.

In the temperature range 450–530°C β-FeSi$_2$ grows from the monosilicide [1.139]. It originates from FeSi via a nucleation-controlled process as can be concluded from hillock-like growth emerging in the films [1.138, 1.204]. Further growth of the disilicide is dominantly diffusion-controlled. It is characterized by an activation energy in the range 1.5–2.3 eV (see Table 1.38), which is higher than that for the monosilicide formation. Comparison of the growth rates presented in Fig. 1.33 confirms that the disilicide grows more slowly with respect to the disilicide. The experimental data for the parabolic rates of the β-FeSi$_2$ growth are generally represented by the fit $U = 0.133\exp(-1.96 \text{ eV}/k_BT) \text{ cm}^2/\text{s}$.

In connection with the type of growth kinetics one should note the data of Amesz et al. [1.204]. They observed the lateral growth of β-FeSi$_2$ on Si(001) in the temperature range 625–700°C to follow a linear time dependence with an activation energy of 1.35 eV. At the moment there is no other data indicating linear kinetics for the β-FeSi$_2$ growth. Thus, it seems reasonable to consider this result to be attributed to the influence of impurities or to size effects.

In the temperature range 550–650°C similar growth rates of monosilicide and disilicide result in two phase composition, e.g. FeSi and β-FeSi$_2$, of the synthesized silicide films. These can be preserved upon annealing up to 750°C [1.205]. It is indeed pronounced when polycrystalline silicon is used as a substrate. In this case grain boundary segregated impurities may retard the process.

Higher temperature processing of thin film iron/silicon structures performed at above 900°C resulted in the β-FeSi$_2$ to α-FeSi$_2$ transformation [1.135, 1.136, 1.205, 1.206]. Impurity contamination and oxygen in particular are considered to control this process. Oxygen atoms stabilize the β-phase, which results in a need for higher temperature for the atomic rearrangement.

In the case of rapid thermal processing of thin film structures [1.202, 1.204, 1.206] the general phase sequence in silicidation, e.g. Fe/Si → FeSi → FeSi$_2$, is preserved, while the related temperatures are 100–150°C higher.

Structural peculiarities of the iron silicides synthesized during solid state reactions in thin film structures are discussed in Sect. 2.2.

Ru-Si system. Only one silicide phase Ru$_2$Si$_3$ was observed when metal ruthenium film on a silicon substrate was annealed in the temperature range from 375°C to 1000°C [1.140, 1.207, 1.208]. The metal film is completely converted into the silicide after processing at 485–500°C. The growth of the silicide layer is controlled by the diffusion mechanism. In the temperature range 375–450°C an activation energy of 1.8 eV and parabolic pre-exponential factor of about 1.4 cm^2/s [1.140] characterize the process. Silicon atoms were observed in the thin film experiment with tungsten marker to be the most mobile species during the silicide formation [1.209].

The behavior of the film which contained rhodium (about 33 at %) and traces of osmium is considerably different from that of the pure ruthenium one [1.140].

Two phases: RuSi upon furnace processing at 475°C for 90 min and below or Ru_2Si_3 after longer times at that temperature or with higher temperatures were observed. Studies of the kinetics of RuSi formation were found to be diffusion-controlled with an activation energy between 2.25 and 2.6 eV. The growth rate in the form $U = 82\exp(-2.2 \text{ eV}/k_BT)$ cm^2/s covers the experimental data obtained for this phase in the temperature range 400–475°C. The transformation from RuSi to Ru_2Si_3 occurs in a very short time interval since the heat treatment at 475°C for 90 min is insufficient to cause the complete formation of RuSi, yet only 30 min more allows not only the complete formation of RuSi but further the complete transition to Ru_2Si_3. A sample annealed at 485°C for 60 min had a silicide layer composed of only Ru_2Si_3. Its rough surface is indicative of a nucleation process, presumably through the initial formation of RuSi. Both the rough appearance and the rapidity of the transformation from RuSi to Ru_2Si_3 cause one to classify the transformation with nucleation-controlled silicidation.

Os-Si system. An investigation of the silicidation kinetics in this system is hardly realized in the classical metal-on-silicon thin film structures. The problem is connected with poor adhesion of the metal to silicon. Osmium films thicker than 30 nm peel off during subsequent thermal processing.

The kinetic data available have been obtained for the metal films of 3 nm [1.210] and 30 nm [1.140] and for the thicker films (150 nm) deposited onto a ruthenium sublayer [1.140]. In both cases, Os_2Si_3 was observed to be formed first. It appeared at 600°C in the very thin metal/silicon structures and at 450°C in the combined osmium/ruthenium/silicon ones. Growth of this phase is diffusion-limited. In the temperature range 450–525°C it is characterized by an activation energy of 1.8 eV [1.140]. Transformation of Os_2Si_3 into $OsSi_2$ takes place at 730–750°C, being completed at the highest temperature. Meanwhile, oxygen contamination of the metal film preserves the multiphase silicide layer even after annealing at 1000°C [1.210].

Silicon atoms are the most mobile species during Os_2Si_3 and $OsSi_2$ formation. There is, however, an indication of metal atom motion in the temperature range 750–1000°C [1.140]. This seems to be not surprising as the network of silicon atoms surrounding osmium in $OsSi_2$ is relatively open.

Ir-Si system. According to the phase diagram, as many as six or even more silicides may be formed in this system. However, only three distinct phases, namely IrSi, $IrSi_{1.75}$ and $IrSi_3$, were identified [1.43, 1.44] during silicide formation in the metal thin film on silicon structures. The phase indicated in the experiments as $IrSi_{1.75}$ within experimental uncertainty corresponds to stoichiometric Ir_3Si_5, which may also contain an excess of silicon atoms.

The monosilicide IrSi nucleates first at 300–400°C [1.47, 1.142]. The single-phase growth proceeds up to 500°C. It is diffusion-controlled with an activation energy of about 1.9 eV [1.43]. In the range 500–950°C $IrSi_x$ is observed. The stoichiometry varies from $x = 1.5$ to $x = 1.75$. The semiconducting monoclinic phase with $x \sim 1.6$ (Ir_3Si_5) is dominant.

After processing at 500°C unreacted metal, IrSi and $IrSi_{1.75}$ are detected to occur simultaneously in the film [1.43, 1.44]. Preferential growth of $IrSi_{1.75}$ takes place in the temperature range 550–900°C. It forms in layers more or less parallel to the original Me/Si interface like IrSi. There have been difficulties in obtaining specific kinetic information for this phase, since below 600°C its growth occurs after the development of an IrSi layer, whereas at higher temperatures it proceeds so rapidly that detailed observations could not be made. Nevertheless, diffusion-limited kinetics with an activation energy of about 2.1 eV has been concluded. A dominant motion of silicon atoms forms both IrSi and Ir_3Si_5, as was observed in thin film experiments with rhodium and cobalt markers [1.209].

Above 950°C $IrSi_3$ nucleates and grows in the system. This proceeds in a very narrow temperature range. In the vicinity of 1000°C its formation is completed [1.43, 1.211]. It presents some interesting features since $IrSi_3$ was found to grow more or less throughout the thickness of the film rather than in well defined layers parallel to the original metal/silicon interface [1.43]. $IrSi_3$ grows via a mechanism which is quite distinct from that observed for the two other silicide phases. This compound forms by nucleation at certain points in the film by a heterogeneous process. The growth occurs without observable composition gradients through the thickness of the film. It requires rapid redistribution of at least one of the two components through the whole film thickness, which is expected to proceed via grain boundary diffusion and relatively slower intragrain transformation of a lower silicide into $IrSi_3$. The activation energy was presumed to be of the order of 3 eV.

When silicide formation was performed in the thin film mixture of iridium and silicon [1.68], Ir_4Si_5 and Ir_3Si_5 were detected as intermediate phases between the monosilicide and the silicon-richest phase $IrSi_3$ [1.68].

Rapid thermal processing of iridium films on monocrystalline silicon investigated by Rodriguez et al. [1.143, 1.212, 1.213] generally confirms the above peculiarities in the silicidation kinetics. However, there is a difference in the temperature ranges for particular phase formation presented in the publications. In the earlier studies [1.143, 1.212] only IrSi was detected to grow in the samples annealed at 400°C for times shorter than 45 s. For longer processing times the silicide with the composition of $IrSi_{1.75}$ was observed at the IrSi/Si interface. Both silicides and unreacted iridium were present simultaneously upon processing for 1 s at 500°C. Single-phase silicide films consisting of $IrSi_{1.75}$ appeared after processing for the time increased to 10 s at 500–650°C. The data published later in [1.122] show that the monosilicide is growing intensively at 575°C. The subsequent $IrSi_{1.75}$ phase becomes the only one upon processing at 675°C after 720 s. These correlate better to the results from the conventional furnace annealing as far as one can expect that shorter heat processing time should be compensated by an increased temperature.

The growth of IrSi and $IrSi_{1.75}$ follows the parabolic law [1.212, 1.213] indicating formation of these phases to be diffusion-controlled. The estimated silicidation rate is from three to five orders higher than that observed in conventional furnace experiments. We have calculated activation energies and pre-exponential factors for the growth of IrSi and $IrSi_{1.75}$ using their thickness measured in [1.212] after processing at 400 and 450°C. They are 2.0 eV/1.2×10^3 cm²/s and 4.93 eV/5.7×10^{23} cm²/s, respectively. Meanwhile, activation energies of 1.3 and 3.9 eV for the growth of these phases were extracted in the later experiments [1.213]. The activation energy of the IrSi growth correlates well with that (1.9 eV) evaluated in the samples subjected to conventional long-term annealing. As for the growth parameters for the $IrSi_{1.75}$ phase, they appear to be incredibly high. While additional experiments are needed for their reexamination, there is no doubt that silicidation during rapid thermal processing is enhanced. The most reasonable explanation of this in the case of iridium silicides is the dramatically reduced influence of oxygen on the silicidation process which was achieved by annealing in vacuum ($\sim 1.3 \times 10^{-3}$ Pa) for very short times. The phase sequence in the silicide formation remains unchanged.

1.3.5 Ternary Silicides

The kinetics of silicidation in ternary systems is to a great extent dependent on a particular structure of the sample. There are two principal possibilities for solid-state-diffusion-induced synthesis of ternaries in thin films. The first supposes the use of metallic alloy or mechanical mixture of the components deposited on a silicon substrate. In this type of structures the three components involved in ternary compound formation contact each other from the very beginning of the silicidation process. The first phase to nucleate will be either intermetallic Me1-Me2 compound, Me1-Si or Me2-Si silicide. Reaction between the metals will be favorable whenever their heat of mixing is greater than the heat of formation ΔH_f of the first nucleating silicide phase. The threshold is estimated to be at $\Delta H_f = -11$ kcal/g [1.49] as calculated from the theory of Miedema et al. [1.88] accounting for the difference in work function and the change in electron density at a boundary of the Wigner–Seitz atomic cells. Subsequent phase growth in the structure is regulated by kinetic factors.

The second approach is based on the use of metallic bilayers on silicon. Intermetallic compound formation at the Me1/Me2 interface and silicidation at the Me1/Si interface first proceeds in such structures independently. Further phase evolution is considered to be controlled by atomic diffusion across the compound formed and by an interfacial reaction barrier connected with the necessary rearrangement of atoms at the interface [1.49, 1.214]. Thus, the phase to grow becomes that with the lowest effective interface reaction barrier. A second compound does not appear until this first phase has reached a critical thickness. This means that due to kinetic instabilities or nucleation difficulties, equilibrium phases may form only sequentially in contrast to the possibility of the simultaneous growth in alloyed or mixed samples.

In both approaches the ternary phase is produced via reactions of metal with the silicide or of silicon with the intermetallic compound. Diffusion of species through a growing compound and the rate of reaction at the phase interface compete to control the process. As for the majority of silicon-rich binary silicides, silicon is the most mobile species responsible for a single-phase ternary silicide growth [1.76]. However, the experimental kinetic data, which are available only for the W-Ti-Si system (see Table 1.37), demonstrate a linear growth rate indicating reaction-controlled processes as dominant. It is not typical for pure binary systems.

Thermal processing of $W_{0.7}Ti_{0.3}$ thin film on monocrystalline silicon was carried out by Harris et al. [1.74] to produce the ternary silicide $W_{0.7}Ti_{0.3}Si_2$. At 700–750°C linear kinetics with an activation energy of 4.5 eV was observed. The pre-exponential factor $V_0 = 1.7 \times 10^{14}$ cm/s is estimated from the experimental data presented in the reference for Si(111) substrate. The mixed metal films deposited on Si(111) react slower than those on Si(110), and on Si(100) the reaction is the fastest. Even in vacuum (9.3×10^{-5} Pa) the silicidation rate in the part of the samples unprotected from the ambient is a factor of two lower than that in the protected one. This is related to the oxygen contamination, which is definitely detected in the unprotected regions.

Similar results were obtained by Nava et al. [1.75] who synthesized $W_{0.76}Ti_{0.24}Si_2$ from the $W_{0.76}Ti_{0.24}$ alloy film on PtSi sublayer on Si(100) in the overlapping temperature range. The ternary compound grows linearly with time, but with lower activation energy, e.g. 3.3 eV, as deduced from the experimental data. Oxygen in the concentration range up to 5% was also detected in the structures. Comparison of the kinetics and activation energies for the growth of binary and ternary tungsten silicides collected in Table 1.37 confirms that the ternaries are formed in a reaction-controlled process. Higher activation energies may be explained by lower rates of a silicon reaction with the alloy than with the pure components. Even a sublayer of PtSi can not contribute much in the silicidation rate as far as silicon diffusion through PtSi is characterized by an activation energy of 1.4–1.6 eV [1.4] which is different from that for the tungsten silicide growth. Moreover, oxygen at the interface undoubtedly enhances reaction-controlled regularities and increases the activation energy.

Silicide phase mixing and separation were analyzed in experiments with Mo/Cr and W/Cr bilayer metal films on silicon [1.215]. In the bilayer Mo(65 nm)/Cr(95 nm)/Si(100) structures $CrSi_2$ and $MoSi_2$ form upon processing at 550°C for 60 min. Significant interdiffusion of the metal atoms does not occur until 800°C, while near coincidence of the RBS spectra for the samples annealed at 900 and 1000°C implies that the equilibrium requires temperatures as high as these. For the samples annealed at 700°C the amount of compound in the solution is about constant, approximately 1 mol % $MoSi_2$ in $CrSi_2$ and 0.5 mol % $CrSi_2$ in $MoSi_2$. At 1000°C the respective fractions are 8.5 and 5.5 mol %. All $CrSi_2$ has regular hexagonal structure, while the structure of $MoSi_2$ is expected to change from hexagonal after annealing at 700°C to pure tetragonal after 1000°C annealing.

Identical regularities are observed for W(51 nm)/Cr(92 nm)/Si(100). Two phases, which are $CrSi_2$ and WSi_2, intermix at 800°C and above. Below 1000°C very little WSi_2 is found in $CrSi_2$. After annealing at 1000°C the amount of WSi_2 in $CrSi_2$ is about 10 mol %. The amount of $CrSi_2$ in WSi_2 increases from about 5 mol % after 700°C to 25 mol % after processing at 1000°C.

Diffusion mixing of molybdenum and tungsten atoms in $CrSi_2$ occurs at remarkably high temperatures (900–1000°C). This might be due to the slow diffusion of large atoms in a matrix where the host atoms are smaller.

In the Fe-Si system, addition of Co or Ni at concentrations below 12 at % does not change the formation and stability of different iron based silicides [1.114]. The first stages of the synthesis are initiated at 450°C. They result in the formation of the monosilicides Fe(Co)Si, Fe(Ni)Si. No evidence of solid metal solutions in silicon and Fe_3Si related phases was observed.

Summarizing what is known for kinetics of ternary semiconducting silicide formation, one can easily conclude that while there are some acceptable theoretical viewpoints on the regularities of the processes involved, but their experimental investigation is far from being completed.

1.4 Thermal Oxidation of Silicides

Thin films of all studied semiconducting and metallic silicides on a silicon substrate oxidize and form SiO_2 on their surface when they are annealed in an oxidizing ambient. *A priori* one could expect formation of a mixture of silicon and metal oxides, but thermodynamic analysis based on the comparison of heat of silicon and metal thermal oxidation in silicides performed by Bartur [1.216] and Strydom *et al.* [1.217] proves that all transition metal silicides, except metallic $ZrSi_2$ and $HfSi_2$, have a preference to form SiO_2. However, traces of metal oxides registered during oxidation of $TiSi_2$, $NbSi_2$, and $TaSi_2$ reveal that kinetics of the proceeding processes may play a role too.

Formation of a metal-free SiO_2 has been confirmed in numerous experiments with metallic silicides (see [1.218]). The data for semiconducting silicides, in general, are identical, while they are limited by the results obtained in the study of thermal oxidation of $CrSi_2$ [1.171, 1.217, 1.219–1.221], $Mn_{11}Si_{19}$ [1.220, 1.222], $ReSi_2$ [1.220], β-$FeSi_2$ [1.222], Ru_2Si_3 [1.220, 1.223, 1.224], Ir_3Si_5 [1.220, 1.224]. In addition, results for metallic tetragonal $MoSi_2$ [1.220, 1.225, 1.226] and WSi_2 [1.220, 1.225, 1.227] are also useful to estimate the oxidation behavior of the semiconducting hexagonal phases of these disilicides.

Tracer experiments using radioactive silicon [1.217] and inert markers [1.222, 1.228] show that SiO_2 formation takes place at the oxide/silicide interface, silicon is very mobile in silicides during oxidation, and the process is mainly controlled by a slower oxidant diffusion through the growing SiO_2 layer. As for the role of metal atoms, the experimental observations of metal-free SiO_2 formation and integrity of the oxidized silicide on silicon substrates suggest two possible modes of their behavior during the oxidation. These are fast diffusion of metal atoms from the oxide/silicide interface towards the silicide/silicon interface and restoration of the silicide composition there and/or formation and subsequent vaporization of volatile metal oxides.

A small amount of metal oxide, as has been detected on $CrSi_2$ [1.217], can be formed only during the low temperature initial stages of oxidation when silicon is not mobile enough in the silicide. At high temperatures (900–1100°C) the metals are eventually lost, presumably in the form of volatile oxides [1.226, 1.227]. Moreover, if the pressure is lower than the dissociation pressure of the metal oxide there will be no metal oxidation. Thus, even SiO_2 formation is not observed on semiconducting Ru_2Si_3 and metallic $CoSi_2$ oxidized at 750°C at a pressure of about 1.3×10^{-4} Pa [1.223], which is due to the volatilization of the nascent product.

However, both processes, which are fast diffusion of metal atoms and vaporization of volatile metal oxides, result in the dominant role of silicon and oxidant behavior in the oxide layer growth. Concequently, the oxide growth may be described by processes separated into four major distinct steps:

1. Diffusion of oxygen-containing molecules through the SiO_2 layer to the oxide/silicide interface.
2. Silicon oxidation at the oxide/silicide interface.
3. Transport of silicon and metal through the silicide.
4. Reactions at the silicide/substrate interface.

The first two steps are, in general, identical for an oxide grown on silicides and on silicon. At the same time, the third and fourth ones, being much faster than those, provide an unlimited source of silicon for oxidation at the oxide/silicide interface. This makes it useful to analyze the kinetics of the silicide thermal oxidation with the linear-parabolic equation first proposed by Deal and Grove for silicon [1.229]. Then, the specific behavior of silicon and metal atoms in the oxidizing semiconducting silicides can be discussed.

The derivation of the linear-parabolic kinetics for oxidation of silicides is in agreement with the experimental observations [1.216, 1.220, 1.225] and theoretical approaches developed [1.230]. Thus, the oxide growth obeys the equation

$$d^2 + (U/V)d = U(t + \tau^*), \qquad (1.26)$$

where d is the oxide thickness, V and U are, respectively, linear and parabolic rates, and τ^* represents the time shift corresponding to the presence of the initial oxide layer before oxidation. The rates and the time shift depend on the type of the oxidizing ambient (dry oxygen or wet oxidation), temperature, and pressure in the oxidation chamber. As most oxidation processes are performed at atmospheric pressure, the first two factors appear to be of primary importance. The temperature dependence of linear and parabolic rates is usually presented in the Arrhenius form (see (1.17)) with corresponding pre-exponential factors and activation energies.

There was another attempt to describe silicide oxidation combining thermodynamic and kinetic approaches [1.231, 1.232]. This tries to account for variations in the chemical activity of the material being oxidized, as a result of the oxidation process itself. Unfortunately, practical application of the derived equations to a quantitative analysis requires specific information about diffusion in silicides and about the correlation between silicon concentration and activity coefficient, which is not available today. Moreover, the linear-parabolic kinetics appear to be a close case to the mathematics proposed.

The available experimental data characterizing linear and parabolic rates of thermal SiO_2 growth on semiconducting silicides and on silicon are summarized in Table 1.39. Their absolute values calculated with these data are shown in Figs. 1.34 and 1.35. For monocrystalline silicon both linear and parabolic rate constants exhibit a break in their activation energies at about 950°C [1.233]. At higher temperatures the activation energy for the linear rate becomes 1–1.5 eV lower, while for the parabolic one it appears to be higher by the same value.

Table 1.39. Growth parameters of SiO_2 on silicon and silicides on silicon at atmospheric pressure

Oxidized material	Temperature range (°C)	Linear rate		Parabolic rate		Refs
		Pre-exponential factor[a], V_0 (cm/s)	Activation energy, E_a (eV)	Pre-exponential factor[a], U_0 (cm²/s)	Activation energy, E_a (eV)	
dry oxygen oxidation						
Si(111)	700 – 1200	0.175	1.99	2.38×10^{-9}	1.24	1.229[b]
	800 – 1000	2.22×10^{-2c}	1.74	2.23×10^{-7c}	1.71	1.233
Si(100)	800 – 1000	1.23×10^{-2c}	1.76	2.83×10^{-5c}	2.22	1.233
	780 – 1000	3.78×10^{-3}	1.6	5.83×10^{-6}	2.1	1.235
Si(110)	800 – 1000	0.788^c	2.10	6.22×10^{-8}	1.63	1.233
CrSi$_2$	950 – 1100	-	-	1.0×10^{-9c}	1.3^c	1.217
	750 – 800	4.92×10^{-6}	0.815^a	2.57×10^{-6}	1.89^a	1.220
MoSi$_2$[d]	808 – 1104	-	-	1.69×10^{-7c}	1.6	1.236
	900 – 1100	8.58×10^{-2}	1.9	3.28×10^{-8}	1.6	1.226
WSi$_2$[d]	750 – 800	4.35×10^{-5}	1.02^a	8.57×10^{-6}	2.07^a	1.220
Ru$_2$Si$_3$	750 – 800	1.67×10^{5}	3.08^a	2.63×10^{-12}	0.767^a	1.220
Ir$_3$Si$_5$	750 – 800	9.02	2.15^a	1.0×10^{-10}	1.09^a	1.220
wet oxidation						
Si(111)	920 – 1200	2.08	1.96	1.11×10^{-9}	0.79	1.229
	800 – 1000	7.02×10^{-2}	1.65^a	2.83×10^{-8}	1.15^a	1.225
	917 – 1098	34.7	2.25^a	5.88×10^{-10}	0.70	1.237
Si(100)	800 – 1000	5.68×10^{-2}	1.64^a	1.21×10^{-7}	1.35^a	1.225
Si-poly	800 – 1000	9.23×10^{-2}	1.66^a	1.21×10^{-7}	1.35^a	1.225
CrSi$_2$	850 – 1100	-	-	1.6×10^{-7c}	1.4	1.217
MoSi$_2$[d]	800 – 1000	-	-	8.06×10^{-8c}	1.3	1.236
	800 – 1000	1.85×10^{-4}	0.81^a	$\sim 2.20 \times 10^{-8}$	$\sim 1.10^a$	1.225
WSi$_2$[d]	800 – 1000	9.77×10^{-4}	0.98^a	1.51×10^{-7}	1.35^a	1.225

[a] Calculated from the experimental data presented in the reference,
[b] $\tau^* = 1.2 \times 10^{-5} \exp(2.02 \text{ eV}/k_B T)$ s has been calculated from the experimental data presented in the reference, [c] Presented in the reference, [d] Tetragonal metallic phase.

The effect has been attributed to a viscous flow which occurs in SiO_2 at the above mentioned temperature [1.234]. This is why in the analysis and use of the data presented one should be careful with regard to the temperature range in which they have been obtained.

In general, silicides oxidize at a higher rate than pure silicon. The data collected in Table 1.39 show that the activation energies for the linear rate for the silicide oxidation are much lower than those for the silicon oxidation. The unexpectedly high activation energies extracted for Ru$_2$Si$_3$ (3.08 eV) and Ir$_3$Si$_5$ (2.15 eV) can be considered to be a result of the extremely narrow temperature range (750–800°C) of the experimental investigation performed enhancing the role of the experimental uncertainty of the measured oxide thickness.

The linear rates of metallic silicide oxidation are about ten times greater than that of pure silicon, but for semiconducting ones it has been found to be intermediate between the values for silicon and the metallic silicides.

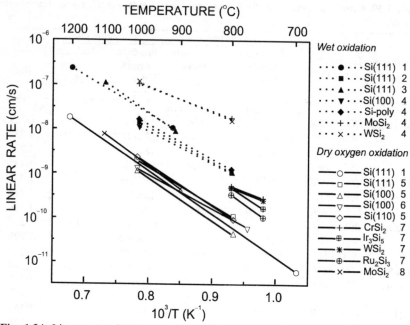

Fig. 1.34. Linear rate of SiO$_2$ growth on silicon and silicides on silicon at atmospheric pressure as a function of temperature: 1-[1.229], 2-[1.225], 3-[1.237], 4-[1.225], 5-[1.233], 6-[1.235], 7-[1.220], 8-[1.226]

As a first hypothesis it is assumed [1.224] that the increased rate of oxide formation (referring to the linear term) over metallic silicides results from the lack of the *sp* hybridization which is a characteristic of the covalent bonding of pure silicon. Actually, formation of silicon–oxygen bonds depends on the possibility of charge transfer between the surface atoms and the reacting species [1.238]. This is less probable when the material is an insulator, but it is greatly enhanced when the density of electrons at the Fermi surface is sufficiently high, as in metallic materials.

Charge transfer to molecules will eventually result in the population of antibonding orbitals, which weaken the chemical bonds. The rate of oxide formation is enhanced over metallic silicides, as compared to semiconducting ones or silicon, because of the ability to sustain localized variations of the electron density, i.e. to form a new chemical bond, which seems to be a characteristic of metallic surfaces.

The parabolic rates and their activation energies are about the same for oxidation of both silicides and silicon. For dry oxidation the data obtained by various studies clearly fit the process with one activation energy. Even the data for wet oxidation can be taken as essentially indicative of one process only. The first order conclusion is that the diffusion process through the oxide during oxidation is the same for all the silicides.

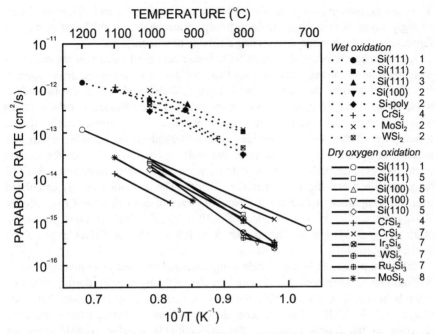

Fig. 1.35. Parabolic rate of SiO₂ growth on silicon and silicides on silicon at atmospheric pressure as a function of temperature: 1-[1.229], 2-[1.225], 3-[1.237], 4-[1.217], 5-[1.233], 6-[1.235], 7-[1.220], 8-[1.226]

This fact implies that the equilibrium concentration of the oxidant in the oxide and its diffusion coefficient through the oxide are the same regardless of the substrate (mono-, polycrystalline silicon or silicide). The comparison confirms the rule: during oxidation of a silicide, which has been formed by a diffusion-controlled process, the dominant moving species are the same as in the silicidation. This is physically plausible. In the silicide formation or oxidation we deal with the same bulk and hence with the same relative magnitude for diffusivities.

If, at a certain temperature, the silicide formation (which usually occurs at lower temperatures than oxidation for comparable silicon consumption rates) has already become diffusion-controlled, the reaction rate at the silicon/silicide interface during oxidation is not expected to influence the process much. Hence, we infer similar boundary conditions generated by the chemical reactions at the interfaces. Thus, similar bulk and interface properties result in similar atomic fluxes in both cases.

Diffusion of the oxidant through SiO₂ proceeds in the same way for all silicides. The observed difference in the oxide thickness is therefore caused by dissimilarities in the linear (early) stage of the oxidation process. It is difficult to discuss these dissimilarities quantitatively because the data available are scarce.

The same activation energy for the parabolic rate of silicides and silicon oxidation strongly suggests that the controlling transport mechanism is diffusion of oxygen regardless of whether one is oxidizing a silicide or silicon.

Kinetic measurements show that SiO_2 formation on silicides is some orders of magnitude faster during wet oxidation than that during oxidation in dry oxygen. It is well illustrated by the data plotted in Figs. 1.34 and 1.35. The linear rate of the oxide growth over silicides depends on the nature of the silicides. For the silicides with a strong metallic character, it is one or two orders of magnitude higher than it is for pure silicon. For the silicides with a marked semiconducting nature the reaction at the silicide/oxide interface becomes much slower. The corresponding linear rate assumes values intermediate between those which have been obtained for pure silicon and those for metallic silicides. The activation energy calculated from the experimental data for the parabolic rate of oxidation for $CrSi_2$, $MoSi_2$, WSi_2 (also $TaSi_2$) is in the range of 1.10–1.35 eV, which is close to silicon (1.35 eV) [1.225]. For the linear rate it is 0.8–1.0 eV for disilicides and about 1.65 eV for silicon.

Diffusion processes in the silicide being oxidized proceed in an unusual way. It is important to note here that in silicides the diffusing species during oxidation were believed to correlate with the moving species in the silicide formation [1.216, 1.228, 1.230]. Thus, silicon diffusion from a silicide/silicon substrate interface to the reaction zone at the oxide/silicide interface should dominate during oxidation of disilicides and close in composition silicon-rich silicides. With respect to thermodynamics the diffusion of silicon in one direction and that of metal in the opposite direction are perfectly equivalent. Nevertheless, there are reliable experimental observations of parallel diffusion of both metal and silicon atoms from the oxide/silicide to the silicide/silicon interface. These clearly indicate that chemical potential of the atoms in the oxidation zone can be dramatically modified resulting in the change of the direction of the driving force for the diffusion. Such a parallel diffusion was observed with inert markers in $CrSi_2$ [1.219], $Mn_{11}Si_{19}$ and $FeSi_2$ [1.222]. The same general trends were also detected in the typically metallic $CoSi_2$ and $PdSi$ [1.228], $TiSi_2$ [1.222], $NiSi_2$ [1.222, 1.228], $NbSi_2$ [1.222], VSi_2 [1.219]. Concequently, both metal and silicon diffusion towards the silicon substrate are concluded to be responsible for a metal-free SiO_2 growth on top of the silicides regardless of the predominant moving species during silicide formation. About 25% of the silicon atoms available for oxidation are involved in this process [1.219]. The observed dual parallel diffusion is explained by the very nature of the oxidation process, which modifies the chemical potential of the atoms at the surface being oxidized.

Reactions at the silicide/substrate interface depend to a great extent on the nature of the substrate material. The traditionally used silicon actively responds to the diffusion processes in the silicide. The silicide/silicon interface absorbs silicon and metal atoms coming from the oxide/silicide interface or releases silicon for diffusion to the oxide/silicide interface. In both cases the reactions taking place there preserve the composition of the silicide. When the oxidized silicide is on a

SiO_2 covered substrate, silicon is consumed only from the silicide, thus reducing its composition to a metal-rich phase [1.217, 1.227].

In conclusion, properties of the oxides thermally grown on semiconducting silicides are close to SiO_2 on pure silicon. Their dielectric constants (3.1–3.7) and densities (2.14–2.27 g/cm^3) are consistent with those (3.8–3.9 and 2.18–2.29 g/cm^3, respectively) of SiO_2 thermally grown on silicon [1.239]. Buffered HF etch rates of such oxides [1.220] indicate that those formed on the rapidly oxidizing silicides are less dense than the oxides on the intermediate and slowly oxidizing ones. At the same time, the dielectric strength (up to 0.3×10^{-6} V/cm) is much lower than that of SiO_2 on silicon ((6.9–9.0)$\times 10^{-6}$ V/cm) and varies considerably. Additional experimental investigations of optical and electrical properties of the oxides thermally grown onto silicides are of practical importance for their application in semiconductor devices and integrated circuits.

2 Thin Film Silicide Formation

Victor E. Borisenko

Belarusian State University of Informatics and Radioelectronics,
Minsk, Belarus

CONTENTS

Fabrication of thin film silicides exploits practically the whole variety of techniques developed in microelectronics. Three main groups of these techniques can be technologically distinguished on the basis of the dominant physical and chemical processes involved. These are: diffusion synthesis, ion beam synthesis, and atomic and molecular deposition. The detailed classification of the techniques and schematic presentation of the thin film structures produced are given in Fig. 2.1. Atomic diffusion, ion implantation, and surface reactions of deposited atoms are considered as the dominant processes used for the silicide formation within the selected technological groups. The silicon substrate at which a silicide film is expected to be formed is shown in the scheme to have both film free and silicon dioxide covered regions. This represents a typical surface structure of an integrated circuit or discrete semiconductor device where silicon dioxide film is used to protect and electrically isolate selected regions of the silicon substrate.

The main technological steps in the fabrication of thin film silicides include deposition of silicide forming components and their thermal processing for silicidation. In the group of atomic and molecular beam deposition, these two steps are realized within one technological cycle by maintaining substrates during the component deposition at an elevated temperature sufficient for silicidation.

The variety of the techniques developed supposes pure metal, mixed metal+silicon, or one component enriched silicide films to be deposited at the first step.

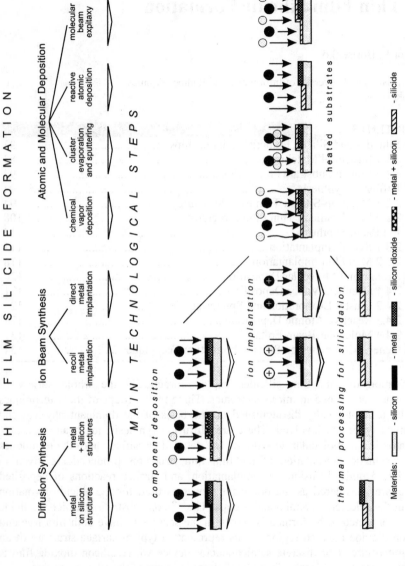

Fig. 2.1. Technological methods used for fabrication of thin film silicides and the structures produced

Subsequent or superposed thermal processing initiates and completes synthesis of stoichiometric silicide phases in the films. Rapid thermal processing regimes, involving heating cycles of a few tens of seconds, show great promise in carrying out such a treatment [2.1]. The short duration of the anneals eliminates or considerably reduces contamination and oxidation-related problems.

Practical application of different techniques for formation of thin film semiconducting silicides is illustrated by Table 2.1 summarizing examples from basic and applied studies. It is evident that not all of the technological approaches developed for silicides have been used for semiconducting ones. Meanwhile, there is no universal technology appropriate for all of them. Diffusion synthesis seems to be the best candidate for this, but one should remember that in addition to stoichiometry we need a particular crystalline structure and impurity composition of the semiconductors. Polycrystalline and epitaxial monocrystalline films differ in electronic properties and in applications. Dramatic changes in the properties can be achieved by doping or ternary compound formation. Flexible regulation of these factors is usually impossible within one approach. This makes it imperative to analyze peculiarities of the technological approaches within a selected group. This is done in the sections of this chapter. With regard to the prospects of epitaxial silicide films in silicon-based optoelectronics and nanoelectronics, we start the analysis with the theoretical consideration of epitaxial relationships between monocrystalline silicon and semiconducting silicides, while related experimental examples are given and discussed in the following sections.

2.1   Silicides and Silicon Epitaxial Relationships

Two main factors are considered to determine the quality of an epitaxial growth. These are structural matching of the substrate and the growing film and atomic perfection of the substrate surface during deposition. They are analyzed here for semiconducting silicides and silicon.

2.1.1   Structural Matching

The geometrical concept of the lattice match for any pair of crystal lattices in any given crystal direction has been proposed by Zur et al. [2.2, 2.3]. It allows a reconstruction of the interface with the periodicity of the common unit cell. Lattice match in this context means that the two-dimensional interface possesses translation symmetry that is compatible with that of the bulk on both sides of the interface. Such compatibility enables local structure of the interface to repeat periodically over large distances.

The model covers only sharp, single-crystalline interfaces with a possible periodic reconstruction. Polycrystalline interfaces, stepped ones and interfaces in which the atomic reconstruction is not periodic are not covered.

Table 2.1. Experimentally demonstrated applicability of different technological approaches to form polycrystalline (poly) and epitaxial (mono) thin film semiconducting silicides

TECHNOLOGICAL APPROACH	II group metals		VI group metals			VII group metals				VIII group metals			
	Mg_2Si	$BaSi_2$	$CrSi_2$ (ternaries)	$MoSi_2$	WSi_2 (ternaries)	$MnSi_{2-x}$ (ternaries)	$ReSi_{1.75}$ (ternaries)	$\beta\text{-}FeSi_2$ (ternaries)	Ru_2Si_3	$OsSi$	Os_2Si_3	$OsSi_2$	Ir_3Si_5
DIFFUSION SYNTHESIS													
Diffusion in metal-on-silicon thin film structures	poly mono		poly mono (poly)	poly mono	mono (poly)	poly mono	poly mono (poly)	poly mono (poly)	poly mono		poly	poly mono	poly
Diffusion in metal+silicon thin film structures			poly	poly	poly			poly mono (poly)				poly	
ION BEAM SYNTHESIS													
Recoil implantation			poly	poly	poly								
Metal ion implantation			mono					poly mono (poly)					
ATOMIC AND MOLECULAR DEPOSITION													
Chemical vapor deposition			poly				poly	poly mono					
Cluster evaporation and sputtering								poly mono (poly)					
Reactive atomic deposition	mono	mono	mono			mono	mono (mono)	poly mono (mono)					
Molecular beam epitaxy	poly	mono	mono					mono (mono)					

Local chemistry at the interface is also not considered. The lattice match, as defined by Zur *et al.* [2.2, 2.3], is characterized by two parameters, which are a mismatch and a dimension of the common unit cell. In the general case a common unit cell can be represented by a parallelogram, whose sides are a and b, with an acute angle α between them. These three parameters, namely a_s, b_s and α_s for the substrate, may be slightly different for the corresponding a_f, b_f and α_f of the deposited material. This difference determines the mismatch:

$$\eta = \max\left(\left|\frac{a_s - a_f}{a_s}\right|, \left|\frac{b_s - b_f}{b_s}\right|, \left|\frac{\alpha_s - \alpha_f}{\alpha_s}\right|\right). \tag{2.1}$$

The area of the common parallelogram, i.e.,

$$s = ab\sin\alpha \tag{2.2}$$

is the other parameter relevant to the lattice match. Actually, one can calculate two areas corresponding to the substrate and the film, but in the case of epitaxial growth they should be very close to each other. For every choice of maximal area and mismatch, the number of possible matches is finite. Most of them for binary semiconducting silicides are presented in Table 2.2.

Table 2.2. Calculated geometrical lattice matching of semiconducting silicides to silicon

Silicide	Matching faces silicide//silicon	Orientations silicide//silicon	Common unit cell area (nm²)	Lattice mismatch (%)	Refs
Mg_2Si	(111)//(111)		0.177	1.9	2.4
$BaSi_2$	(100)//(111)	[001]//[110]	0.771	1.6	2.5
$CrSi_2$	(0001)//(100)	[1$\bar{1}$00]//[011]	1.19	0.9	2.6
	(11$\bar{2}$0)//(100)	[1$\bar{1}$00]//[011]	1.47	0.6	
	(11$\bar{2}$0)//(100)	[1$\bar{1}$00]//[011]	1.47	0.6	
	($\bar{1}\bar{1}$22)//(001)	[1$\bar{1}$00]//[1$\bar{1}$0]	0.55	0.95	2.7
	(0001)//(110)	[1$\bar{1}$00]//[1$\bar{1}$0]	0.85	1.7	2.6
	(10$\bar{1}$0)//(110)	[$\bar{1}$2$\bar{1}\bar{1}$]//[1$\bar{1}$0]	0.85	1.0	
	($\bar{2}$114)//(011)	[01$\bar{1}$0]//[01$\bar{1}$]	0.84	0.41	2.7
	(0001)//(111)[a]	[01$\bar{1}$0]//[1$\bar{1}$0][a]	0.51	0.1	2.6
		[10$\bar{1}$0]//[10$\bar{1}$][a]	0.51	0.13	2.7
		[10$\bar{1}$0]//[1$\bar{1}$0][a]	0.51	0.1	2.8
		[10$\bar{1}$0]//[110][a]	0.51	0.17	2.9
		[10$\bar{1}$0]//[110][a]	0.51	0.3	2.10
	(0001)//(111)[b]	[11$\bar{2}$0]//[1$\bar{1}$0][b]	0.17	14	2.8

Silicide	Matching faces silicide//silicon	Orientations silicide//silicon	Common unit cell area (nm²)	Lattice mismatch (%)	Refs
CrSi₂		[112̄0]//[110]ᵇ	0.17	15.3	2.10
	(0001)//(111)	[451̄0]//[11̄0]	1.19	1.7	2.6
	(101̄1)//(111)	[033̄1̄]//[231]	0.99	2.2	
	(101̄1)//(111)	[21̄1̄1̄]//[011̄]	1.32	3.0	
	(0001)//(111)	[112̄0]//[110]	6.12	1.2	2.10
	(111)//(111)	[101̄0]//[110]	0.51	1.05	2.9
MoSi₂ (hexagonal)	(101̄0)//(110)	[2̄42̄1̄]//[11̄0]	1.21	1.9	
	(101̄0)//(110)	[0001]//[11̄2]	0.60	2.1	
	(101̄1)//(111)	[0331]//[231]	1.06	2.6	2.6
	(0001)//(001)	[21̄1̄0]//[11̄0]	1.84	4.04	2.11
	(1̄21̄2)//(001)	[12̄13]//[010]	0.84	2.89	
	(0001)//(111)	[11̄00]//[11̄0]	0.55	4.04	
WSi₂ (hexagonal)	(0001)//(001)	[21̄1̄0]//[11̄0]	1.84	4.04	2.11
	(1̄21̄2)//(001)	[12̄13]//[010]	0.84	3.01	
	(0001)//(111)	[11̄00]//[11̄0]	0.55	4.04	
MnSi₁.₇₅ (Mn₄Si₇)	(001)//(100)	[100]//[001]	0.30	1.4	2.6
ReSi₂ᶜ	(110)//(100)	[001]//[011]	1.19	1.0	2.6
	(110)//(100)	[33̄1̄]//[011]	1.19	0.8	
	(100)//(100)	[001]//[011]	1.20	2.1	
	(100)//(001)	[010]//⟨110⟩	1.20	1.8	2.12
	(110)//(110)	[001]//[11̄0]	0.85	1.8	2.6
	(110)//(110)	[331]//[11̄0]	0.85	1.6	
	(110)//(111)	[001]//[11̄0]	0.51	0.2	2.6
β-FeSi₂	(100)//(001)	[010]//⟨110⟩	0.59	2.1	2.13
	(100)//(001)ᵃ	[010]//⟨110⟩ᵃ	0.31	1.8	2.14
	(100)//(001)ᵃ	[010]//⟨110⟩ᵃ	0.61	1.5	2.15
		[001]//⟨110⟩ᵃ	0.61	2.0	
	(100)//(001)ᵇ	[010]//⟨100⟩ᵇ	1.22	4.0	2.14, 2.15
	(100)//(001)ᵇ	[001]//⟨100⟩ᵇ	1.22	4.0	2.15
	(101)//(111)ᵃ	[101]//[11̄0]ᵃ	0.98	1.45	2.16
		[101]//[1̄01]ᵃ			
		[101]//[01̄1]ᵃ			

Silicide	Matching faces silicide//silicon	Orientations silicide//silicon	Common unit cell area (nm²)	Lattice mismatch (%)	Refs
β-FeSi₂	(110)//(111)[b]	[101]//[$\bar{1}$10][b]	0.98	2.0	2.16
		[101]//[$\bar{1}$01][b]			
		[101]//[0$\bar{1}$1][b]			
Ru₂Si₃	(100)//(100)	[001]//[001]	1.48	1.8	
	(100)//(110)	[001]//[001]	1.48	1.8	2.6
OsSi	(001)//(100)	[110]//[013]	0.70	2.5	
cubic with	(001)//(100)	[230]//[001]	1.14	1.7	
a=0.2963 nm	(001)//(100)	[100]//[011]	1.40	2.8	
	(001)//(100)	[100]//[012]	1.40	2.5	
	(001)//(100)	[110]//[013]	1.40	2.5	
	(001)//(100)	[140]//[012]	1.49	0.6	
	(223)//(100)	[1$\bar{1}$0]//[013]	1.45	2.5	
	(014)//(100)	[100]//[012]	1.45	2.5	
	(124)//(100)	[23$\bar{1}$]//[001]	1.21	2.0	
	(001)//(110)	[120]//[1$\bar{1}$2]	1.49	2.8	
	(023)//(110)	[032]//[001]	1.27	2.8	
	(234)//(110)	[20$\bar{1}$]//[1$\bar{1}$2]	1.42	2.3	
	(001)//(111)	[210]//[1$\bar{2}$1]	1.40	2.2	
	(012)//(111)	[02$\bar{1}$]//[2$\bar{1}$$\bar{1}$]	0.79	2.8	
	(234)//(111)	[20$\bar{1}$]//[121]	1.42	2.3	2.6
OsSi	(001)//(100)	[120]//[001]	1.12	2.7	
cubic with	(011)//(110)	[01$\bar{1}$]//[1$\bar{1}$2]	0.63	0.5	
a=0.4729 nm	(011)//(111)	[01$\bar{1}$]//[2$\bar{1}$$\bar{1}$]	1.27	1.5	
	(111)//(111)	[10$\bar{1}$]//[1$\bar{2}$1]	0.39	0.5	2.6

[a] The epitaxial relationships identified as A-type, [b] The epitaxial relationships identified as B-type, [c] For the smallest unit cell area. For structure comments see the text.

A small lattice mismatch is a favorable prerequisite for epitaxial layer growth. However, note that the lattice match depends on temperature and the criterion for selection does not include any chemical effects at the interface.

For good quality epitaxial layers, the lattice mismatch between the silicide and the silicon substrate is typically less than 3% [2.17]. At a certain mismatch a pseudomorphic epitaxial growth can take place up to a critical thickness determined by the ratio of the interatomic forces in the growing layer, substrate and across the interface. This has been confirmed by the analysis performed by Markov and Milchev [2.18] on the basis of the anharmonicity in the interatomic forces in epitaxial structures. Moreover, it is concluded that pseudomorphic growth should be observed at larger mismatch when the interatomic distance of the overgrowth is smaller than that of the substrate, as compared with the reverse case (the interatomic distance of the overgrowth is larger than that of the substrate). The lattice mismatch also plays a prominent role in determining the mode of growth [2.19]. The larger the mismatch, the greater is the tendency for island-like growth and vice versa.

A good geometrical lattice match is never the only sufficient condition for a high quality epitaxy. Chemistry of the interface always plays a major role as far as particular epitaxial growth is concerned. Nevertheless, an estimated poor match indicates that high quality epitaxial growth in the system is hardly possible. Available geometrical details of lattice matching of semiconducting silicides and silicon are analyzed below. Note that not all of the semiconducting silicides have been carefully analyzed in this respect. The silicides with a clear tendency to epitaxial growth have been mostly covered.

Mg_2Si has, similar to silicon, a cubic structure but with a larger lattice parameter. For $a = 0.6351$ nm a side of the unit cell on the $Mg_2Si(111)1\times1$ surface is 0.449 nm, that is about 1.3% longer than 0.4434 nm, which is $2/3\sqrt{3}$ times the surface lattice parameter of the ideal $Si(111)1\times1$ surface (0.384 nm). Due to the small lattice mismatch high quality epitaxial growth of Mg_2Si onto monocrystalline silicon appears to be possible [2.4, 2.20].

$BaSi_2$. The lattice parameters for orthorhombic $BaSi_2$ suggest good structural matching on $Si(111)$ [2.5]. The c parameter (1.158 nm) is almost identical to three times the [110] spacing for silicon (1.152 nm). The b parameter (0.680 nm) is also very close to three times the [112] spacing for silicon (0.669 nm). Thus, epitaxial growth with $BaSi_2(100)//Si(111)$ and $BaSi_2\langle001\rangle//Si\langle110\rangle$ as well as related variants can be expected. The arrangement of atoms in the silicide (100) plane is shown in Fig. 2.2.

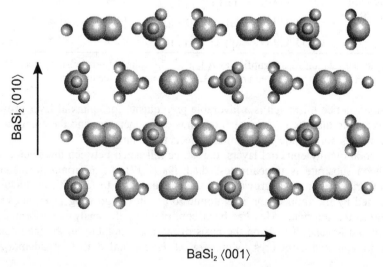

BaSi₂ ⟨010⟩

BaSi₂ ⟨001⟩

Fig. 2.2. A one-unit-cell projection of orthorhombic $BaSi_2$ [2.5]

Rows of barium atoms are seen as bridging elements between the isolated silicon tetrahedra. Stacking of these rows in (100) planes forms (010) planar channels through which silicon or barium can diffuse to react at the surface.

CrSi$_2$. Two principal epitaxial orientation relationships have been actually identified for CrSi$_2$ on Si(111) substrates [2.8–2.10, 2.21]: A-type, which has CrSi$_2$(0001)//Si(111) with CrSi$_2$[10$\bar{1}$0]//Si[1$\bar{1}$0], and B-type, which has CrSi$_2$(0001)//Si(111) with CrSi$_2$[11$\bar{2}$0]//Si[1$\bar{1}$0].

The major hexagonal axis of the CrSi$_2$ coincides with the Si(111) surface normal for A-type and B-type. These are related by a 30° rotation about that axis. This is illustrated in Fig.2.3. In the case of the A-type epitaxial relationships the common unit area is 0.511 nm^2 while there is scatter in the estimated mismatch from –0.1% to –0.3% (see Table 2.2). For the B-type relationships, the area is only 0.17 nm^2, while the mismatch is +15.3% [2.10]. The A-type lattice matching is so excellent in theory that it seems surprising that any B-type domains occur [2.10]. The understanding of this phenomenon is strengthened with the observation that the unit cell of the reconstructed Si{111}(7×7) surface is a possible common unit cell pertaining to the B-type epitaxial relationship, and that for this definition of the B-type common cell there is a small negative mismatch.

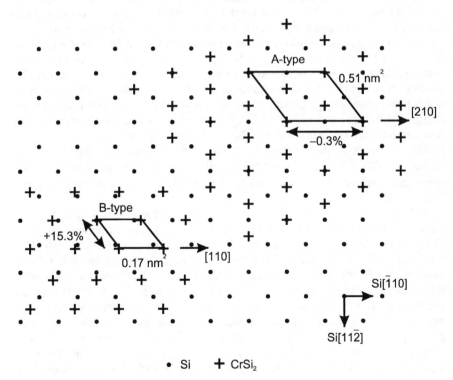

Fig. 2.3. Possible common unit cells of A-type and B-type azimuthal orientations for CrSi$_2$(100) on the Si(111) [2.10]

Steps on the Si(111) surface directly control the A-type or B-type relationships [2.22]. When the step width corresponds to the coincidence distance between silicon and silicide planes, the creation energy of the misfit dislocation associated with the B-orientation is reduced by a factor of 3 and the probability of developing the related orientation is substantially increased. This is the case for the B-type orientation: an angle of 7.5° gives an atomic step width of 2.3 nm, which corresponds to the unit cell of the reconstructed Si{111}(7×7), and the distance between misfit dislocations, as discussed by Mahan et al. [2.10].

MoSi$_2$, WSi$_2$. Hexagonal MoSi$_2$ and WSi$_2$ have almost identical lattice parameters and close thermal expansion coefficients. This results in their identical geometrical lattice match to silicon, as is clear from the data collected in Table 2.2. While the calculated mismatch reaches 4.04% at room temperature and increases up to 4.65% at 1100°C, epitaxial growth of both hexagonal phases is experimentally observed [2.11, 2.23].

ReSi$_{1.75}$(ReSi$_2$). Epitaxial relationships of this semiconducting silicide to silicon were theoretically analyzed expecting it to have ReSi$_2$ stoichiometry and tetragonal (MoSi$_2$-type) [2.6, 2.24] or orthorhombic structure [2.12, 2.25, 2.26]. In fact, the confirmed triclinic lattice of semiconducting ReSi$_{1.75}$ phase is very close to the body-centered orthorhombic lattice of this compound with $a = 0.3128$ nm, $b = 0.3144$ nm, $c = 0.7677$ nm (see Table 1.1), which in its form closely resembles the well known tetragonal MoSi$_2$-type structure. This was used in the analysis of its possible epitaxial relationships to monocrystalline silicon performed by Bai et al. [2.25] and Mahan et al. [2.12]. Fig. 2.4 illustrates the matching faces for both structures. Since the difference between a and b is only ~0.5%, the silicide structure was assumed for convenience to be tetragonal with $a = 0.313$ nm. It is evident that two rotations with mutually perpendicular azimuthal alignments ReSi$_2$[010]//Si[011] and ReSi$_2$[001]//Si[011] can be realized because the Si(100) face has fourfold symmetry and the ReSi$_2$(100) face has only twofold symmetry [2.25]. Moreover, the preferred epitaxial relationship is found to be ReSi$_2$(100)//Si(001) with ReSi$_2$[010]//Si⟨110⟩ [2.12]. This provides a room temperature mismatch of +1.8% along ReSi$_2$[010] but only of −0.03% along ReSi$_2$[001].

β-FeSi$_2$. Orthorhombic β-FeSi$_2$ acceptably matches Si(001) and Si(111) substrates. For β-FeSi$_2$(100) on Si(001) there are two possible common unit cells [2.13–2.15]. These are shown in Fig. 2.5. The matching faces of these two cells are identical, β-FeSi$_2$(100)//Si(001), but the azimuthal orientations differ by a rotation of 45° about the substrate surface normal. These are the cells with β-FeSi$_2$[010]//Si⟨110⟩ identified as A-type and with β-FeSi$_2$[010]//Si⟨100⟩ identified as B-type with the area of 0.61 and 1.22 nm^2, respectively. Each type has two variants corresponding to the orientation along β-FeSi$_2$[010] and β-FeSi$_2$[001] axes. The A-type looks more favorable as it has smaller mismatch, +2% against −4% for the B-type, to the substrate. Geometrical matching of β-FeSi$_2$ to Si(111) is also favorable for an epitaxial growth. Two crystallographic planes of this disilicide, the (101) and (110) planes, match the (111) surface of silicon [2.16].

ReSi$_2$(100)//Si(001) with ReSi$_2$[010]//Si⟨110⟩

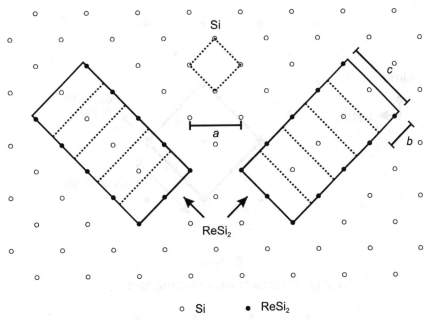

o Si   • ReSi$_2$

Fig. 2.4. Predicted common unit cell for ReSi$_2$(100) on the Si(001) face, together with the primitive unit cell of each material [2.12]

They have very close rectangular unit cells of 1.259×0.779 nm^2 and 1.257×0.783 nm^2, respectively, with acceptably low lattice mismatches (1.45 and 2.0%) along the [$\bar{1}$10] axis of silicon. Other equivalent azimuthal orientations presented in Table 2.2 are also possible.

OsSi. Analysis of geometrical matching between this monosilicide and silicon has been performed for two types of its cubic lattice, namely for the FeSi-type structure with $a = 0.4729$ nm and for the CsCl-type structure with $a = 0.2963$ nm [2.6]. While only the first one (with the FeSi-type structure) was reported to have semiconducting properties [2.27], it is useful to keep in mind the data presented in Table 2.2 for both of them, as this semiconducting silicide remains poorly investigated and new clear results can be expected. In the case of the FeSi-type structure, the best matching is calculated for OsSi(111)//Si(111) with OsSi[10$\bar{1}$]//Si[1$\bar{2}$1]. This is characterized by the common unit cell area of 0.39 nm^2 and 0.5% lattice mismatch. The monosilicide with the CsCl-type structure should provide the best epitaxy on (100) silicon substrates demonstrating OsSi(001)//Si(100) with OsSi[140]//Si[012].

OsSi$_2$. Only small changes of atomic positions in orthorhombic OsSi$_2$ are needed to obtain an atomic arrangement corresponding to that of the cubic CaF$_2$-type structure [2.28].

A - type

FeSi$_2$(100)//Si(001) with FeSi$_2$[010]//Si⟨110⟩

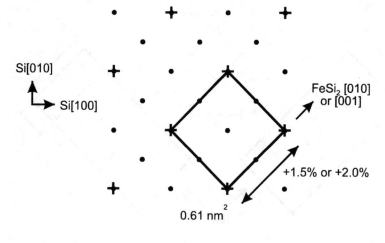

B - type

FeSi$_2$(100)//Si(001) with FeSi$_2$[010]//Si⟨100⟩

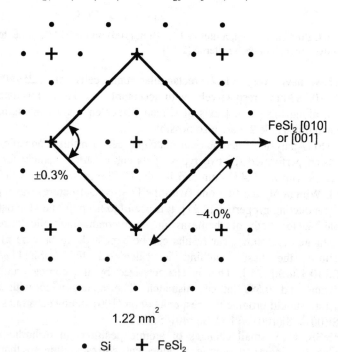

Fig. 2.5. Lattice matching of β-FeSi$_2$ to Si(001) [2.15]

The distortion of the OsSi$_2$ lattice from that structure type is reflected in the ratio between the length of the diagonal in the *bc* plane and the length of the *a* axis. This ratio is 1.14 showing that the deviation from the cubic structure can be interpreted as an expansion in the *bc* plane and a contraction in the direction of the *a* axis. As for the absolute value of the above estimated mismatch, it seems that perfect epitaxial growth of the orthorhombic OsSi$_2$ on silicon is not possible. An intermediate template layer is needed.

In general, lattice adjustment of a silicide to a silicon substrate can be achieved not only with an intermediate template layer but also with atomic step profiling of the substrate and by formation of an appropriate intermediate ternary or quaternary silicide with the best fitted lattice parameters.

2.1.2 Silicon Substrate Preparation

Perfect epitaxial growth needs an atomically clean surface of a silicon substrate and high purity components for deposition. A typical substrate preparation procedure includes cleaning in the buffered oxide etch (10/1 NH$_4$F/HF) followed by vacuum *in situ* ion beam milling and/or high temperature annealing. As far as the above chemical solution dissolves silicon dioxide, it can be employed only for silicon substrates without any oxide film used for pattering on the surface.

One of the ways to clean the silicon surface in vacuum is heating of the substrate to a temperature of about 800°C [2.29]. Under this condition, surface silicon dioxide reacts with underlying silicon to form silicon monoxide (SiO), which vaporizes and leaves an oxide-free surface. This is the so-called Shiraki clean after its inventor.

In situ "silicon-beam clean" [2.13, 2.30] is widely employed for preparation of atomically perfect silicon surfaces. It consists of the exposure of the surface to a silicon flux corresponding to a deposition rate of 0.06 nm/s for 120 s at 800°C [2.31]. The substrate temperature may be reduced to 700°C [2.13], while the silicon deposition rate has to be maintained at 0.02 nm/s for 250 s.

In addition, high vacuum heating at 900–1000°C was shown [2.32] to eliminate the influence of carbon-related precipitates on the crystal perfection of a silicon surface. However, carbon contamination is actually much more difficult to remove. It may significantly inhibit step propagation on Si(111) during the oxide desorption. A silicon buffer layer of about 100 nm grown *in situ* by molecular beam epitaxy at 700°C solves the problem [2.33]. A subsequent annealing at 830°C for 2 min removes the waviness present right after the growth. The step structure becomes regular. Another approach is vacuum annealing at 400°C followed by "silicon-beam clean" at 800°C [2.34] or vacuum annealing at 1100°C [2.35]. The above cleaning techniques result in a sharp 7×7 reconstructed pattern on Si(111) wafers and the two-domain 2×1 pattern on Si(001) ones as indicated by reflection high energy electron diffraction (RHEED). The purity of the components used for deposition is usually in the range of 99.9–99.999%.

2.2 Diffusion Synthesis

Two types of structure are used for formation of thin film silicides by solid state diffusion synthesis. These are pure metal films on a silicon substrate and films composed of intermixed silicon and metal atoms, as is schematically shown in Fig. 2.1. The films are deposited at room temperature by evaporation or sputtering. In the first case, the metal/silicon interface coincides with the interface between the metal film and the silicon substrate. Silicidation starts there during subsequent thermal processing. Silicon from the substrate is consumed by the growing silicide. No silicidation occurs on the substrate regions covered by silicon dioxide. In the second case, both silicide forming components are simultaneously deposited onto the substrate in amounts corresponding to or close to the stoichiometric ratio. Subsequent annealing provides uniform diffusion mixing of the components and their chemical interaction. The silicide is formed inside the film all over the substrate independently of the composition of the sublayers. Specific features of the above approaches and technological regimes employed for different semiconducting silicides are discussed below.

2.2.1 Metal-on-Silicon Thin Film Structures

Solid-state-diffusion synthesis of silicides in metal-on-silicon (metal/silicon) thin film structures inevitably brings about net volumetric changes related to silicon consumption from the substrate. Table 2.3 shows the thickness of the silicon consumed and the thickness of the silicide formed in units of the as-deposited metal thickness.

Table 2.3. Volumetric changes during silicide formation [2.36]

Metal	Silicide formed	Thickness of the consumed silicon normalized to the thickness of the as-deposited metal (d_{Si}/d_{Me})	Thickness of the silicide formed normalized to the thickness of the as-deposited metal (d_{sil}/d_{Me})
Mg	Mg_2Si	0.43[a]	1.38[a]
Ba	$BaSi_2$	0.64[a]	1.40[a]
Cr	$CrSi_2$	3.33	3.00
Mo	$MoSi_2$	2.57	2.60
W	WSi_2	2.52	2.58
Mn	$MnSi_{1.73}$ ($Mn_{11}Si_{19}$)	2.83	2.73
Re	$ReSi_2$	2.72	2.56
	$ReSi_{1.75}$	2.38[a]	2.49[a]
Fe	α-$FeSi_2$	3.64	3.36
	β-$FeSi_2$	3.39	3.20
Ru	Ru_2Si_3	2.21	2.53
Os	OsSi	1.43	1.89
	Os_2Si_3	2.15	5.00
	$OsSi_2$	2.87	3.07
Ir	Ir_3Si_5	2.37[a]	2.77[a]

[a] Calculated by V.E. Borisenko.

These are calculated from the atomic density of the respective elements in the silicide phases [2.36]. The volume change was assumed to be completely accommodated by a change of the thickness. Depending on the silicide composition, its final surface can appear below ($d_{Si}/d_{Me} > d_{sil}/d_{Me}$) or above ($d_{Si}/d_{Me} < d_{sil}/d_{Me}$) the original metal/silicon interface. The total volumetric change always results in a shrinkage, i.e. $d_{sil}/(d_{Me}+d_{Si}) < 1$. Both polycrystalline and epitaxial monocrystalline silicide films may be formed by this technique. The latter case is usually called solid state epitaxy.

The silicidation process is very critical on the ambient and temperature–time regimes of post-deposition thermal processing. Most of the metals chemically react with oxygen and nitrogen. These reactions compete with the silicide synthesis and can even suppress it when high temperature processing (about 1000°C and above) is employed for refractory metal silicide formation. Dry oxygen free argon or helium [2.37, 2.38], reducing hydrogen/argon mixture [2.39], forming gas (85% N_2, 15% H_2) [2.40, 2.41], or vacuum [2.42] ambient as well as protective films help to solve this problem. For processing at low temperatures, an aluminum film may be used as a protective cover [2.43]. Silicon protective films [2.44–2.47] or silicon dioxide ones [2.46, 2.48, 2.49] better suit the technology. They are efficient over the whole temperature range of practical interest. The negative influence of the residual ambient impurities may also be reduced in rapid thermal processing cycles lasting from a few to tens of seconds [2.1].

Thin film metal/silicon structures are classical for an investigation of silicidation kinetics. Details of the phase transformations in such structures were discussed in Sect. 1.3. Here we focus on technological regimes proposed for producing only semiconducting phases. A summary based on the published data is given in Table 2.4. The indicated thermal processing regimes suppose formation of these phases to be completed. One-step and two-step annealing schemes are used. The latter has been found to be effective in formation of epitaxial silicide films [2.1]. Semiconducting $BaSi_2$ and $OsSi$ films appear not to be possible as no example of their formation by diffusion synthesis in metal/silicon structures has been published.

Mg₂Si. Silicidation proceeds rather rapidly in magnesium/silicon thin film structures at relatively low temperatures. Thus at 275°C, 400 nm magnesium film is observed to be converted into the silicide within 70 min [2.50], while at 300°C, 30 min was enough to produce the silicide from the 300 nm metal film [2.43]. In the latter case, silicide columns matched to the monocrystalline silicon substrate were formed in the film.

There has been an attempt to fabricate Mg_2Si by laser melting ($\lambda = 1.06$ μm, 35 ns pulse) of magnesium film on silicon [2.81]. The synthesized layer contained both silicide and silicon crystallites.

CrSi₂. Formation of $CrSi_2$ films by diffusion synthesis in thin film structures meets practically no problems because this phase is the dominantly growing compound in the system.

Processing temperatures from the range 450–600°C can be used for formation of uniform single-phase disilicide polycrystalline films (see Table 2.4). A particular choice of annealing temperature and time depends on the thickness of the film desired.

Table 2.4. Conditions of diffusion synthesis of semiconducting silicides in metal-on-silicon thin film structures and crystallinity of the silicide produced

Silicide formed	Substrate	Thickness of as-deposited films (nm)	Thermal processing regime	Crystalline structure of the silicide	Refs
Mg_2Si	Si(100)	300	300°C/30 min	$\langle 100 \rangle$ oriented columns of 40–80 nm diameter	2.43
	Si	400 +500 nm of Al	275°C/70 min	no information	2.50
$CrSi_2$	Si(100)	100	450°C/120 min	polycrystalline	2.51
	Si(100)	110	475°C/30 min	polycrystalline (grains of ~10 nm)	2.52
	Si(111)	100–400	500–600°C/30 min	polycrystalline (grains of 20 nm)	2.53
	Si-am (200–400 nm) on SiO_2	50–100	900°C/60 min	polycrystalline (grains with preferred orientation along c axis)	2.54
	Si(001)	50	450°C/60 min	polycrystalline with epitaxial grains (100–150 nm): $CrSi_2(001)//Si(001)$ with $CrSi_2(110)//Si(110)$	2.55
	Si(111)	30 +20 nm of Si	1000–1100°C/60 min	epitaxial: $CrSi_2 (2\bar{2}0)//Si(111)$, $CrSi_2 (22\bar{4}0)//Si (22\bar{4})$, $CrSi_2 (20\bar{2}0)//Si (20\bar{2})$, $CrSi_2 [1\bar{2}13]//Si[101]$	2.56
	Si(001)	30 +20 nm of Si	1000–1100°C/60 min	epitaxial:$CrSi_2 (2\bar{2}00)//Si (2\bar{2}0)$ and $CrSi_2 [\bar{2}243]//Si[001]$	2.7
$Cr_{1-x}V_xSi_2$	Si(100)	270[a]	1200°C	polycrystalline	2.57
$MoSi_2$	Si(111)	30 +20 nm of Si	600–700°C/60 min 1000–1100°C/60 min	polycrystalline (grains of 17–27 nm) mixture of hexagonal (h) and tetragonal (t) epitaxial grains h-$MoSi_2$: $MoSi_2 (20\bar{2}0)//Si (20\bar{2})$ and $MoSi_2[0001]//Si[111]$ t-$MoSi_2$: $MoSi_2(004)//Si (\bar{2}02)$ and $MoSi_2[110]//Si[111]$	2.23
	Si(001), Si(111), Si-am	5–10	400–690°C/30 min	polycrystalline	2.58

Silicide formed	Substrate	Thickness of as-deposited films (nm)	Thermal processing regime	Crystalline structure of the silicide	Refs

WSi_2	Si(001), Si(111)	30 +30 nm of Si	600°C/60 min +1100°C/60 min	mixture of hexagonal (h) and tetragonal (t) epitaxial grains on Si(001)
				h-WSi$_2$: WSi$_2$ (20$\overline{2}$0)//Si(2$\overline{2}$0) and WSi$_2$[0001]//Si[001]
				WSi$_2$(2$\overline{1}$1̄2)//Si(2$\overline{2}$0) and WSi$_2$[$\overline{2}$42̄3]//Si[001]
				t-WSi$_2$: WSi$_2$(004̄)//Si(2$\overline{2}$0) and WSi$_2$[110]//Si[001]
				WSi$_2$(112̄)//Si(2$\overline{2}$0) and WSi$_2$[111]//Si[001]
				WSi$_2$(004)//Si(220) and WSi$_2$[100]//Si[001]
				on Si(111)
				h-WSi$_2$: WSi$_2$(20$\overline{2}$0)//Si(20$\overline{2}$) and WSi$_2$[0001]//Si[111]
				t-WSi$_2$: WSi$_2$(004)//Si(20$\overline{2}$) and WSi$_2$[110]//Si[111]
				WSi$_2$(112̄)//Si(20$\overline{2}$) and WSi$_2$[111]//Si[111]
					2.11

$W_{0.7}Ti_{0.3}Si_2$	Si(100), Si(110), Si(100)	100–150 of $W_{0.7}Ti_{0.3}$	725–750°C/45 min	polycrystalline with the hexagonal C40 structure	2.59
$W_{0.76}Ti_{0.24}Si_2$	PtSi(120 nm) on Si(100)	20 of $W_{0.7}Ti_{0.3}$	700–735°C/ 30–210 min	polycrystalline with the hexagonal C40 structure	2.60
$MnSi_{1.7}$ [b]	Si(100)	240	600°C/30 min	polycrystalline	2.61
	Si(100), Si-poly (500 nm) on SiO$_2$	140–550[c]	1000°C/60 min	polycrystalline	2.62
	Si(111)	30	600–700°C/60 min	polycrystalline (grains of 210–270 nm)	2.44
	Si(100)	300	540°C/60 min	polycrystalline	2.63

Silicide formed	Substrate	Thickness of as-deposited films (nm)	Thermal processing regime	Crystalline structure of the silicide	Refs
MnSi$_{1.7}$[b]	Si(111)	30	300°C/60 min +1050°C/60 min	epitaxial: MnSi$_{1.7}$(220)//Si($2\bar{2}0$) and MnSi$_{1.7}$[$1\bar{1}0$]//Si[111]	2.44
	Si(100)	30	300°C/60 min +1100°C/60 min	epitaxial: MnSi$_{1.7}$(100)//Si(400) and MnSi$_{1.7}$[001]//Si[001]	2.44
	Si-am (300 nm) on SiO$_2$	100 +15 nm of Si	450°C/60 min +850°C/60 min	no information	2.64
ReSi$_2$ (ReSi$_{1.75}$)	Si(111), Si(100), Si-poly (500 nm) on SiO$_2$	12–30[c] +25 nm of Si	900°C/60 min	polycrystalline	2.45
	Si(111), Si(001)	41	800–1000°C/30 min 1100°C/30 min or 500°C/60 min +1100°C/30 min	polycrystalline epitaxial on Si(111): ReSi$_2$(002)//Si($\bar{2}02$) and ReSi$_2$[110]//Si[111] on Si(001): ReSi$_2$($\bar{1}10$)//Si($2\bar{2}0$) and ReSi$_2$[110]//Si[001] ReSi$_2$($\bar{1}1\bar{2}$)//Si(220) and ReSi$_2$[110]//Si[001]	2.24
Re$_{1-x}$Mo$_x$Si$_2$	Si(111)	3–12[c]	750–850°C/15 min	epitaxial: ReSi$_2$[001]//Si[$1\bar{1}0$] and ReSi$_2$[$3\bar{3}1$]//Si[$01\bar{1}$]	2.65
	Si(100)	100[a]	1200°C	polycrystalline	2.57
β-FeSi$_2$	Si(001), Si(111)	30 +60 nm of SiO$_2$ or +30 nm of Si	700°C/60 min	polycrystalline (grains of ~160 nm)	2.46

Silicide formed	Substrate	Thickness of as-deposited films (nm)	Thermal processing regime	Crystalline structure of the silicide	Refs
β-FeSi$_2$	Si(100), Si-poly Si(111), Si-am	150	900°C/120 min	polycrystalline with a (202)/(220) grain texture	2.66
		24–47^c +30 nm of Si	625°C/60 min	polycrystalline	2.67
	Si(111)	40 +120 nm of SiO$_2$	650°C/60 min	polycrystalline (grains of ~220 nm)	2.48
	Si(100)	150	850°C/120 min	polycrystalline	2.68
	Si(100), Si(111)	24, 130, 160 +50, 70, 90 nm of Si 70, 100, 140 +80, 115, 160 nm of SiO$_x$	600°C/120 min	polycrystalline	2.47
	Si(111)	120	750°C/10 s	polycrystalline with a (202)/(220) grain texture (grains of ~20 nm)	2.49
	Si(001)	2.5 +60 nm of SiO$_2$	670°C/10 min	epitaxial: β-FeSi$_2$(100)//Si(001) with β-FeSi$_2$[010]//Si[011], β-FeSi$_2$[001]//Si[011]	2.69
	Si(111)	4	~520°C	epitaxial	2.16
				A-type: β-FeSi$_2$(101)//Si(111) with β-FeSi$_2$[101]//Si[0$\bar{1}$1] or β-FeSi$_2$[101]//Si[$\bar{1}$01] or β-FeSi$_2$[101]//Si[$\bar{1}$10]	
				B-type: β-FeSi$_2$(110)//Si(111) with β-FeSi$_2$[001]//Si[0$\bar{1}$1] or β-FeSi$_2$[001]//Si[$\bar{1}$01] or β-FeSi$_2$[001]//Si[$\bar{1}$10]	

Silicide formed	Substrate	Thickness of as-deposited films (nm)	Thermal processing regime	Crystalline structure of the silicide	Refs
β-FeSi$_2$	Si(111) Si(001)	1.4^c ~0.4	620°C/5 min 400°C/10 min	epitaxial: identical to that observed in [2.16] epitaxial islands: β-FeSi$_2$[010]//Si⟨110⟩ and β-FeSi$_2$[010]//Si⟨100⟩	2.70 2.71
	Si(100), Si$_{0.59}$Ge$_{0.41}$ on Si(100)	0.387	550°C/30–60 s	epitaxial islands: β-FeSi$_2$[001]//Si⟨110⟩ and β-FeSi$_2$[010]//Si⟨110⟩	2.72

β-Fe(Co)Si$_2$
β-Fe(Ni)Si$_2$
	Si(001)
	Si(111)
		1
		40, 80 of Fe+Co or Fe+Ni alloy
			300°C
			700°C/60 min
				epitaxial: β-FeSi$_2$(100)//Si(001) with β-FeSi$_2$[010]//Si⟨100⟩
				polycrystalline with the texture:
				β-FeSi$_2$(110)//Si(111) and β-FeSi$_2$(101)//Si(111)
					2.14
					2.73

Ru$_2$Si$_3$	Si(100), Si(111) Si(111) Si(100) Si(111)	90 200 17	500°C 600°C/30 min 625°C/60 min 300°C/60 min +1000°C/30 min	no information no information polycrystalline epitaxial: Ru$_2$Si$_3$(402)//Si(0$\bar{2}$2) and Ru$_2$Si$_3$[010]//Si[111]	2.74 2.41 2.75 2.76, 2.77
Os$_2$Si$_3$	Si(111) Si(100) Si(100)	3 30 150 on 50 nm of Ru	750–1000°C/15 min 600–700°C/60 min 550–730°C/60 min	polycrystalline, mixture of OsSi$_2$ and Os$_2$Si$_3$ polycrystalline polycrystalline	2.39 2.75 2.75
OsSi$_2$	Si(111) Si(100)	3 30	750–1000°C/15 min 800–1000°C/60 min	polycrystalline, mixture of OsSi$_2$ and Os$_2$Si$_3$ polycrystalline	2.39 2.75

Silicide formed	Substrate	Thickness of as-deposited films (nm)	Thermal processing regime	Crystalline structure of the silicide	Refs
$OsSi_2$	Si(100)	150	750–1000°C/60 min	polycrystalline	2.75
	Si(111)	3 on 50 nm of Ru	200°C/15 min +1000°C/15 min	epitaxial: $OsSi_2(040)//Si(2\overline{2}0)$ and $OsSi_2[10\overline{2}]//Si[111]$	2.39
Ir_3Si_5	Si-mono	200	700°C/60 min	polycrystalline	2.78
	Si(100)	200	500–950°C/30 min– 128 h	polycrystalline, IrSi and $IrSi_3$ are present	2.40
	Si(111)	240	450°C/30 s, 500°C/1 s	polycrystalline, Ir and IrSi are present	2.42
	Si(111)	45	500–650°C/10 s	polycrystalline	2.79
	Si(111)	160	675°C/720 s	polycrystalline	2.80

[a] Final thickness of the silicide, [b] Expected stoichiometric phase is $Mn_{15}Si_{26}$, [c] Calculated from the silicide thickness presented in the reference.

These can be *a priori* estimated from the growth kinetic parameters presented in Table 1.35 accounting for the comments made in Sect. 1.3 or directly reproduced from Table 2.4. Note that in order to convert relatively thick metal films (up to 400 nm) into the disilicide, higher annealing temperatures from the mentioned range, i.e. 500–600°C, and times of about 30 min are needed [2.53]. If an as-deposited metal film is heavily contaminated with oxygen (3–5 at %), much higher temperature annealing should be used, for example 900°C/60 min, to obtain a uniform composition of the disilicide film on SiO_2 [2.54].

Local formation of $CrSi_2$ in a window of an oxide mask is characterized by its noticeable lateral growth on the top of the mask. It can be reduced and even completely suppressed by arsenic ion implantation (1–2 at %), but not using bombardment by silicon ions [2.82].

The hexagonal lattice of $CrSi_2$ acceptably matches the silicon crystalline structure. It provides good opportunities for an epitaxial growth of the disilicide onto silicon monocrystalline substrates. Epitaxial grains of 100–150 nm in size were registered already after processing at 450°C for 60 min of the metal film deposited onto the atomically clean Si(001) surface and *in situ* annealed in a high vacuum chamber at 1.3×10^{-6} Pa [2.55]. The grains had $CrSi_2(001)//Si(001)$ with $CrSi_2(110)//Si(110)$.

The best epitaxy is obtained on Si(111) and Si(001) for samples heated to 300 or 400°C during metal film deposition and subsequently annealed at 1000–1100°C for 60 min [2.7, 2.56]. Epitaxial islands of 1–5 μm size cover most of the surface. The epitaxial relationships observed are listed in Table 2.4. For samples annealed at 1000°C without the benefit of substrate heating during metal deposition, only a small fraction of $CrSi_2$ grains, about 0.5 μm in size, become epitaxial. Two-step annealing of room temperature deposited films at 300°C for 60 min plus 1000°C for 60 min results in the formation of epitaxial grains only of 1–2 μm in size.

On Si(001), epitaxial regions are about 1 μm in size after annealing at 1000–1100°C for 60 min [2.7]. They cover 30% of the silicon substrate. Only a small fraction of $CrSi_2$ grains exhibited epitaxial orientation relationships on a Si(011) substrate. They are only 0.5–1 μm in size. In conclusion, the quality of the $CrSi_2$ epitaxy in terms of the size, extent of silicon coverage and the regularity of the interfacial dislocations of the epitaxial regions is best on Si(111) and poorest on Si(011).

Some attempts have been made to produce $CrSi_2$ by pulse electron beam (50 ns) or laser (30 ns) melting of the metal film on silicon substrates [2.83–2.85]. Complete reaction between the metal and silicon needs fine adjustment of the energy deposited in the pulse, which is a function of the metal film thickness. The disilicide film synthesized is polycrystalline consisting of nanocrystals. Grains of such a small size are formed due to the very high quenching rates usually following nanosecond pulse anneals.

$Cr_{1-x}V_xSi_2$. Diffusion synthesis was also successfully employed by Long and Mahan [2.57] to form ternary $Cr_{1-x}V_xSi_2$ thin films. Binary metallic Cr-V alloy sputter deposited onto monocrystalline silicon substrates from composite targets was used. The alloy composition was varied to provide x ranging between 0 and 1.

Conversion of the film properties from semiconducting to metallic was noted with increase of vanadium concentration.

MoSi₂. Formation of thin films consisting of the hexagonal $MoSi_2$ phase is achieved in two temperature ranges which are 400–700°C and 1000–1100°C (see Table 2.4).

The lower temperature annealing is used to produce this phase in a polycrystalline form [2.23, 2.58]. In very thin molybdenum-on-silicon structures the hexagonal $MoSi_2$ is synthesized already at 400°C [2.58]. Its growth proceeds intensively with increase of the annealing temperature up to 700°C, as has been recorded with Raman spectroscopy of the annealed samples. This is illustrated with the typical data presented in Fig 2.6 obtained for the thickness of the as-deposited molybdenum films of 5–10 nm on Si(111).

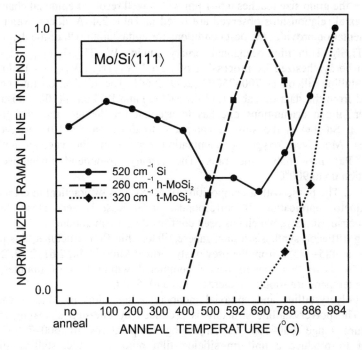

Fig. 2.6. Dependence of Raman line intensity on anneal temperature for Mo/Si(111) samples [2.58]

The hexagonal phase is clearly evident in the temperature range 500–700°C. Grains of hexagonal $MoSi_2$ are doubled in size with increase of the annealing temperature in the range 600–700°C [2.23]. Both hexagonal and tetragonal phases are detected after an anneal at 790°C. Mostly identical results are observed for the molybdenum films on Si(100), polycrystalline and amorphous silicon but with a small shift of the temperature limits. Annealing at 1000–1100°C provides epitaxial growth of both phases. The epitaxial regions observed were less than 1 μm in size. The best epitaxy was found in the samples annealed at 1050°C.

WSi$_2$. Hexagonal WSi$_2$ was found to be formed epitaxially only in the mixture with the tetragonal phase of this disilicide [2.11] when diffusion synthesis in thin film structures is employed. This occurs in the temperature range 600–1100°C in both one- and two-step annealing cycles. Technological details are given in Table 2.4.

Hexagonal grains of the disilicide appear on the Si(001) substrate upon one-step processing at 600°C for 60 min when pure metal is still present in the film. Both hexagonal and tetragonal grains, along with the unreacted metal, are observed after processing at 700°C. In the samples annealed at and above 800°C the metal film completely transformed into WSi$_2$ with the tetragonal phase prevailing over the hexagonal one. The grain size increases from 100 to 250 nm when the annealing temperature is increased from 800 to 1000°C. The two crystalline phase composed films are also formed upon two-step processing. In the latter case the grain size reaches 0.5–1 μm with a well resolved epitaxial character. The epitaxial relationships observed are listed in Table 2.4. A silicon wafer with (001) orientation provides the best conditions for epitaxial growth of the disilicide.

W$_{1-x}$Ti$_x$Si$_2$. Thin film hexagonal ternary silicide W$_{0.7}$Ti$_{0.3}$Si$_2$ (W$_{0.76}$Ti$_{0.24}$Si$_2$) may also be synthesized by processing of mixed W$_{0.7}$Ti$_{0.3}$ (W$_{0.76}$Ti$_{0.24}$) films on monocrystalline silicon at 700–750°C [2.59, 2.60]. The silicidation rate can be estimated from the data of Table 1.37 for Si(111) and PtSi on Si(100) substrates. For other silicon orientations one has to account for the fact that the growth process on Si(111) is the slowest compared to those on Si(110) and Si(100) substrates. Moreover, oxygen contamination even from the vacuum ambient (9.3×10^{-5} Pa) may halve the rate. The ternary compound maintains this composition up to 900°C.

MnSi$_{2-x}$. The phase with a composition of MnSi$_{1.7}$ is of practical interest for a semiconductor application. It corresponds, or at least is very close, to the stoichiometric Mn$_{15}$Si$_{26}$, which has been confirmed as a semiconductor.

During diffusion synthesis in manganese/silicon thin film structures, this phase originates at 485–525°C from the previously formed MnSi [2.44, 2.61, 2.63, 2.86]. In order to have this transformation completed within tens of minutes, the annealing temperature should be increased to 540–600°C.

Both polycrystalline and epitaxial monocrystalline films of MnSi$_{1.7}$ can be formed. The polycrystalline structure is produced by one-step annealing in the temperature range 600–1000°C. Annealing temperatures of 600–700°C are sufficient to produce a uniform silicide film onto a monocrystalline silicon substrate [2.44, 2.61]. The uniformity can be improved during two-step thermal processing supposing formation of MnSi first at 400°C and its subsequent conversion into MnSi$_{1.7}$ by annealing at 600°C [2.86]. When the metal film was deposited onto a polycrystalline silicon layer isolated from the substrate by thermal silicon dioxide, the annealing temperature may be as high as 1000°C [2.62] in order to obtain the best composition uniformity of the silicide. This can be explained by an influence of impurities segregated at grain boundaries in the polycrystalline silicon in the silicidation process.

Two-step thermal processing with the first anneal at 300°C followed by the second step anneal at 1000–1100°C was demonstrated in [2.44] to provide epitaxial growth of $MnSi_{1.7}$ on Si(100) and Si(111) substrates. The epitaxial relationships observed are presented in Table 2.4. The epitaxial regions were found to be about 500 nm in size on average.

$ReSi_{1.75}$ ($ReSi_2$). Thermal processing of metallic rhenium films on monocrystalline silicon substrates provides both polycrystalline and epitaxial growth of the related semiconducting silicide. Here, we will preserve indication of the highest rhenium silicide, $ReSi_2$ or $ReSi_{1.75}$, as was done in the original papers, while in most studies $ReSi_{1.75}$ was actually formed and investigated as was discussed in Sect. 1.1.

Polycrystalline $ReSi_2$ starts growing on (111) and (001) silicon substrates at 600°C during one-step annealing [2.24]. After processing at 800°C no pure metal is detected in the film. An increase of the annealing temperature to 1000°C results in grain growth characterized by the increase of their average size from about 30 nm to 310–340 nm. Epitaxial grains with a mean size of 500–700 nm appeared after annealing at 1100°C on both substrates. The epitaxial relationships observed are listed in Table 2.4. The epitaxial grains cover about 70% of the substrate surfaces. The silicon surface coverage, average grain size, and fraction of epitaxy are found to increase to 90%, 900 nm, and 50%, respectively, when two-step annealing (500°C/60 min + 1100°C/30 min) is performed. These characteristics of the epitaxial regions do not differ much for (111) and (001) substrates.

When fabricating the epitaxial semiconducting rhenium silicide, one should be careful with the use of very thin metal films. Experiments with metal films of 3–5 nm [2.65], demonstrate that the best crystalline quality is achieved after annealing at 750°C, while the films formed at 900°C and above can be broken, possibly by islanding. In the silicide produced from both thin (3–5 nm) and thick (~12 nm) metal films $ReSi_2[110]$ coincides with Si[111]. Other epitaxial relationships are presented in Table 2.4. There are two of these. In thick silicide films $ReSi_2[001]//Si[1\bar{1}0]$ is observed. The second one is detected in very thin films, and attributed to the silicon/silicide interface. It is obtained by a 30° rotation of the first around the [110] $ReSi_2$ axis. This orientation relation is $ReSi_2[3\bar{3}1]//Si[01\bar{1}]$.

Polycrystalline films with well resolved semiconducting electronic and optical properties are formed by annealing of rhenium films on monocrystalline and polycrystalline silicon at 900°C for 60 min [2.45]. In order to prevent oxidation of the metal layer on transfer from the deposition chamber to the furnace, the sample surface is protected by *in situ* deposited silicon.

$Re_{1-x}Mo_xSi_2$. Ternary $Re_{1-x}Mo_xSi_2$ silicide thin films were demonstrated by Long and Mahan [2.57] to be synthesized in Re/Mo alloy sputter-deposited onto monocrystalline silicon substrate. Variation of the target composition used for the alloy deposition provides regulation of x between 0 and 1. Conversion of the film properties from semiconducting to metallic is noted with increase of the molybdenum concentration.

β-$FeSi_2$. Thin films of this silicide can be synthesized in both polycrystalline and epitaxial forms. The technological regimes tested and the results obtained are presented in Table 2.4. While they are characterized by wide ranges of processing temperatures and times, some general features may be summarized as follows.

First, an annealing temperature as high as 900°C seems to be the upper limit for β-FeSi$_2$ formation. Above this temperature the semiconducting β-phase transforms into metallic α-FeSi$_2$. The lowest temperature is dependent on the thickness of the as-deposited metal, crystalline structure of the substrate, and impurities in the thin film structure and in the annealing ambient.

Polycrystalline films are usually synthesized when the thickness of the deposited iron exceeds 5 nm. This can be done on both polycrystalline and monocrystalline silicon substrates. Annealing at 600–700°C for 1–2 h is sufficient to transform 20–50 nm of the metal into the disilicide [2.46–2.48, 2.67]. Below this range FeSi remains in the silicide film. The metal films thicker than 100 nm need the annealing temperature to be increased to 750–850°C for this [2.49, 2.68]. A processing temperature as high as 900°C (120 min) has to be used for synthesis of single-phase polycrystalline β-FeSi$_2$ from an oxygen or other impurity contaminated iron films [2.66, 2.87]. The surface roughness of the synthesized films can be regulated by a capping layer. It is significantly decreased by amorphous silicon, SiO$_x$ ($x \approx 1$), SiO$_2$, or Si$_3$N$_4$ deposited before annealing.

Epitaxial β-FeSi$_2$ films may be grown only to a thickness limited to a few tens of nanometers. Variation of crystalline structures of the FeSi precursor as well as of the disilicide itself tends to become dominant in the competition in epitaxial relationships . Substrate orientation and a template layer, if it is used, have a great influence on the process.

On Si(111), an epitaxial β-FeSi$_2$ film of 13 nm thickness (from a 4 nm metal film) is formed at about 580°C [2.16]. When the metal thickness is reduced to a few monolayers, the processing temperature may be lowered to 540°C [2.88]. Two types of epitaxial zones are observed, with pinholes between them. The relevant orientation relationships are presented in Table 2.4. The disilicide films are stable up to 670°C where they break into islands, clearing large silicon substrate areas. An identical result was observed in [2.70, 2.89] for 4.5 nm of β-FeSi$_2$ synthesized by thermal processing at 620°C for 5 min. When the as-deposited iron was ~0.3 nm thick, epitaxial γ-FeSi$_2$ (cubic) was found to grow on Si(111) at 570°C [2.90]. Meanwhile, 1.5 nm of the metal transformed into FeSi at 520°C and then into β-FeSi$_2$ upon annealing above 620°C.

On Si(001), very thin metal films (~0.4 nm) form epitaxial islands of β-FeSi$_2$ upon processing at 400–550°C [2.71, 2.72]. A significant improvement of the epitaxial growth has been shown to be achieved by the use of a template or strained (for example SiGe) layers to better match the disilicide parameters.

Atomic scale steps on Si(001) are also found to promote solid phase epitaxial growth of β-FeSi$_2$ [2.69]. The steps were formed on a Si(001) vicinal substrate cut 5° off the crystallographic plane in the [011] direction. The thickness of the metal film and annealing regime are presented in Table 2.4. The general epitaxial relation observed is β-FeSi$_2$(100)//Si(001). Two azimuthal orientations deduced by LEED are β-FeSi$_2$[010]//Si[011] and β-FeSi$_2$[001]//Si[011] with a lattice mismatch of 1.5% and 2%, respectively.

Processing at high temperatures, which are typically in the range 900–1000°C, provides formation of a mixture of β- and α-phase epitaxial grains of 1–2 μm size on both Si(111) and Si(001) substrates [2.46, 2.91]. Their orientation relationships are β-FeSi$_2$(220)//Si(111), β-FeSi$_2$(044)//Si(004) and α-FeSi$_2$(100)//Si(100), α-FeSi$_2$(010)//Si(010), respectively.

β-Fe(Co)Si$_2$ and β-Fe(Ni)Si$_2$. Annealing of Fe-Co and Fe-Ni alloy films on Si(111) provides silicidation starting from 450°C [2.73] when the concentration of Co or Ni in the alloy is 10 at % and lower. The disilicides β-Fe(Co)Si$_2$ and β-Fe(Ni)Si$_2$ appear upon processing at 700–800°C for 60 min. They are polycrystalline with a slight texture corresponding to typical relationships on the Si(111) substrate: β-FeSi$_2$(110)//Si(111) and β-FeSi$_2$(101)//Si(111). Annealing at 900°C results in grains of the α-phase in the silicide layer.

Ru$_2$Si$_3$. As far as this silicide is the only phase formed in the structures composed of pure thin metal film on silicon (see Sect. 1.3), there is a variety of the appropriate thermal processing regimes to fabricate it. The annealing temperature can be chosen in the range from 500 to almost 1000°C [2.41, 2.74, 2.75]. The lower limit is related to the need to have silicidation completed within the whole film thickness during a real processing time. The upper one preserves this phase to be orthorhombic with semiconducting properties. The regimes used for preparation single-phase Ru$_2$Si$_3$ films are listed in Table 2.4.

Epitaxial orthorhombic Ru$_2$Si$_3$ is formed by one- and two-step annealing of the metal films chemically electroless deposited onto Si(111) [2.76, 2.77]. After one-step annealing at 600–800°C, mixtures of three polycrystalline phases of Ru$_2$Si, RuSi, and Ru$_2$Si$_3$ are found. The grain size is about 10 nm. In the case of annealing at higher temperatures, 900–1100°C, localized epitaxial ruthenium silicides are formed. Ru$_2$Si$_3$ is dominant among them. Single-phase epitaxial Ru$_2$Si$_3$ is identified only after two-step annealing. The regime employed and epitaxial relationships observed are reported in Table 2.4. The epitaxial grains are 2–10 μm in size.

Os$_2$Si$_3$ and OsSi$_2$. Pure metallic osmium films have poor adhesion to silicon substrates. Above a thickness of 30 nm they fail by peeling off during subsequent heat processing. In order to overcome this limit, a sublayer of ruthenium (50 nm) was proposed [2.75] to be used for experiments with thicker films.

Very thin osmium films (of about 3 nm) are proposed to be formed by electroless deposition in the water solution of osmium tetraoxide, ammonia citrate, and sodium hypophosphite [2.39] and then convert to silicides by thermal processing at 600–1000°C in the reducing atmosphere containing 10% of hydrogen in argon. Both polycrystalline and epitaxial monocrystalline films of OsSi$_2$ were fabricated. The metal completely transforms into silicides after processing at 750°C for 15 min. The silicide film formed contains a mixture of polycrystalline Os$_2$Si$_3$ and OsSi$_2$. An increase of the annealing temperature up to 1000°C results only in the grain growth from about 20 nm to 300 nm, while the two silicide phases coexisted in the film. The multiphase silicide film formation is explained by the influence of oxygen penetrated into the metal. At the same time, the proposed two-step annealing scheme including thermal processing at 200°C for 15 min and then at 1000°C for 15 min provides formation of single-phase OsSi$_2$ epitaxial film. The epitaxial relationships observed are listed in Table 2.4.

The thicker osmium film on silicon substrate demonstrates another behavior [2.75]. Thermal processing of 30 nm metal film on Si(100) at 400°C for 60 min causes only small changes in the structure. An anneal at 600°C results in the formation of Os_2Si_3, the same being true for an anneal at 700°C. At 800°C, one forms $OsSi_2$, a phase which is also obtained at 1000°C.

Much thicker osmium silicide layers can be fabricated when a ruthenium adhesive interlayer 50 nm of thick is used [2.75]. For 150 nm of osmium deposited onto the adhesive layer, annealing at low temperatures, about 500°C, leads to the formation of Os_2Si_3. Whereas $OsSi_2$ is synthesized at higher temperatures, in agreement with the observation for the thin osmium films. Ru_2Si_3 appears almost immediately at 500°C, whereupon Os_2Si_3 grows in well-behaved layers. Formation of Os_2Si_3 occurs in lapsed time, which is much less than 60 min for annealing temperatures above 550°C. Yet the resulting Os_2Si_3 remains unchanged after processing at 730°C for 60 min. After annealing just 20°C above this temperature, i.e. at 750°C, $OsSi_2$ is formed through the whole thickness of the film. The Ru_2Si_3 phase, which is the terminal phase for the Ru-Si system, remains in that form. At higher temperatures there are no changes. The ruthenium interlayer or that of Ru_2Si_3 does not significantly affect the behavior of the top osmium-containing film.

Ir_3Si_5. Formation of thin films consisting only of this silicide meets difficulties as it is an intermediate phase between IrSi and $IrSi_3$, which accompany silicidation in iridium/silicon film structures. It grows from the IrSi precursor in the temperature range 500–950°C when conventional furnace annealing is used [2.40, 2.92]. At higher temperatures, $IrSi_3$ proceeds rapidly consuming the Ir_3Si_5 layer formed. For practical purposes, a temperature range as narrow as 550–600°C may be recommended in order to produce silicide films preferably containing the Ir_3Si_5 phase within annealing for some hours.

Rapid thermal processing of iridium/silicon thin film structures in vacuum was observed in [2.42, 2.79] to produce $IrSi_{1.75}$ upon annealing at 450–500°C for seconds. This phase within the experimental uncertainty can be identified as Ir_3Si_5. Processing at 500°C for 1 s still leaves unreacted metal and IrSi in the silicide layer. They disappear when the processing time is increased to 10 s as well as after 10 s annealing at 550–650°C [2.79]. Nevertheless, some grains of IrSi can exist in the silicide layers synthesized from the metal films thicker than 100 nm [2.80]. In order to obtain single $IrSi_{1.75}$ phase film, an annealing temperature as high as 675°C and processing time of the order of 720 s are recommended. The above temperatures reported to be acceptable for iridium silicide formation by rapid thermal processing are about 100°C lower than those which can be expected from conventional furnace experiments for heating within seconds. Such a large difference has no reasonable explanation. At the same time one should keep in mind the possible underestimation of the real wafer temperature monitored in the apparatus for rapid thermal processing.

2.2.2   Metal+Silicon Thin Film Structures

Mixed metal and silicon films are formed by coevaporation or cosputtering in a vacuum chamber [2.93]. Substrates are maintained at room temperature or a little higher (for improved film adhesion to the substrate and reduced mechanical stresses in the thin film structure). Thus, no uniform silicidation within the codeposited films takes place. While the film composition usually corresponds to the stoichiometry of the desired silicide, it is only a mixture of the components at the atomic scale. The as-deposited films are usually amorphous. Subsequent thermal processing is needed to initiate appropriate chemical reactions and diffusion intermixing. As far as codeposited films contain both silicide-forming components, the substrate material is not so important for silicidation. The technique can be used for fabrication of silicide films on different substrates including silicon dioxide and silicon nitride covered ones (see Fig. 2.1.).

Codeposited films are usually formed with an excess of one of the components with respect to the phase composition desired. During subsequent thermal processing lack of silicon is compensated by its diffusion from the substrate. Excess silicon atoms diffuse to the substrate interface where they condense in a form of epitaxial layers onto monocrystalline surfaces [2.94] or precipitate at oxide ones.

In the case of coevaporation, both components are simultaneously evaporated from the two high purity elemental sources and allowed to condense on the substrate. The vapor sources are generally electron-gun-heated metal and silicon charges in crucibles. For each source a deposition rate is established so that a desired silicon-to-metal ratio results during evaporation.

Cosputtering is performed in vacuum chambers supplied with two or more independently powered sputtering units. The targets mounted in these units can be made of the pure components or of their composition prepared by powder sintering or alloying. The composed targets may also be used in a single-target sputtering deposition process. The actual ratio of the components in the films deposited by cosputtering is very sensitive to the sputtering gas pressure in the chamber and residual gas impurities in it. The sputtering gas is usually argon and the pressure is around 0.65 Pa. The relatively high gas pressure in the chamber results in scattering of the sputtered species. Their mean free path becomes shorter than the distance between the substrate and the sputtering unit. Thus, even small fluctuations in the pressure will change the ratio between the components arriving at the substrate. The sputtering gas impurities, especially oxygen and water vapor, also significantly affect the ratio by modifying sticking conditions on the substrate surface. One should be careful with monitoring of the above mentioned parameters in order to obtain reproducible composition and an appropriate quality of the deposited films.

Unlike cosputtering, the deposits produced by coevaporation have higher purity because typical vacuum conditions during evaporation are characterized by a residual gas pressure of about 1.3×10^{-5} Pa or lower. The peculiarities observed for particular semiconducting silicides formed with coevaporation and cosputtering techniques are discussed below. Table 2.5 presents some details of the experimental experience accumulated.

Table 2.5. Conditions of diffusion synthesis of semiconducting silicides in metal+silicon thin film structures and crystallinity of the silicide produced

Silicide formed	Substrate	As-deposited films composition	thickness (nm)	Thermal processing regime	Crystalline structure of the silicide film	Refs
CrSi$_2$	SiO$_2$ on Si	CrSi$_2$	40, 60, 90	375–425°C/ 20–30 min	polycrystalline	2.95
MoSi$_2$	Si-poly (200 nm) on SiO$_2$	MoSi$_x$[a]	200	600°C/60 min	polycrystalline	2.96
WSi$_2$	Si-poly (150 nm) on SiO$_2$	WSi$_{2.0}$	~150	550–650°C/30 min	polycrystalline	2.94
	SiO$_2$ on Si	WSi$_{1.8}$–WSi$_{2.2}$	200	420–455°C	polycrystalline	2.97, 2.98
ReSi$_2$ (ReSi$_{1.75}$)	Si(100), SiO$_2$ on Si	ReSi$_{2.2}$	200 70	1000°C/60 min	polycrystalline	2.38
	SiO$_2$ on Si	ReSi$_{1.75}$	140	800°C	polycrystalline	2.99
β-FeSi$_2$	sapphire, glass	FeSi$_{2.1–2.2}$	500–600	600°C/30 min	polycrystalline	2.100
	Al$_2$O$_3$-ceramic, sapphire	FeSi$_{2.1–2.2}$	600	870°C/60 min	polycrystalline (grains of ~15 μm)	2.101
	Si$_{0.6}$Ge$_{0.4}$(100)	FeSi$_2$	~80	700°C/60 s	polycrystalline (grains of 5–10 μm)	2.102
	Si(100)	FeSi$_2$	210	700°C/60 min or 900°C/30 s	polycrystalline (grains of 130–210 nm)	2.103
	Si(100)	FeSi$_x$[b]	82 +30 nm of Si	600°C/60 min or 800°C/30 s	polycrystalline	2.104, 2.105
	Si(111)	FeSi$_2$	100–420	800°C/120 min	polycrystalline (grains of 30–80 nm) with ⟨202⟩ texture	2.106

Silicide formed	Substrate	As-deposited films		Thermal processing regime	Crystalline structure of the silicide film	Refs
		composition	thickness (nm)			
β-FeSi$_2$	Si(100), Si$_{0.59}$Ge$_{0.41}$(100)	FeSi$_2$	3–18	560°C/15 s	epitaxial: β-FeSi$_2$[010] and [001]//Si$\langle110\rangle$	2.107
	Si(111)	FeSi$_2$	2–8	500–550°C	epitaxial: β-FeSi$_2$(101)//Si(111) with β-FeSi$_2$[010]//Si$\langle0\bar{1}1\rangle$ β-FeSi$_2$(001)//Si(111) with β-FeSi$_2$[100]//Si$\langle11\bar{2}\rangle$ β-FeSi$_2$(001)//Si(111) with β-FeSi$_2$[100]//Si$\langle1\bar{1}0\rangle$	2.108
	Si(111)	FeSi$_{2.2}$^{c}	200–300	700°C/60 min	polycrystalline	2.109
β-Fe$_{1-x}$Co$_x$Si$_2$	CoSi$_2$ template on Si(111)	Fe$_{1-x}$Co$_x$Si$_2$	10	650°C	epitaxial	2.110
β-Fe$_{0.9}$Mn$_{0.1}$Si$_2$	Si(100)	Fe$_{0.9}$Mn$_{0.1}$Si$_2$	500	900°C/1–120 min	polycrystalline	2.111
Ir$_3$Si$_5$	Si(100)	IrSi$_{-1.6}$	145	600°C/300 min	no information	2.112
	SiO$_2$ on Si	Si/IrSi=0.7	185, 200	650–680°C/30 min	polycrystalline (grains of ~65 nm)	2.113, 2.114

^{a} The film is composed of alternating 0.5 nm of Mo and 1 nm of Si layers, ^{b} The film is composed of 47 periods including 0.5 nm of Si and 1.3 nm of Fe, ^{c} Composition of the target from which the film was deposited by RF sputtering.

CrSi$_2$. A stoichiometric CrSi$_2$ target can be used to fabricate disilicide films by RF sputtering [2.95]. The as-deposited films are amorphous. They start to crystallize around 300°C. Optimal annealing corresponds to 375–425°C for 20–30 min. This provides formation of a polycrystalline film with a stable electrical resistivity.

MoSi$_2$. Thin films of polycrystalline hexagonal MoSi$_2$ may be fabricated at relatively low temperatures. Namely annealing at 600°C for 60 min in forming gas was used in [2.96] to produce this phase in the layered structure consisting of 0.5 nm of molybdenum and 1 nm of silicon at a total thickness of 200 nm on a polycrystalline silicon layer. The silicide film formed was 186 nm thick.

WSi$_2$. Formation of polycrystalline films consisting of semiconducting hexagonal WSi$_2$ is performed with the use of the pure component mixture electron beam coevaporated onto polycrystalline silicon (150 nm) [2.94] or oxidized (500 nm of SiO$_2$) silicon wafers [2.97, 2.98]. Synthesis of this phase starts at 300–360°C. It is completed in the temperature range 420–500°C when conventional furnace annealing [2.94] or continuous heating with the rate of 6°C/min [2.97, 2.98] is used. Further increase of the annealing temperature to 615°C and above results in the hexagonal-to-tetragonal phase transformation. The mixture of as-deposited components corresponding to the stoichiometric WSi$_2$ is reported to have the lowest activation energy for the hexagonal phase formation. The hexagonal phase produced in this case is a highly strained crystalline structure.

ReSi$_{1.75}$ (ReSi$_2$). A nominal composition ReSi$_{2.3}$ was deposited by coevaporation from the two elemental sources at room temperature onto monocrystalline silicon in order to obtain the semiconducting phase which was believed to be ReSi$_2$ [2.38]. Annealing at 1000°C for 60 min in purified helium provides this transformation. The excess silicon epitaxially condenses on the monocrystalline surface or precipitates at the oxide interface.

Magnetron cosputtering from pure component targets has been shown to be appropriate for fabrication of ReSi$_{1.75}$ polycrystalline films onto oxidized unheated silicon wafers [2.99]. The structure of the as-deposited films is x-ray amorphous. Annealing at 800°C leads to formation of the crystalline silicide phase with the composition of ReSi$_{1.75}$ and monoclinic structure observed by Gottlieb *et al.* [2.115]. Annealing at lower temperature (700°C) is insufficient for the crystal structure formation, leaving vacancies in the Si$_3$ and Si$_4$ atom positions. A Si-O(N) layer with a thickness of 3 nm was identified by XPS at the surface of the annealed samples.

β-FeSi$_2$. Codeposition of the pure iron and silicon onto sapphire, Al$_2$O$_3$-ceramic, or glass substrates held at a temperature below 250°C provides formation of completely amorphous films [2.100, 2.101]. Polycrystalline β-FeSi$_2$ appears upon processing at 400°C for tens of hours. Meanwhile, an annealing time of 30 min seems to be appropriate to complete the synthesis at 600°C [2.100]. In order to obtain large grains (15–30 μm) in the film, annealing temperatures as high as 870°C should be used [2.101]. A slight excess of silicon corresponding to the composition range of FeSi$_{2.1-2.2}$ is advised to obtain optimal semiconducting properties [2.101, 2.109]. This is believed to be due to a reduced heteroepitaxial lattice misfit with the silicon substrate related to a large amount of silicon interstitials and silicon precipitates.

Polycrystalline β-FeSi$_2$ was also formed from a codeposited stoichiometric mixture of the components on the relaxed Si$_{0.6}$Ge$_{0.4}$(100) substrate [2.102]. Rapid thermal processing at 600–700°C for 60 s produces grains of 5–10 μm in size.

Films deposited by cosputtering are often contaminated with oxygen. Its concentration can reach 10^{17} atoms/cm^2 [2.106]. Synthesis of β-FeSi$_2$ in such films usually needs annealing temperatures as high as 800°C (see Table 2.5). They are typically polycrystalline.

Thin epitaxial β-FeSi$_2$ is found to grow on Si(100) from a codeposited iron and silicon mixture when a template layer or relaxed Si$_{0.59}$Ge$_{0.41}$ is used [2.107]. The template is formed by deposition of 0.387 nm of iron and subsequent rapid thermal annealing at 560°C for 15 s. Epitaxial films as thick as 3–18 nm are then formed by identical heat treatment of the codeposited overlayer. The orientation relationships observed are typical for the β-FeSi$_2$ epitaxy on Si(100). They are presented in Table 2.5. While the films are formed on properly lattice-matched substrates, islanding still takes place. This is attributed to the peculiar structure of the β-FeSi$_2$ unit cell providing formation of surface antiphase domains.

High perfection thin epitaxial films consisting of about 1 μm grains are also formed on Si(111) substrates with a template layer [2.108]. The template is prepared by room temperature deposition of 2 nm of iron, then 7 nm of FeSi$_2$ mixture and subsequent annealing at 450–500°C. By using a slightly iron-rich FeSi$_2$ mixture for the further epitaxy, β-FeSi$_2$ with (001) and (100) orientation is formed. Related azimuthal orientation relations obtained are listed in Table 2.5.

β-Fe$_{1-x}$Co$_x$Si$_2$. Ternary epitaxial Fe$_{1-x}$Co$_x$Si$_2$ with the orthorhombic β-FeSi$_2$ structure was observed by Long et al. [2.110] to be formed in a narrow range of the cobalt concentration limited by $0 \leq x \leq 0.15$. Processing at 650°C of a Fe+Co+Si mixed 10 nm film deposited at room temperature on the CoSi$_2$ template layer on Si(111) substrate is used. The films formed are reported to be multidomain with the epitaxial relationships typical for β-FeSi$_2$.

β-Fe$_{0.9}$Mn$_{0.1}$Si$_2$. Films of this ternary silicide are formed on Si(100) by electron beam evaporation of a FeSi$_2$ ingot added with about 10% of Mn [2.111]. After annealing at 900°C they demonstrate semiconducting properties with p-type conductivity. The films are polycrystalline with flat interfaces. Moreover, similar results can be obtained when Mn is implanted (400 keV/1.6×10^{16} ion/cm^2) in an already formed β-FeSi$_2$ film and annealing at 850°C is performed.

Ir$_3$Si$_5$. Electron-beam coevaporation of iridium and silicon was tested in [2.112] for fabrication of this iridium silicide phase. In order to obtain a specific film composition, the deposition process should be controlled by adjusting the individual deposition rates. A uniform IrSi$_{-1.6}$ film is formed after annealing of the corresponding iridium and silicon mixture at 600°C. While no data on the structure of the film is presented in the paper, one can expect it to be polycrystalline.

Deposition of metal+silicon films by magnetron sputtering from two targets is also appropriate for fabrication of Ir$_3$Si$_5$ containing films [2.113, 2.114]. The first target is prepared by component sintering at concentrations corresponding to IrSi$_{1.75}$. The films deposited during sputtering only this target have a much larger concentration of iridium than in the target material. Consequently they contain the IrSi phase. In order to obtain higher silicon content in the films, simultaneous sputtering of a pure silicon target is performed. Fluxes of the particles from the targets are regulated by the electric power applied to the sputtering units.

After thermal processing most of the films consist of two or even more phases, but single-phase Ir_3Si_5 may also be synthesized. The power ratio providing $N(Si)/N(IrSi) = 0.7$ and annealing temperature in the range 650–680°C are the best for fabrication such single-phase films onto oxidized silicon wafers. The films are polycrystalline. The initial grain size (of about 65 nm in the film of 200 nm thickness) remains unchanged after processing up to 1000°C, demonstrating the high thermal stability of the film material.

2.3 Ion Beam Synthesis

There are two principal approaches to silicide formation with the use of ion implantation. The first employs the effect of ion beam induced atomic mixing of the components at a metal/silicon interface of a thin film structure subjected to ion bombardment. Scattering of implanted ions from the metal atoms in the film results in some of the atoms obtaining kinetic energy sufficient for penetration into the silicon substrate. Implantation of these recoil atoms into silicon produces an intermixed layer at the interface where silicides nucleate and grow first or during subsequent thermal processing. The most efficient component mixing is achieved when the projected range of the ions in the metal is close to the film thickness. From the technological point of view such an approach can be considered as a modification of the diffusion synthesis in metal/silicon thin film structures discussed in the previous section. This helps to suppress the influence of an interfacial oxide and impurities located there on the silicidation process.

Another approach to silicide formation based on the use of ion beams supposes direct implantation of metal ions into silicon substrates. It gained attention in the late 1980s with the progress in generation of intense beams of metal ions. Various transition metals were implanted at very high doses ($>10^{17}$ ions/cm^2) with energies sufficient to produce buried layers (see Mantl [2.116]). During the implantation, silicon targets are maintained at an elevated temperature which is not lower than 350°C. Direct silicide synthesis is observed under these conditions while subsequent annealing is often employed in order to improve uniformity and crystalline quality of the silicides.

A uniform silicide layer is formed if the dose of implanted ions exceeds a threshold value, which is a function of the ion energy for any particular metal [2.116]. In this case nearly all the implanted metal atoms will be found in the synthesized layer after high temperature annealing. Depending on the ion energy, surface or buried layers can be produced. Sputtering of implanted atoms is usually negligible, except for very high doses at low energies where surface layers are formed. Therefore, the thickness of the synthesized layer d depends linearly on the ion dose Φ

$$d = \Phi/N, \qquad (2.3)$$

where N denotes the atomic density of the metal in the silicide.

Most of the investigations performed with direct ion beam synthesis were focused on the formation of epitaxial buried $CoSi_2$ and $NiSi_2$, which have metallic properties. As for semiconducting silicides, only $CrSi_2$ and β-$FeSi_2$ were attempted by this method.

Available examples of applications of both recoil implantation and direct metal ion implantation for fabrication of semiconducting silicides are presented and discussed below.

2.3.1 Recoil Implantation

The recoil implantation technique is quite limited in examples for semiconducting silicides. The only ones tested are $CrSi_2$, $MoSi_2$, and WSi_2. However, no principal restrictions can be found for other silicides as is illustrated by experiments with metallic type silicides [2.1]. A summary of the technological regimes employed is given in Table 2.6.

$CrSi_2$. Formation of uniform stoichiometric $CrSi_2$ is detected at the metal/silicon interface implanted with 45 keV Cr^+ ions at the temperature of 250°C and above [2.117]. Other available details are presented in Table 2.6. As the same result is also observed after 300 keV Xe^+ irradiation, it is concluded that formation of the disilicide phase does not depend on the bombarding species.

$MoSi_2$. The hexagonal $MoSi_2$ was shown to be produced by ion beam induced mixing in thin film molybdenum-on-silicon structures [2.118]. The molybdenum film was 20 nm thick. The samples were implanted with 200 keV arsenic ions. This corresponds to a projected range of 50 nm and straggling of 24 nm in the metal. There was no special heating of the target during implantation but the temperature induced was estimated to be about 350°C. For other details the reader is referred to Table 2.6.

The hexagonal phase is registered already in the as-implanted samples. Postimplantation annealing at 800°C changes the preferred orientation of the crystallites but not the type of lattice structure.

The hexagonal phase transforms into the tetragonal one during annealing at higher temperatures. Thus, 800°C is considered as a temperature limit for the hexagonal phase formation. It is more than 100°C higher than that for thin film structures produced by solid state diffusion synthesis [2.23, 2.96].

WSi_2. Thin film hexagonal WSi_2 can be synthesized with the use of recoil implantation identically to $MoSi_2$ [2.118]. A tungsten film 40 nm thick on monocrystalline silicon substrate was implanted with 300 keV arsenic ions. It corresponds to a projected range of 54 nm and straggling of 22 nm in the metal. The sample temperature was raised to about 350°C during implantation. The hexagonal phase was detected in the as-implanted samples. It withstood annealing at 650°C but converted into the tetragonal one during processing at 700°C.

Table 2.6. Conditions of ion beam synthesis of semiconducting silicides and crystallinity of the silicide produced

Silicide formed	Substrate	Thickness of as-deposited metal film (nm)	Implantation				Postimplantation annealing	Crystalline structure of the silicide film	Refs
			ions	energy (keV)	dose (ions/cm²)	temperature (°C)			
CrSi$_2$	Si(100)	20	Cr$^+$	45	1.5×10^{16}	250–300	-	no information	2.117
MoSi$_2$	Si(100)	20	As$^+$	200	2×10^{16}	$\sim$350^a	800°C/30 min	polycrystalline	2.118
WSi$_2$	Si(100)	40	As$^+$	300	2×10^{16}	$\sim$350^a	650°C/30 min	polycrystalline	2.118
CrSi$_2$	Si(111)	-	^{52}Cr$^+$	200	3×10^{17}	450	1100°C/30 min or 1100°C/40 s	epitaxial: CrSi$_2$(0001)//Si(111) CrSi$_2$(000$\bar{1}$)//Si(111)	2.119
β-FeSi$_2$	Si(100)	-	^{56}Fe$^+$	140 +80 +50	2.13×10^{17} $+1.00\times10^{17}$ $+5.75\times10^{16}$	350	600°C/60 min+ 915°C/120 min or 600°C/60 min+ 1100°C/1 min+ 850°C/18 h	polycrystalline	2.120–2.123
	Si(001)	-	Fe$^+$	200	4×10^{17}	350	900°C/18 h	polycrystalline (grains of ~160 nm)	2.124, 2.125
	Si(111)	-	Fe$^+$	200	$(1-3)\times10^{17}$	350	1150°C/10 s+ 800°C/17 h	epitaxial: β-FeSi$_2$(010)//Si(111) with β-FeSi$_2$[001]//Si(110), β-FeSi$_2$[100]//Si(110)	2.126–2.129
	Si(111)	-	Fe$^+$	450	$(6-10)\times10^{17}$	320	600°C/30 min+ 900°C/30–150 min	epitaxial: identical to that observed in [2.16] – see Table 2.4	2.130, 2.131

Silicide formed	Substrate	Thickness of as-deposited metal film (nm)	Implantation				Postimplantation annealing	Crystalline structure of the silicide film	Refs
			ions	energy (keV)	dose (ions/cm²) Φ^b	temperature (°C)			
β-FeSi$_2$	Si(001)	-	Fe$^+$	50, 50+100, 50+100+180	Φ^b	RT	900°C/120 s or 800°C/30 min	epitaxial: β-FeSi$_2$(100)//Si(001) with β-FeSi$_2$[010]//Si[110], β-FeSi$_2$[200]//Si[002]	2.132
	Si(100)	-	Fe^{+n}	40^c	(4–6)×10^{17}	RT–660°C	-	epitaxial: β-FeSi$_2$(110)//Si(100) with β-FeSi$_2$[001]//Si$\langle$011$\rangle$	2.133
	Si(100)	-	^{56}Fe$^+$	2000	1×10^{18}	350	800–900°C / 30 min	polycrystalline (grains of ~5 μm)	2.134
	Si(111)	-	Fe$^+$	120	4×10^{17}	300	750°C/60 min	epitaxial: β-FeSi$_2$(101)//Si(111), β-FeSi$_2$(100)//Si(100), and β-FeSi$_2$[010]//Si[01$\overline{1}$]	2.135
β-Fe$_{1-x}$Co$_x$Si$_2$	Si(100)	-	Fe$^+$ + Co$^+$	300, 180	4×10^{17}, (0.5–1)×10^{17}	350, 350	600°C/60 min+ 850°C/30 min, 600°C/60 min+ 850°C/30 min	polycrystalline (grains of 0.2–1 μm) with β-Fe$_{1-x}$Co$_x$Si$_2$[220]//Si[001] preferred orientation	2.136, 2.137
	Si(100)	-	Co$^+$ + Fe$^+$	200, 200	(1–75)×10^{15}, 4×10^{17}	350, 350	850°C/18 h, -	no structural data	2.138

^{a} This temperature was reached during implantation, ^{b} The dose of implantation was chosen to have iron concentration between 1 and 28 at %, ^{c} It is an extracted voltage in kV providing the ion beam consisting of 40 keV Fe$^+$-ions (31%), 80 keV Fe^{+2}-ions (64%), 120 keV Fe^{+3}-ions (5%).

2.3.2   Metal Ion Implantation

Direct synthesis of silicides by metal ion implantation supposes "stoichiometric" doses of this component, which typically are 10^{17}–10^{18} ions/cm^2, to be introduced into a silicon target. This has become possible with the appearance of high current sources of metal ions. The method has been mainly employed for formation of polycrystalline and epitaxial β-FeSi$_2$, while CrSi$_2$ was also demonstrated to be produced using this approach.

CrSi$_2$. Stoichiometric epitaxially aligned buried CrSi$_2$ layers were formed by implantation of 200 keV metal ions into Si(111) [2.119]. The implantation was performed at 350–470°C in order to promote dynamic annealing, while 450°C was mentioned to be the most effective. Other details can be found in Table 2.6. After annealing at 1100°C Si(60 nm)/CrSi$_2$(100 nm)/Si(substrate) structure was formed. The buried CrSi$_2$ layer is highly oriented with its (0001) and (000$\bar{1}$) planes parallel to the (111) of the silicon substrate. It is characterized by a minimum channeling yield of probing high energy He$^+$ ions (χ_{min}) of less than 20%. An improved alignment with $\chi_{min} < 7\%$ is achieved when chromium ions are implanted at current densities above 50 μA/cm^2. There are cracks observed in the buried disilicide and silicon overlayer. Their number is reduced in the samples subjected to rapid thermal processing at 1100°C for 40 s instead of the conventional furnace annealing.

β-FeSi$_2$. Iron implantation into silicon at elevated temperature, typically at 320–350°C results directly in the silicide phase formation. There are usually ε-FeSi, β-FeSi$_2$, and α-FeSi$_2$ [2.132, 2.139, 2.140]. Ion beam synthesis of FeSi$_2$ with a particular crystalline structure has been shown to be annealing regime and iron concentration dependent [2.132]. Ion-beam-induced recrystallization of the implanted samples involves epitaxial growth of the disilicide in the phase sequence of γ-FeSi$_2$, α-FeSi$_2$, and β-FeSi$_2$ with increasing iron concentration along the implantation profile. Up to the concentration of about 9 at %, the only phase present is cubic γ-FeSi$_2$. Between 11 and 18 at % γ-FeSi$_2$ and α-FeSi$_2$ coexist. Finally, at concentrations above 21 at % all the three phases form. This sequential phase formation is explained by lattice match between the different phases and therefore depends on the interfacial energy. Precipitates of cubic γ-FeSi$_2$ can also be synthesized by a room temperature iron ion implantation followed by 500 keV silicon ion-beam-induced epitaxial crystallization at 320°C [2.141]. This is also observed in the samples implanted at 350°C and annealed at up to 600°C [2.142]. The γ-phase matches best with the silicon matrix.

The thermally induced recrystallization produces the β-FeSi$_2$ phase. This can be obtained by conventional furnace annealing of implanted samples at 520°C and also via rapid thermal processing at 900°C for 120 s [2.132]. In both cases iron atoms segregate toward the surface while ion-beam-induced recrystallization substantially preserves the as-implanted profile.

In the temperature range 700–900°C the γ-FeSi$_2$ and α-FeSi$_2$ phases are metastable with respect to β-FeSi$_2$. The γ → β transformation starts at 700°C and occurs via an Ostwald ripening mechanism. It is completed at around 900°C. In addition, annealing at 800°C for 30 min induces almost complete transformation of γ-FeSi$_2$ and α-FeSi$_2$ into β-FeSi$_2$. The epitaxial relationships observed on the Si(001) substrate are listed in Table 2.6.

Triple-energy implantation of iron ions into Si(100) and Si(111) has been performed to synthesize β-FeSi$_2$ [2.120–2.123]. Implantation and subsequent annealing regimes are presented in Table 2.6. Single-phase layers are observed only on Si(100) substrates two-step annealed at 600°C and then at 915°C, and three-step annealing, at 600°C then at 1100°C and finally at 850–900°C. In the latter case the synthesized film demonstrated direct-gap semiconductor ($E_g = 0.88$ eV) properties. Residual defect densities were in the range of 10^{17}–10^{18} cm^{-3}. The α-FeSi$_2$ phase coexists with β-FeSi$_2$ when Si(111) is used as a substrate. This takes place even in the as-implanted samples. The phase transformation β → α occurs upon second step annealing at 930°C. Nonepitaxial growth of the disilicide was revealed from 2 MeV He$^+$ ion channeling analysis.

Annealing in the temperature range 750–940°C of 200 keV iron implanted Si(111) provides synthesis of buried β-FeSi$_2$ layers, but these layers are not continuous [2.126–2.129]. An alternative annealing scheme has been developed supposing fabrication of α-FeSi$_2$ first and then its transformation into β-FeSi$_2$. The first annealing step is performed at 1150°C for 10 s and the second at 800°C for 17 h. The implantation and annealing conditions employed (see Table 2.6) ensure formation of the disilicide containing about 17% of iron sites to be vacant upon the first high temperature annealing. This seems to make easier the transformation from α to β phase. The buried layer formed was 95 nm thick. The epitaxial quality of the buried structure is characterized by the minimum yield of channeling He$^+$ ions of 12% for upper silicon and 54% for buried β-FeSi$_2$.

Another example of ion beam synthesis of epitaxial β-FeSi$_2$ is given in [2.130, 2.131] where 450 keV iron ions were implanted in Si(111) at about 320°C and subsequently annealed at 600°C for 30 min plus at 900°C for 30–150 min. The buried layers formed consisted of about 5 μm blocks with epitaxial relationships corresponding to (101) and/or (110) of β-FeSi$_2$ parallel to (111) of silicon. In Si(001), they are only 0.5 μm in size. Implantation with a dose of 6×10^{17} ions/cm^2 causes formation of a buried layer consisting of grains with lateral dimensions of approximately 5 μm. After implantation of 1×10^{18} ions/cm^2 a surface layer is formed.

A nonseparated iron ion beam was used by Liu et al. [2.133] for direct synthesis of β-FeSi$_2$ during implantation. The extracted voltage was 40 kV. The ion beam included Fe$^+$, Fe^{+2}, and Fe^{+3} ions with equivalent energies and ratios are mentioned in Table 2.6. The ion current density was 152 μA/cm^2, which was high enough to heat the silicon substrate from room temperature up to 530–660°C during the implantation. The silicide films fabricated are about 60 nm thick. The epitaxial β-FeSi$_2$ formed in these regimes demonstrates no visible microstructure changes upon additional annealing at 800°C for 8 h.

Implantation of 2 MeV iron ions with the dose of 1×10^{18} ions/cm^2 and subsequent annealing at 800–900°C provides formation of buried β-FeSi$_2$ 200 nm thick about 1.5 μm below the surface [2.134]. An annealing temperature of 900°C is reported to initiate the transformation of the β-FeSi$_2$ phase into α-FeSi$_2$.

A surprising result was observed by Jin *et al.* [2.135] in Si(111) implanted with 120 keV iron ions at 300°C. A new type of orthorhombic FeSi$_2$ was detected. Although the lattice parameters of the new phase are the same as those of the known β-FeSi$_2$, its point group and space group were determined to be *mmm* and *Pbca*, respectively. Annealing at 750°C for 60 min transforms only a minority of the silicide particles into β-FeSi$_2$, while the primitive FeSi$_2$ remains the dominant phase. Some additional experiments are necessary in order to clarify the point.

β-Fe$_{1-x}$Co$_x$Si$_2$. Dual metal ion implantation is shown by Pankin *et al.* [2.136], and Harry *et al.* [2.143] to be an appropriate technique for ternary semiconducting silicide formation on the basis of β-FeSi$_2$. Their crystalline structure is largely determined by the first implanted metal and subsequent annealing regimes. When cobalt is implanted first, the ternary phase has cubic fluorite structure and phase separation takes place in the surface layers [2.143, 2.144].

Single-phase buried β-Fe$_{1-x}$Co$_x$Si$_2$ is successfully synthesized in Si(100) with x varied from 0 to 0.2 [2.136, 2.137]. A procedure including iron implantation followed by an intermediate annealing to produce iron disilicide, then cobalt implantation and subsequent annealing was found to allow formation of this ternary silicide with the best uniformity in the phase composition and crystalline structure. The appropriate regimes are presented in Table 2.6. The typical thickness of the silicide layers is 120–150 nm. Two temperatures of the intermediate annealing can be employed providing formation of β-FeSi$_2$ at 850°C or α-FeSi$_2$ at 1050°C first. In the latter case cobalt implantation is performed as the temperature reduced to 180°C for partial amorphization of the silicide formed. This enhances the α- to β-phase transformation during the subsequent annealing. A reduced defect density in the top silicon layer and inside the silicide grains but more iron silicide precipitates distinguish this approach from that when β-FeSi$_2$ is formed during the intermediate annealing. Band gap tuning between 0.84 and 0.70 eV is achieved by the above mentioned variation of the cobalt content.

2.4   Atomic and Molecular Deposition

This group of methods is based on almost individual deposition of atoms or molecules of silicide forming components, which are metal and silicon, on a substrate. Substrates are maintained at an elevated temperature during the deposition in order to initiate diffusion intermixing of the components on the surface and stimulate appropriate chemical reactions yielding silicide formation.

The components can come to the substrate surface from gaseous compounds dissociating at the heated surface as takes place during chemical vapor deposition, in the form of atomic or molecular clusters, as can be achieved with laser ablation, or from atomic/molecular beams of pure components employed in reaction deposition and molecular beam epitaxy techniques (see Fig. 2.1). These technological approaches are mostly at the beginning of practical application to thin film semiconducting silicide formation. Their prospects and effectiveness have been demonstrated with β-FeSi$_2$, where epitaxial and polycrystalline films were produced by all the mentioned techniques. General regularities, the technological regimes used, and the results obtained for this semiconducting silicide as well as for others, when available, are discussed in this section.

2.4.1 Chemical Vapor Deposition

Chemical vapor deposition (CVD) of silicides supposes silicon and metal species to be produced simultaneously from the gaseous phase [2.145]. Gas pressure in the reaction chamber and substrate temperature are the most important factors controlling the process. The silicide forming components come into the reaction chamber in the form of gaseous or vapor compounds. Their decomposition and release of metal and silicon atoms may be performed only at the elevated temperature. This can take place directly on the heated substrate or in the gaseous phase just at the substrate surface. In the latter case the silicon and metal species produced in the gaseous phase condense on the substrate and uniform coverage of the patterned substrates can be achieved. Meanwhile, surface-controlled decomposition reactions provide selective deposition of the components onto silicon in the windows in a silicon dioxide mask.

The gas ambient in the chamber is mainly heated by convection from the substrate and substrate holder. This makes the ambient temperature determined by the gas pressure, thus regulating the choice between uniform film deposition at atmospheric pressure and selective substrate surface-controlled component condensation at lower pressures. This necessitates separate analysis of the results obtained in CVD at atmospheric pressure, low pressure chemical vapor deposition (LPCVD) usually performed at 133–1330 Pa, and chemical beam epitaxy (CBE). The latter is also called gas-source molecular beam epitaxy as it is realized in ultrahigh vacuum chambers of the machines for molecular beam epitaxy rearranged to operate with reactive gases at pressures of the order of 1.3×10^{-4} Pa.

High purity gases and *in situ* produced vapor compounds are used as precursors of silicon and metal atoms, and are listed in Table 2.7. Silanes provide an appropriate choice for silicon. The choice of metal precursor gases is, in general, limited to carbonyls, while high vapor pressure halides (mainly fluorides and chlorides) possess a variety of possibilities. *In situ* metal chlorination can be used to obtain such precursors as has been shown for chromium [2.146], tungsten and tantalum [2.147, 2.148], rhenium [2.149], and iron [2.150, 2.151].

In the case of tungsten chlorination, the product is principally composed of WCl$_4$ (tetrachloride) and WCl$_2$ (dichloride). The latter is totally evaporated above 880°C [2.147]. Gaseous ReCl$_5$ is predominantly formed during rhenium chlorination at above 200°C [2.149]. A chlorination temperature of 527°C is used.

Table 2.7. Precursor gases and vaporizing compounds appropriate for chemical vapor deposition of silicides*

Compound formulae	Name	Melting point (°C)	Boiling point (°C)	Decomposition temperature (°C)	Refs
SiH_4	silane	−185	−111.9		2.152
Si_2H_6	disilane	−129.3	−14.4		2.152
		−132.5	−14.5		2.153
Si_3H_8	trisilane	−117.4	52.9		2.153
Si_4H_{10}	tetrasilane	−108	84.3		2.153
SiH_2Cl_2	dichlorosilane	−122	8.3		2.153
$SiCl_4$	tetrachlorosilane	−70	57.57		2.153
Si_2Cl_6	hexachlorodisilane	−1	145		2.153
GeH_4	germane	−165	−88.5	350	2.153
Ge_2H_6	digermane	−109	29	215	2.153
Ge_3H_8	trigermane	−105.6	110.5	195	2.153
$GeCl_4$	germanium tetracloride	−49.5	84		2.153
$Ge(CH_3)_4$	germanium tetramethyl	−88	43.6		2.154
$Cr(CO)_6$	chromium hexacarbonyl	110		110	2.153
$CrCl_2$	chromium dichloride	824			2.153
$Mo(CO)_6$	molybdenum hexacarbonyl		156.4	150	2.153
MoF_6	molybdenum hexafluoride	17.5	35		2.153
$MoCl_5$	molybdenum pentachloride	194	268		2.153
$W(CO)_6$	tungsten hexacarbonyl	~150	175	~150	2.153
WF_6	tungsten hexafluoride	2.0	17.2		2.155
		2.5	17.5		2.153
WCl_6	tungsten hexachloride	275	346.7		2.153
$Re(CO)_5$	rhenium pentacarbonyl	250		250	2.153
ReF_4	rhenium fluoride	124.5	500	500	2.153
ReF_6	rhenium hexafluoride	18.8	47.6		2.153
$ReCl_4$	rhenium tetrachloride		500		2.153
$Fe(CO)_5$	iron pentacarbonyl	−21	102.8		2.153
$FeCl_3$	iron chloride	306	315	315	2.153
$Ru(CO)_5$	ruthenium pentacarbonyl	−22			2.153
RuF_5	ruthenium pentachloride	101	250		2.153
OsF_6	osmium hexafluoride	32.1	45.9		2.153
$Ir_2(CO)_8$	iridium carbonyl	160 (subl)			2.153
IrF_6	iridium hexafluoride	44.4	53		2.153

*Some other chemicals and their properties can be found elsewhere [2.153, 2.156].

Generation of $FeCl_3$ may be performed with high-purity iron wire heated to above 300°C through which Cl_2 gas diluted in Ar passes [2.150]. Iron chlorinated compounds ($FeCl$, $FeCl_2$, $FeCl_3$, Fe_2Cl_4, Fe_2Cl_6) are solids with a high vapor pressure above 300°C. $FeCl_3$ can be produced in an open system under continuous Cl_2 flow in a carrier gas through a heated iron wire. It is one of the most volatile compound among the iron chlorides [2.151].

Chemical vapor deposition at atmospheric pressure

CVD of silicides at atmospheric pressure is very attractive from the point of view of uniform film formation and step coverage. It is characterized by controlled substrate consumption, thus there is almost a total absence of dopant redistribution, a minimum number of technological steps improving throughput and reliability, and current availability of industrial single wafer CVD reactors. Among the silicides of interest, $MoSi_2$ [2.157], WSi_2 [2.147, 2.148, 2.159], and β-$FeSi_2$ [2.160] were investigated by atmospheric pressure CVD.

$MoSi_2$ and WSi_2. In the case of these disilicides, studies were focused on the formation of their tetragonal phases, which are known to have metallic properties. Semiconducting hexagonal $MoSi_2$ is definitely observed only in the beginning of the silicidation process when molybdenum film is subjected to thermal processing in SiH_4/H_2 (10/90) gas mixture at 600–850°C [2.157]. Regarding the deposition temperature, one could expect the appearance of the hexagonal WSi_2 phase when this disilicide is deposited from WF_6/Si_2H_6 or $SiH_4/H_2/N_2$ at 290–300°C [2.158] and from WF_6/SiH_2Cl_2 at 550–600°C [2.159, 2.161]. The latter deposition conditions ensure formation of inherently cleaner films as compared to the use of a WF_6/SiH_4 mixture. The silicon carrier in the form of SiH_2Cl_2 has lower reactivity with WF_6 than SiH_4, which results in better step coverage and lower fluorine incorporation into the films. The as-deposited films consist of both hexagonal and tetragonal grains.

β-$FeSi_2$. The results obtained for β-$FeSi_2$ appear to be optimistic. This semiconducting disilicide is successfully deposited from $Fe(CO)_5$ and Si_2H_6 at 1% dilution in pure hydrogen [2.160]. Formation of epitaxial β-$FeSi_2$ films on Si(111) includes: (1) *in situ* silicon surface deoxidation before deposition by annealing the samples at 1040°C for 5 min in a hydrogen flow, (2) deposition of a 5 nm sticking layer at 540°C and annealing in a hydrogen atmosphere at 760–800°C for 5 min in order to improve the material quality and provide β-$FeSi_2$ formation, (3) silicide film growth at 540°C with a rate of 1.2 nm/min up to a typical thickness of 100 nm, (4) annealing at 760–800°C for 30 min to increase the crystalline quality of the film.

Without steps (2) and (4) the deposited silicide film consists of $\in$-FeSi at a Si/Fe ratio as low as 0.5 or α-$FeSi_2$ at a ratio between 1 and 4. The sticking layer formed at step (2) is composed of embedded β-$FeSi_2$ crystallites with the epitaxial relationships β-$FeSi_2(101)$ or $(110)//Si(111)$. Meanwhile, there are α-$FeSi_2$ crystallites at the surface. The subsequent silicide film growth at step (3) occurs homoepitaxially for both phases which leads to the need for high temperature annealing (step (4)) for transformation of the mixed α- and β-phase film into a continuous β-$FeSi_2$ one. The epitaxial grains formed are several 100 nm in size, while in general the film remains polycrystalline with some preferential orientations besides the epitaxial ones. Oxygen at a level of about 10% and a few per cent of carbon are detected all over the silicide film before final annealing. After annealing the impurities locate only at the film surface.

Low pressure chemical vapor deposition

LPCVD processes possess both uniform and selective silicide deposition. This was tested for $MoSi_2$ [2.162], WSi_2 [2.148, 2.163, 2.164], $ReSi_{1.75}$ [2.149], β-$FeSi_2$ [2.150, 2.151, 2.165–2.167] films.

$MoSi_2$ and WSi_2. Identically to the experiments on CVD of $MoSi_2$ and WSi_2, most attention was paid to obtaining metallic tetragonal phases of these disilicides. This is why the molybdenum silicide films deposited from $MoCl_5/SiH_4$ with Ar, H_2 or N_2 gas carriers at 520–800°C [2.162] and tungsten ones deposited at 450–700°C from a mixture of *in situ* produced WCl_4 with SiH_4, H_2, Ar at a total pressure of 133 Pa [2.148], as well as RF plasma-enhanced deposited from WF_6 (100%) and SiH_4 (20% in He) at 230°C [2.163] were subsequently annealed at 700–1000°C to accomplish the tetragonal phase formation.

$ReSi_{1.75}$. Polycrystalline thin films of this semiconducting silicide can be deposited onto oxidized silicon substrates from the gaseous phase composed of SiH_4, *in situ* generated $ReCl_5$, H_2, and Ar at the total pressure of 800 Pa [2.149]. While the deposition temperature may be varied in the range 650–800°C, below 700°C weakly crystallized material composed of metallic rhenium and Re_5Si_3 is produced. Above 700°C pure crystallized $ReSi_{1.75}$ with some excess of silicon or SiO_2 presumably located at grain boundaries is formed. Oxygen in an amount of 5–10% is detected by NRA. It is assumed to be in the form of ReO_2 and/or ReO_3. Regardless of the oxygen content, the temperature dependence of the resistivity of the films demonstrates their semiconducting behavior.

β-$FeSi_2$. The polycrystalline structure of the films deposited onto oxidized silicon substrates is typical for β-$FeSi_2$ [2.151]. The deposition can be carried out at 900°C from SiH_4 and the metal precursor obtained by *in situ* chlorination of the heated iron wire when the total pressure in the reaction chamber is 800 Pa. The oxygen content recorded with NRA and AES is around 1% but there is no detectable contamination from carbon and chlorine.

Deposition of β-$FeSi_2$ onto patterned oxidized silicon substrates can provide selective epitaxial growth of the silicide on silicon in the windows of the silicon dioxide mask [2.150, 2.165, 2.167]. A gas mixture composed of SiH_4, H_2, He, and $FeCl_3$ (diluted in Ar) coming from a separate line is used for this. Predeposition thermal processing of the substrates at 1000°C for 30 s in H_2 at 133 Pa are important to clean the silicon surface in a native oxide.

At a deposition temperature in the range 750–850°C epitaxial granular structured β-$FeSi_2$ films 50–190 nm thick are selectively formed on Si(111). Silicon substrates are used for the disilicide synthesis. About 130 nm of the substrate is consumed during formation of 190 nm of the disilicide. In order to minimize the substrate consumption the concentration of SiH_4 should be increased until the nucleation process starts over the silicon dioxide mask. This will result in the loss of the selectivity in the deposition process. The same is observed with the use of hydrogen as a dilutent or carrier gas. Only $SiH_4/He/FeCl_3/Ar$ compositions provide selective deposition. The minimum line width achieved is 500 nm.

The same deposition technique was used by Berbezier *et al.* [2.166] to investigate epitaxial relationships obtained for β-$FeSi_2$ but on uncovered Si(111) and Si(001) substrates. It has been observed that: (i) annealing of the deposited films has no influence on the crystalline quality of β-FeSi$_2$ formed, (ii) deposition with an approximate stoichiometric composition greatly increases the flatness of the interface. The epitaxial relationship detected on Si(111) is β-FeSi$_2$(220)//Si(111). Domains with different azimuthal orientations related to rotation around [111] and [110] silicon axis exist in the films. These three domains are always present. On Si(001), the only epitaxial relationships found are β-FeSi$_2$(100)//Si(001) with β-FeSi$_2$[010] or [001]//Si[110]. Thus, the same could be expected in selectively deposited β-FeSi$_2$ on silicon structures.

Chemical beam epitaxy

CBE is performed in MBE chambers retrofitted with gas sources. Each gas is injected into the chamber separately and focused onto the heated silicon substrate. This method was applied to the formation only of β-FeSi$_2$ epitaxial films. Precursor gases used are SiH$_4$ or Si$_2$H$_6$ and Fe(CO)$_5$.

Surface kinetics and the silicide phase formed are controlled by the gas source used (SiH$_4$ or Si$_2$H$_6$), substrate temperature, and its crystallographic orientation [2.168]. Thermal cleaning of monocrystalline silicon substrates is performed at 820°C for 15 min in order to remove a native oxide layer from the surface. The relevant temperature range for the disilicide growth is 450–550°C. At higher temperatures the films are very rough with a rough interface to the substrate.

The use of SiH$_4$ induces the formation of a very thin cubic γ-FeSi$_2$ strained layer first. It remains stabilized at the interface during subsequent growth of the semiconducting β-FeSi$_2$ phase [2.168–2.171]. Below 500°C there is no semiconducting phase formation. The smoothest films are found to grow at 550°C with the pressure ratio of SiH$_4$ to Fe(CO)$_5$ in the range 1/2–2/5. There is a clear tendency to greater roughness the more the pressure ratio deviates from 1/2. The temperature dependence of the growth rate is illustrated in Fig. 2.7. The growth rate is 0.20 nm/min on Si(111) and 0.25 nm/min on Si(100) at the optimal deposition temperature of 550°C. With increasing substrate temperature to 700°C the growth rate is doubled.

On Si(111), the films grown at 550°C demonstrate epitaxial relationships β-FeSi$_2$(100)//Si(111). Above this temperature there is a transition from two- to three-dimensional growth. This results in an increasing roughness of the silicide films as the temperature is increased. At 700°C the epitaxial relationships are identical to that observed during solid state epitaxy: β-FeSi$_2$(101) or (110)//Si(111).

No γ-FeSi$_2$ has been observed. On Si(100), the epitaxial relationship with the substrate is β-FeSi$_2$(100)//Si(100) in the whole temperature range. Growth of the grains with different azimuthal orientations takes place. There are three of them rotated by 120° on Si(111) and two azimuthal directions having an angle of 45° to each other on Si(100). The grains are 0.1–1 μm in size. They have a very sharp interface with the substrate. No traces of carbon incorporation within the sensitivity of the *in situ* techniques employed have been detected.

When the silicon supply at the surface is increased by using Si$_2$H$_6$ [2.168, 2.172–2.174], the tetragonal nonstoichiometric α-FeSi$_2$ is stabilized by the interface and a corresponding film grows to a thickness of several tens of nanometers.

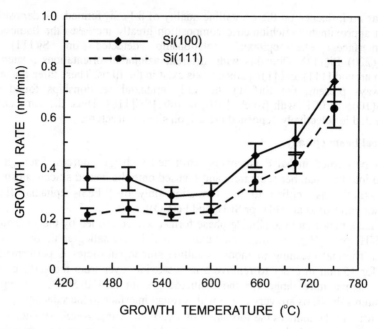

Fig. 2.7. Growth rate of β-FeSi₂ versus substrate temperature on Si(111) and Si(100) during CBE [2.171]

This phase can be transformed into β-FeSi₂ by annealing above 600°C. In the case of Si_2H_6 at its fixed partial pressure of $1.7×10^{-2}$ Pa the growth of α-FeSi₂ at 550°C proceeds almost one order of magnitude faster than with the use of SiH₄. The ratio between deposited iron and silicon atoms is effectively controlled by variation of the $Fe(CO)_5$ partial pressure in the range $(4–22)×10^{-3}$ Pa. As the partial pressure is increased, the film thickness and its surface roughness also increase. The best result is obtained for a pressure $(5–6.5)×10^{-3}$ Pa corresponding to the deposition of stoichiometric strained tetragonal α-FeSi₂. Its grains of size 30–50 nm have apparently random orientations. In order to transform them into the β-phase, annealing at 600–650°C has to be performed.

This has been demonstrated with a film deposited at 400°C and then at 500°C and subsequently annealed at 600°C [2.172]. The film formed consists of 20–50 nm epitaxial grains with the (110) or (101) planes of β-FeSi₂ parallel to Si(111) planes. Much bigger epitaxial grains reaching 1 μm in size are obtained during α to β transformation in the films deposited at 550°C [2.173, 2.174]. Annealing at 650°C provides this transformation preserving the FeSi₂ stoichiometry, but with a slight enrichment of the surface with silicon atoms. Coalescence of numerous fine α-grains into large β-grains accompanies the transformation. Remarkable smoothing of the silicide/silicon interface on an atomic scale also takes place. The lower symmetry of β-FeSi₂ in comparison with the relevant silicon surfaces always results in the presence of azimuthal rotated grains.

In conclusion, the fabrication of micro- and nanostructures by means of selective CBE of β-FeSi$_2$ on SiO$_2$ patterned silicon substrates might open new prospects for improvement of the material quality.

2.4.2 Cluster Evaporation and Sputtering

Semiconducting materials in the form of nanosize clusters demonstrate dramatic changes in their fundamental electronic and optical properties [2.175]. Quantum confinement in such structures inevitably brings about an increase of the energy gap and may change the character of the gap from indirect to direct, as takes place for silicon. Thus, there is an increasing interest in technological approaches providing formation of nanosize structures. These can use techniques based on flash thermal evaporation of materials subjected to pulse laser light irradiation (laser ablation [2.176]) or ion sputtering at high current densities. The deposited energy in both cases is sufficient to make the subsequent evaporation or sputtering processes not only atom-by-atom injection from an irradiated target, but also yielding atomic clusters preserving the stoichiometry of the target material. Their subsequent condensation on a substrate is followed by formation of the film stoichiometrically identical to the target but with the crystalline structure controlled by the substrate temperature. Low substrate temperatures preserve nanoclusters in the condensed film, while a temperature sufficient to initiate surface diffusion may result in an epitaxial growth of the deposited material [2.177]. With respect to semiconducting silicides, only β-FeSi$_2$ [2.178] and ternary β-Fe$_{1-x}$Co$_x$Si$_2$ [2.179] thin films were tested by this technological approach.

β-FeSi$_2$. Pulsed laser ablation in vacuum (below 2.6×10^{-7} Pa) of the disilicide target was used by Olk *et al.* [2.178, 2.180, 2.181] to form β-FeSi$_2$ film onto a Si(111) substrate. The silicon substrate was heated to about 500°C. Light from a KrF eximer laser ($\lambda = 248$ nm) was focused on a spot of ~2 mm^2 on the target. Light energy in the range 300–800 mJ and repetition frequency of 2–10 Hz was employed. This provided deposition of the disilicide at the rate of 0.0026–0.0135 nm/pulse. Films as thick as 10–120 nm were deposited. The films were confirmed by AES to have the correct stoichiometry within 10%. XRD analysis showed β-FeSi$_2$ epitaxial grain formation with β-FeSi$_2$(001)//Si(111) and three variants of aligned axis, e.g. β-FeSi$_2$[010]//Si[$\bar{1}$10], β-FeSi$_2$[010]//Si[10$\bar{1}$], or β-FeSi$_2$[010]//Si[0$\bar{1}$1]. The grains were 50–100 nm in size.

Semiconducting β-FeSi$_2$ films were also deposited by laser ablation onto Si(100) by Kakemoto *et al.* [2.182]. They used an ArF eximer laser ($\lambda = 193$ nm) operating at a power of 80 mJ with the repetition frequency of 5 Hz provided light energy deposition onto β-FeSi$_2$ polycrystalline target of 1 J/cm^2. The films grown at the substrate temperature in the range 550–750°C are polycrystalline with the surface mostly β-FeSi$_2$(220) oriented.

Planar magnetron radio-frequency sputtering was tested by Tsunoda *et al.* [2.180] to form polycrystalline β-FeSi$_2$ films 2–3 μm thick. High purity (99.5%) polycrystalline alumina was used as a substrate. The target was arranged as a silicon disc with iron chips mounted on its surface. An atomic ratio close to the disilicide stoichiometry and substrate temperature above 300°C were found to be necessary for direct synthesis of β-FeSi$_2$ in the depositing film. The quality of the films is acceptable for thermoelectric applications.

β-Fe$_{1-x}$Co$_x$Si$_2$. Sputtering of sintered cobalt doped iron disilicide targets of different compositions was used by Teichert *et al.* [2.179] to produce polycrystalline Fe$_{1-x}$Co$_x$Si$_2$ films on quartz substrates with $0 \leq x \leq 0.05$. A substrate temperature above 500°C was found to be necessary during the deposition in order to allow crystallization of the condensed species into the orthorhombic β-FeSi$_2$ type structure. The grain size is in the range 40–70 nm.

2.4.3   Reactive Atomic Deposition

In this method metal atoms are deposited onto a hot silicon surface. Their intermixing and compound formation takes place readily during the deposition process because of well activated atomic diffusion. Low-rate reactive deposition can result in a concentration-controlled phase selection [2.183]. This makes possible a direct formation of disilicides, thereby skipping metal-rich phases in the formation sequence. In the mechanism this approach is very similar to the solid state phase formation controlled by diffusion through an intermediate diffusion barrier layer, while the stoichiometry of the growing phase is self-regulating.

Mg$_2$Si. During deposition at room temperature magnesium atoms are preferrentially adsorbed on the top sites of silicon adatoms [2.4, 2.20]. The rest occupy the Si(111)7×7 surface with increasing metal deposition. At a metal thickness of 0.3–0.4 ML (monolayer) a two dimensional Mg$_2$Si epitaxial film nucleates on the Si(111). As the film grows, the surface structure transforms from the diffuse (1×1) phase into the $(2/3\sqrt{3} \times 2/3\sqrt{3})R30°$ superstructure. This rotation of the surface structure by 30° is assumed to be induced by the stress arising from the lattice mismatch. After growing to a critical thickness, which is of the order of a few monolayers, a magnesium film is deposited in a disordered form on the epitaxial film. The epitaxial film of critical thickness acts as a barrier for magnesium and silicon diffusion mixing. A temperature increase is needed to overcome this restriction.

BaSi$_2$. Epitaxy of orthorhombic BaSi$_2$ on silicon was experimentally investigated by McKee *et al.* [2.5]. Some monolayers of the silicide were grown at high temperature by exposing the silicon surface to barium. The exact temperature of deposition is not mentioned in their paper. Epitaxial growth on Si(111) and Si(001) was observed. The silicide growth is dominated by the lattice matching that occurs when BaSi$_2\langle001\rangle$ aligns with Si$\langle110\rangle$. On both substrates the silicide formed has its (100) plane parallel to the growth surface. The open channels of the silicon tetrahedra are oriented normal to this plane. Thus, epitaxial relations BaSi$_2$(100)//Si(111) with BaSi$_2\langle001\rangle$//Si$\langle110\rangle$ and BaSi$_2$(100)//Si(001) with BaSi$_2\langle001\rangle$//Si$\langle110\rangle$ are experimentally confirmed.

CrSi$_2$. This disilicide can be directly formed by chromium deposition onto Si(111) kept at 400–500°C [2.9, 2.10, 2.184]. The deposition temperature of

450°C seems to be optimal for high epitaxial quality of the films. Epitaxial growth with the orientation relationships $CrSi_2(0001)//Si(111)$ and $CrSi_2[10\bar{1}0]//Si[10\bar{1}]$ is observed. The crystalline quality improves significantly upon additional annealing at 1000–1200°C. However, it is followed by breaking of the film into islands in films with a thickness in the range from a few to tens of nanometers. A deficit of metal atoms is thought to be responsible for this as soon as continuous films are formed when enough metal is deposited. From perpendicular strain measurements, the critical thickness for pseudomorphic growth is estimated to be equivalent to 2 nm of the deposited chromium.

$ReSi_{1.75}(ReSi_2)$. Epitaxial $ReSi_2(100)/Si(100)$ structures can be fabricated by rhenium deposition onto silicon at 400–1000°C [2.12, 2.25, 2.26, 2.185]. The films of 65–530 nm thickness analyzed with the 2 MeV He^+ ion channeling technique and Bragg–Brentano x-ray diffraction demonstrate the best quality of the epitaxy at 650°C. Thus, the 150 nm thick film is characterized by the minimum channeling yield for rhenium of 2% and for silicon of 14%. It is stressed that a high value for the minimum yield of the light element (Si) in a polyatomic single crystal does not necessarily mean that the sublattice of the light element is disordered. The predominant epitaxial relationship is $ReSi_2(010)//Si(001)$ with $ReSi_2[001]//Si[110]$. As the common unit cell area is 1.2 nm^2, the estimated maximum mismatch is +1.8% [2.26]. Although the films are not single-crystalline, they are epitaxial in the sense that only one specification of the heteroepitaxial relationship pertains and that film regions described by other relationships are virtually nonexistent. However, there are two epitaxial variants. Specifically, the films consist of equal number of distinct but crystallographically equivalent rotation twins, which are vertical columns of the order of 10 nm in diameter. They differ in azimuthal orientation by rotation of 90° about Si[001].

Good epitaxial growth of $ReSi_2$ has been demonstrated on Si(111) [2.65, 2.186]. Rhenium is deposited onto the substrate held at 650°C. Then the samples are annealed at 750°C. The observed epitaxial relationships are identical to those obtained on the films formed by solid phase epitaxy. One can find the details in Sect. 2.2.

$Re_{1-x}Mo_xSi_2$. An optimal deposition temperature of 650°C was used by Vantomme et al. [2.187] to form epitaxial ternary $Re_{1-x}Mo_xSi_2$ films on Si(100) by codeposition of rhenium and molybdenum. The films were 110–260 nm thick. The best crystalline quality is observed in films with the fraction of molybdenum corresponding to $x < 0.33$. This is characterized by the minimum yield of channeled 2.5 MeV He^+ ions of 11–12% for both metals.

$MnSi_{2-x}$. Thin films consisting of epitaxial grains of tetragonal $Mn_{15}Si_{26}$ ($MnSi_{1.73}$) were fabricated by reactive manganese deposition onto Si(111) substrates [2.188]. The substrates were maintained at 1040–1070°C during the deposition and subsequent annealing at 350–800°C was performed. The grains were 4–5 μm in size. They had preferred orientation with their c axis normal to the substrate.

β-$FeSi_2$. In order to form this silicide silicon substrates are maintained at a temperature in the range 350–650°C during iron deposition [2.13, 2.14, 2.189–2.192]. Phase transitions in the silicide film are a function of the film thickness and the temperature [2.193]. In the range of 1–2 nm they are typically: (i) At 350°C, strained $FeSi_2 \rightarrow FeSi$; (ii) at 400°C, strained $FeSi_2 \rightarrow \beta\text{-}FeSi_2$. These transitions have a dynamical character. Semiconducting $\beta\text{-}FeSi_2$ is definitely formed in the temperature range 450–550°C. The films produced on monocrystalline silicon substrates are typically epitaxial with the orientation relationships generally identical to those obtained with solid state epitaxy during diffusion synthesis in metal-on-silicon thin film structures (see Sect. 2.2).

On Si(001) substrates, deposition of 6–10 nm of iron at 470–500°C results in formation of $\beta\text{-}FeSi_2$ with a high degree of epitaxial alignment [2.13, 2.190]. The deposition rate of 0.01 nm/s is used providing formation of 20 nm of $\beta\text{-}FeSi_2$ when 6.3 nm of iron is deposited [2.190, 2.194]. The orientation relationships observed are $\beta\text{-}FeSi_2[010]//Si\langle 110\rangle$ and/or $\beta\text{-}FeSi_2[001]//Si\langle 110\rangle$. Fabrication of epitaxial films as thick as 100 nm needs the deposition temperature to be increased to 550°C [2.16]. In order to fabricate thicker (about 200 nm) $\beta\text{-}FeSi_2$ films the substrate temperature should be 600°C [2.192].

Subsequent annealing of the formed thin (20 nm) $\beta\text{-}FeSi_2$ films at 850°C initiates its aggregation into islands [2.194]. The islands have an average height of about 40 nm due to aggregation, dipping into the silicon substrate so that only one half of the height of these islands is above the substrate surface. When silicon is grown on the top of such structures by molecular beam epitaxy at 750°C, the islands continue aggregation and finally show a spherical form being surrounded by the monocrystalline silicon. The epitaxial relationships between the two materials and their monocrystallinity are preserved.

Bombardment of the silicon surface with low energy (100–650 eV) argon ions during reactive deposition of iron at 600°C has been found by Terrasi et al. [2.195] to reduce the grain size of $\beta\text{-}FeSi_2$ formed and to improve the film morphology as compared to the nonirradiated ones. The film of 130 nm thickness produced in such conditions had epitaxial ($\beta\text{-}FeSi_2(001)//Si(001)$) and randomly oriented grains of about 600 and 230 nm in size, respectively. The bombardment induced increase of the nucleation rate is considered to be responsible for the decrease of the grain size and related reduced roughness of the surface. This mechanism is argued by Barradas et al. [2.196] on the basis of their own experiments performed at 700°C. They also observed that argon ion bombardment during iron deposition improved the roughness of the $\beta\text{-}FeSi_2$ films formed, while reduction of the grain size was not detected. An argon energy of 200 eV and an argon ion to iron atom ratio of 0.15 have been found to be the best for this. The films are polycrystalline, but with a pronounced texture. At the film thickness of 180 nm the grains ranging from 40 to 140 nm in width have preferential orientation $\beta\text{-}FeSi_2(110)//Si(001)$ and $\beta\text{-}FeSi_2(101)//Si(001)$ with a few degrees misorientation.

The higher deposition temperature facilitates silicon diffusion to the surface. Ion bombardment enhances the surface mobility of the atoms and removing of overhanging adatoms. Surface atoms are also pushed into the subsurface layer, filling existing voids. As a result, a consequent improvement of surface morphology takes place.

On Si(111) substrates, epitaxial $\beta\text{-}FeSi_2$ has been observed to form only at 630°C [2.191]. Iron deposition at lower temperatures, 550–580°C, gives a mixture of Fe$_3$Si, Fe$_5$Si$_3$, FeSi and β-FeSi$_2$ phases. Epitaxial relationships β-FeSi$_2$(110)//Si(111) and β-FeSi$_2$(101)//Si(111) are realized.

Stress measurements performed on the structures prepared by reactive iron deposition onto Si(111) at 323°C [2.197] show a two-stage stress evolution in the synthesis of β-FeSi$_2$. The interface formation during the first 0.3 nm of the iron deposition is characterized by a compressive stress. An intermixed Fe-Si layer is supposed to be responsible for this compressive stress. For an increased iron coverage a tensile stress arises in the film reaching 11–18 N/m. This has been attributed to the semiconducting silicide phase formed on the top of the interface layer. The appearance of the tensile stress is qualitatively explained by the decreased atomic volume of silicon in β-FeSi$_2$ (0.0187 nm^3) as compared to the bulk silicon (0.02007 nm^3). Thus, one can expect a decisive role of such a large stress in the solid state reactions and growth mode of the film.

β-FeSi$_{1.92}$Ge$_{0.08}$. Thin film of this ternary was formed by iron deposition onto Si$_{0.93}$Ge$_{0.07}$/Si substrate kept at 450°C [2.198]. After deposition the sample was *in situ* annealed in the high vacuum chamber at 650°C for 30 min. The film thickness was about 120 nm with the gomogenious component distribution within 80 nm inner region. It had fine-grained polycrystalline structure. The distortion of β-FeSi$_2$ lattice caused by replacement of silicon with germanium was too small to be determined with the XRD analysis performed. The optical spectroscopy showed the ternary formed to have semiconducting properties close to β-FeSi$_2$.

β-Fe(Co)Si$_2$ and β-Fe(Ni)Si$_2$. Deposition of metals from Fe-Co and Fe-Ni alloy sources onto Si(111) heated to 650°C leads to the direct formation of β-FeSi$_2$ based epitaxial ternaries [2.73]. Acceptably homogeneous distributions of the third elements are obtained at their concentrations of 9 at % and lower. The epitaxial layers consist of the grains with the usual relationships: β-FeSi$_2$(110)//Si(111) and β-FeSi$_2$(101)//Si(111). Subsequent annealing at 800°C for 60 min increases the grain size.

2.4.4 Molecular Beam Epitaxy

MBE of silicides is realized when metal and silicon atoms from separate sources are simultaneously deposited at a stoichiometric ratio onto a heated silicon substrate. It is performed in a high vacuum chamber at a residual gas pressure lower than 1.3×10^{-7} Pa in order to provide an atomically clean surface for condensation of the silicide forming components. Long range surface diffusion of the atoms results in the compound formation and its epitaxial growth on the monocrystalline substrate. A polycrystalline film can also grow when the silicon crystal and the silicide have a considerable lattice mismatch or the components are deposited onto a silicon dioxide layer.

Frequently, a template layer on the monocrystalline silicon is used to improve the epitaxial growth. In this case a two-step process is carried out. At the first step, only metal is deposited and a well controlled template silicide layer of about 1 nm thick is formed by solid state epitaxy during diffusion synthesis at low temperature. At the second step, the temperature is increased and the thicker epitaxial layer is grown by simultaneous deposition of metal and silicon atoms.

Peculiarities of molecular beam epitaxy of semiconducting silicides are illustrated by the following examples.

Mg₂Si. The condensation coefficient of magnesium on a silicon surface is practically zero in the temperature range 200–500°C [2.34, 2.199]. Meanwhile, the presence of silicon adatoms at the surface stimulates condensation of the impinging magnesium atoms. It is a self-regulating process requiring an excess of magnesium. At a substrate temperature of 200°C a magnesium flux 4–10 times higher than the silicon one is needed to provide formation of stoichiometric Mg₂Si films of 80–600 nm thick. No silicon is consumed from the substrate. The growth of three-dimensional islands controlled by extensive surface diffusion is believed to take place in these conditions. For both (111) and (001) substrate orientations, a (111) texture is always observed [2.34].

CrSi₂. Epitaxial films of this disilicide were formed by coevaporation of the pure components at the stoichiometric ratio onto Si(111) substrates maintained at about 450°C [2.200], 720°C [2.8], 825°C [2.21]. They were about 10, 2 and 210 nm thick for the mentioned deposition temperatures, respectively. The films are not single crystalline, while epitaxial blocks are well defined.

Three different epitaxial relationships have been identified by Fathauer *et al.* [2.8]. These are $CrSi_2(0001)//Si(111)$ with $CrSi_2[10\bar{1}0]//Si[1\bar{1}0]$ (referred to as A-type), $CrSi_2(0001)//Si(111)$, $CrSi_2[11\bar{2}0]//Si[1\bar{1}0]$ (referred to as B-type), and $CrSi_2(1\bar{1}00)//Si(111)$ with $CrSi_2[0001]//Si[11\bar{2}]$ (referred to as C-type). As far as the major hexagonal axis of CrSi₂ coincides with the Si(111) surface normal for A- and B-type, one can note that these types are related to a 30° rotation about the axis. In C-type, the major axis of CrSi₂ lies in the plane of the Si(111) surface and is aligned with ⟨112⟩ directions. Since there are three ⟨112⟩ directions in the (111) plane, correspondingly there can be three variants of C-type orientation relationships. This type of epitaxial relationship seems likely to nucleate under specific conditions at steps on the silicon surface.

After deposition at 825°C epitaxial grains of two types with a size of 1–2 μm are observed [2.21]. They are rotated by 30° with respect to each other around the surface normal corresponding to (1.17×1.17)R0° and (1.17×1.17)R30° LEED reflections in the films. The related epitaxial relationships are: $CrSi_2(0001)//Si(111)$ and $CrSi_2[11\bar{2}0]//Si[10\bar{1}]$(B-type) with a mismatch of ~3.8% and $CrSi_2(0001)//Si(111)$ and $CrSi_2[11\bar{2}0]//Si[11\bar{2}]$(A-type) with a mismatch of ~0.1%. The latter one is the most favorable for perfect epitaxy. Identical relationships were detected in solid phase epitaxial grown epitaxial structures [2.201]. However, rather poor epitaxial quality and unregulated island formation is mentioned.

Very thin CrSi₂ template layers can significantly improve the quality of the subsequent epitaxial growth [2.202, 2.203]. Moreover, preferential growth of the A-type structure may be achieved. High deposition rates (> 10 ML/min, 1 ML = 0.094 nm) and temperatures (500–700°C) promote formation of the A-type CrSi₂ template, which should have a thickness of more than 0.5 nm. For example, the template formation procedure used in [2.203] included five times repeated deposition of 2 nm chromium onto silicon substrate at 500°C followed by annealing at 850°C for 30 s. Then continuous epitaxial film with a great extent of monocrystallinity as thick as 100–200 nm can be formed by molecular beam epitaxy or reactive deposition. A process temperature of 750°C has been found to be optimal for this.

β-FeSi$_2$. Molecular beam epitaxy of this silicide is performed at 450–600°C. On Si(111) maintained at 500–550°C, an epitaxial α-FeSi$_2$ grows first when 10 ML of iron are deposited [2.204]. With increasing thickness at a later stage of the growth, as well as during annealing at temperatures higher than 600°C, an irreversible transition toward β-FeSi$_2$ takes place. A hierarchy in the stability of different epitaxial FeSi$_2$ phases is mostly controlled by the interaction with the silicon substrate as follows

$$\text{cubic-FeSi}_2 \rightarrow \alpha\text{-FeSi}_2 \rightarrow \beta\text{-FeSi}_2. \tag{2.4}$$

Analogous sequential phase transitions were observed during formation of epitaxial iron disilicide in ion implanted samples [2.132] (see Sect. 2.3).

On Si(001) substrates, islands of β-FeSi$_2$ with a high degree of epitaxial alignment are already formed at 450°C [2.13]. At about 500°C continuous deposition without islanding may be realized [2.14]. The A-type orientation relationships, e.g. β-FeSi$_2$(100)//Si(001) with β-FeSi$_2$[010]//Si[110], are dominant in the temperature range 300–550°C. A thin, some monolayers, cubic γ-FeSi$_2$ layer grows first [2.205]. On increasing the silicide thickness, this film is overgrown by A-type β-FeSi$_2$. B-type grains, which have β-FeSi$_2$(100)//Si(001) with β-FeSi$_2$[010]//Si⟨100⟩ appear either at lower or higher temperatures. There could also be grains with other azimuthal relationships: β-FeSi$_2$[040]//Si[220], β-FeSi$_2$[010]//Si[100], β-FeSi$_2$[014] or β-FeSi$_2$[041]//Si[220]. The epitaxial grains reach 1 μm in size.

By using codeposition and turning the flux on and off at ~20 s intervals, epitaxial growth was observed even at temperatures as low as 200°C, but on a template layer. The template preparation includes deposition of 0.1–0.2 nm of pure iron followed by near to stoichiometric coevaporation of iron and silicon for a total of about 0.7 nm of FeSi$_2$ onto the clean substrate kept near room temperature [2.206]. On Si(111), this first step results in the formation of epitaxial Fe$_{0.5}$Si with the defect CsCl structure, whereas on Si(001), the film remains amorphous, as can be seen with RHEED. Annealing at 400–450°C is performed to produce the epitaxial template. On Si(001), the template has predominantly A-type orientation relationships, while B-type grains are also present. In order to increase the thickness of the template layer the above deposition and annealing steps are repeated several times, each time increasing the annealing temperature by 20–50°C. On Si(111), the largest epitaxial β-FeSi$_2$ grains are obtained by annealing slightly iron-rich films with a thickness around 3 nm to a temperature above 600°C. The predominant epitaxial orientation obtained turns out to be β-FeSi$_2$(001)//Si(111) with β-FeSi$_2$[100]//Si⟨11$\bar{2}$⟩.

The templates possess the MBE-growth of thick (>200 nm) epitaxial β-FeSi$_2$ at 650–700°C on both Si(111) and Si(001) substrates [2.206–2.208]. Epitaxial grains reach 0.5 μm in lateral dimensions. The epitaxial relationship is β-FeSi$_2$(101)//Si(111) or β-FeSi$_2$(110)//Si(111).

An appropriate β-FeSi$_2$ template can also be fabricated by iron evaporation at a rate of 0.03 nm/s onto a silicon substrate heated to 480°C with the subsequent annealing at 600°C for 20 min [2.208]. Then, component codeposition is recommended to be performed at 400°C with the total iron and silicon stoichiometrically balanced rate of 0.1 nm/s. The samples should be additionally annealed at 700°C for 2 h in high purity nitrogen in order to complete crystallization. The procedure provides quite a flat interface between β-FeSi$_2$ and the silicon substrate.

β-Fe(Co)Si$_2$ and β-Fe(Ni)Si$_2$. Possibilities of MBE of these ternaries were tested for room temperature and 550°C depositions on Si(111) [2.73]. The concentration of nickel or cobalt additives in the iron source was in the range of 7–10 at %. Room temperature deposited films are found to be amorphous. Polycrystalline β-Fe(Co)Si$_2$ and β-Fe(Ni)Si$_2$ appear only after annealing at 700°C for 16 h.

After deposition at 550°C the silicide films consist of the disilicides with a small amount of strongly textured monosilicides. In this case a post anneal at 700°C for 60 min is necessary to transform them into epitaxial β-Fe(Co)Si$_2$ or β-Fe(Ni)Si$_2$. The codeposition of Fe-Ni alloy and silicon onto Si(111) at 550°C leads to large epitaxial, mirror-like β-Fe(Ni)Si$_2$ films with a homogeneous concentration of nickel.

2.5   General Comments on Silicide Formation Techniques

During the decades of research practically the whole variety of thin film fabrication techniques has been definitely shown to be suitable for semiconducting binary silicides and related ternaries. This is clearly illustrated by Table 2.1 and in particular by the examples with β-FeSi$_2$, for which formation all the techniques have been successively employed. All the other semiconducting silicides, except OsSi, have also been obtained and investigated in thin films. The uncertainty with OsSi mainly relates to the limited number of experiments performed with the Os-Si system in general. In an attempt to form this monosilicide one may meet difficulties related to formation of a single-phase film, as far as the phase diagram of the system has two more relatively stable phases, which are Os$_2$Si$_3$ and OsSi$_2$. Nevertheless, the stoichiometry and cubic structure of OsSi can be technologically stabilized, at least by aluminum or oxygen doping [2.209].

Among the technological approaches tested, **diffusion synthesis** is most frequently used for fabrication of thin films of semiconducting silicides. It is the simplest in realization, quite flexible, and meets the requirements of productivity in the case of industrial application. The restriction of this approach is determined by the undesired impurities penetrating into the films and accumulating at interfaces during deposition and annealing. Oxygen and nitrogen are the most significant such impurities. They retard silicidation, disturb epitaxial growth, and finally modify properties of the semiconductors. Rapid thermal processing at the annealing step could overcome the contamination related problem [2.1]. Silicide growth under conditions of rapid thermal processing proceeds at higher rates than conventional furnace annealing. The reduced oxygen contamination of an annealing ambient in combination with short processing times is responsible for this.

Ion beam synthesis has the most advanced possibilities for composition and impurity control in fabricating films. Its main limitations are low productivity of the equipment currently used for implantation of metal ions and residual defects in the implanted layers, which need high temperature annealing. Nevertheless, it remains one of the best techniques for fabrication of epitaxial layers and nanocrystals of silicides buried into monocrystalline silicon.

The group of methods based on **atomic and molecular deposition** has the best prospects for fine regulation of stoichiometry and crystalline structure of silicides. In particular this relates to chemical vapor deposition and molecular beam epitaxy. Although chemical vapor deposition uses commercially available highly productive equipment, its main restriction is connected with a limited choice of the gaseous metal-containing components and their purity. These are overcome in molecular beam epitaxy. This offers both precise control and many *in situ* monitoring techniques. Its main drawback could be that it is a rather expensive, it is a mostly laboratory technique and most of the results obtained for semiconducting silicides are from fundamental studies rather than practical applications in electronics.

Polycrystalline films of semiconducting silicides meet no problems in fabrication. The situation is quite different for epitaxial films. If the crystal structures of the film and the substrate hinder good matching, epitaxial layers can hardly be grown with conventional deposition techniques. The reason is that epitaxy is ensured by controlled layer-by-layer growth on a heated substrate. This so-called van der Merwe growth is considered to be a standard growth mode of molecular beam epitaxy, various chemical vapor deposition methods, evaporation or sputter deposition onto a heated substrate and, to a certain extent, also laser ablation. The quality and morphology of the epitaxial silicides within the technology employed increase gradually from solid state diffusion synthesis to reactive atomic deposition and then to molecular beam epitaxy.

Lattice mismatch and the difference in thermal expansion coefficient of two materials cause strains in epitaxial structures. Their relaxation generates dislocations and fragmentation of the epitaxial film, which are not desired for any semiconductor device. None of the binary semiconducting silicides is perfectly lattice matched with silicon. This restricts the maximum size of the uniform epitaxial regions of binary semiconducting silicides to some micrometers. Nevertheless, the problem can be resolved with templates, in particular with the use of SiGe alloys or related ternary semiconducting silicides. The latter has to be carefully studied.

While sophisticated techniques of thin film fabrication have been successively adopted for silicides, the research community is paving the way for additional technological facilities. Possibilities to produce epitaxial silicide layers and nanoclusters buried into monocrystalline silicon have been recently brought about by molecular beam allotaxy [2.210]. In this approach, a buried epitaxial layer is formed from precipitates during annealing. The precipitates are embedded within a single crystalline matrix, grown by molecular beam epitaxy. The key point is that the epitaxial growth of the matrix persists from the substrate through the spacing between the inclusions to the overlayer. Then the precipitates coarsen and coalesce into a uniform layer during subsequent high temperature treatment, which is the second processing step. While the allotaxy has been only tested for $CoSi_2$, there is not doubt of its great potential for semiconducting silicides also.

It is important to note that nanosize precipitates grown embedded into single crystalline silicon at the first step of molecular beam allotaxy are also very attractive for an independent application. As a result of quantum confinement one can expect a dramatic change of their electronic and optical properties with respect to the bulk material, but these phenomena still await investigation.

3 Crystal Growth

Günter Behr

Institute of Solid State and Materials Research Dresden
Dresden, Germany

CONTENTS

Single crystals of semiconducting silicides provide unique possibilities in understanding the bulk properties of these compounds and their dependence on crystallographic orientation and composition within the range of homogeneity. Thin films, polycrystalline samples and ceramics often exhibit properties quite different from the bulk behavior. This is mainly due to atomic segregation at grain boundaries, film–substrate interaction and orientation anisotropy of physical properties. Therefore, many efforts have been made to develop methods of single crystal growth adapted for semiconducting silicides. Their main features and details of realization for particular semiconducting silicides are presented in this chapter.

3.1 Methods of Crystal Growth and Material Problems

Binary phase diagrams of the systems forming semiconducting silicides (see Sect. 1.2) show different phases are formed. There are congruent melting compounds on the one hand, peritectically melting compounds on the other hand and phases which are not in equilibrium with the melt. As a result, very different techniques have to be used for the single crystal growth. Each of them provides crystals of a particular size with a certain morphology, chemical purity and crystalline perfection.

Two principal approaches can be used for the growth of single crystals. These are growth from a gaseous or liquid phase. The smallest samples used for a crystal structure analysis are obtained from arc melted bottoms. Their size is of a few tens of micrometer. In order to perform electrical measurements, single crystals are required of some millimeters. Such samples are grown from a gaseous phase by chemical vapor transport (CVT) or from a melt containing appropriate components. The growth from the gaseous phase is carried out with different transport agents in closed silica ampoules. For the growth from a melt, a flux technique, in which zinc and tin are used as solvents of the main components, and conventional Bridgman, Czochralski, zone melting, gradient freeze melting or floating zone methods are employed. A short description of the methods with the details related to semiconducting silicides is given below. Meanwhile, their fundamentals and more details about the methods can be found in the excellent handbooks [3.1, 3.2]. As far as an adequate choice of a growth technique is dependent on the thermodynamic behavior of the silicide forming components at high temperatures we anticipate the presentation with short comments on binary metal-silicon phase diagrams.

3.1.1 Phase Diagrams and Choice of Growth Methods

The methods which can be used to grow single crystals of semiconducting silicides depend strongly on the phase diagrams of the binary systems for melt growth, on the multi-component phase diagrams with possible solvents for flux growth and on the availability of transport agents to perform a successful chemical transport.

Group II metal silicides

Mg_2Si. The semiconducting silicide Mg_2Si is the only compound in the Mg-Si system and melts congruently at 1085°C (Fig. 1.13). Therefore, crystal growth from the melt should be possible but one has to take into account that the boiling point of pure magnesium is very close to this temperature. Thus, one has to take care about considerable evaporation of the component. Growth by chemical vapor transport is hardly possible because of the lack of stable magnesium halides in the gaseous phase.

$BaSi_2$. The phase diagram of the Ba-Si system presented in Fig. 1.14 shows the congruently melting semiconducting phase $BaSi_2$ with a melting point at 1180°C. As the boiling point of pure barium is 1140°C, the same problems will occur as for the Mg_2Si growth. The lack of appropriate transport agents is also a limiting factor for the crystal growth by the chemical vapor transport method.

Group VI metal silicides

CrSi$_2$. As shown in Fig. 1.15 the semiconducting silicide CrSi$_2$ melts congruently at 1490°C. Partial pressures of the elements at that temperature are relatively low. Single crystal growth from the melt is possible. Due to the existence of thermodynamically stable halides of the elements chemical vapor transport is also expected to be possible. Growth from a flux is known from published experimental results.

MoSi$_2$. The semiconducting high-temperature phase β-MoSi$_2$ melts congruently at 2020°C (Fig. 1.16). However, this phase transforms into tetragonal α-MoSi$_2$ at 1900°C. Although the β-phase is metastable at lower temperatures the transformation is not hindered at temperatures higher than 800°C. Therefore, no methods are available to grow single crystals of the β-phase under conditions near the thermodynamic equilibrium. The α-MoSi$_2$ phase was grown from the melt by Tabata and Hirano [3.3] showing that the β-MoSi$_2$ transforms into α-MoSi$_2$ during cooling without any loss of single crystallinity.

WSi$_2$. The β-WSi$_2$ phase melts congruently at about 2324°C (Fig. 1.17) and undergoes a polymorphic phase transformation at about 2013°C. Although the β-phase is metastable at lower temperatures the transformation is not hindered at temperatures higher than 620°C. Therefore, no methods are available to grow and cool single crystals of the β-phase to room temperature.

Group VII metal silicides

MnSi$_{2-x}$ (Mn$_{11}$Si$_{19}$) melts incongruently at 1155°C (Fig. 1.18) and the peritectic temperature is only 5°C above the eutectic temperature. From the moderate deviation of the peritectic concentration from the stoichiometric composition near MnSi$_{1.73}$ single crystal growth from the melt can be realized. Due to the existence of thermodynamically stable halides of the elements chemical vapor transport is also thought to be possible. Flux growth has not been reported yet in the literature. It is difficult to find a solvent metal, which corresponds to the requirements, because manganese forms alloys with tin and zinc, which are often used as solvent metals.

ReSi$_{1.75}$ (ReSi$_2$). This silicide melts congruently at about 1940°C (Fig. 1.19) and does not undergo phase transitions. Because of the high melting point and the reactivity of the melt with crucible materials melt growth should be attempted by cold crucible Czochralski technique or by a floating zone technique using optical heating. The growth from a flux should be tested. Crystal growth by CVT is described in the literature.

Group VIII metal silicides

β-FeSi$_2$ does not coexist with the melt but it forms from the α-phase by the reaction α-FeSi$_2$ → β-FeSi$_2$ + Si at 937°C (Fig. 1.20). Therefore, one might expect that its single crystals can be grown only from a flux and by the CVT technique.

Ru$_2$Si$_3$ melts congruently at 1710°C (Fig. 1.21). Due to the moderate melting temperature melt growth may be successfully applied to form single crystals. A transformation to the α-phase stable below 1690°C has not been reported. The halides have a limited stability, which makes it difficult to find transport agents for crystal growth by CVT. A growth from a flux has not been reported yet.

OsSi, Os$_2$Si$_3$, OsSi$_2$. In this family of semiconducting silicides OsSi and OsSi$_2$ melt incongruently, whereas Os$_2$Si$_3$ melts congruently at 1840°C (see Fig. 1.22). In principle, the growth of the compounds could be possible from the melt. Probably due to a high toxicity of OsO$_4$, which is formed very easily and sublimates at room temperature, only a few efforts have been made to grow the semiconducting silicides of osmium. The successful OsSi$_2$ growth from a PdSi flux was reported by Mason and Müller-Voigt [3.4]. Due to the low stability of osmium halides a successful CVT growth of these phases seems to be unlikely.

Ir$_3$Si$_5$. This semiconducting silicide forms peritectically from a silicon-rich melt (Fig. 1.23) at temperatures of about 1672°C. Single crystal growth from the nonstoichiometric melt could be possible, a flux growth may also be successful. Finding the appropriate conditions will not be easy and, moreover, the material is expensive. Therefore, only a few attempts have been made to grow iridium silicide single crystals. Iridium halides have low stability, hence a successful CVT growth of the silicide looks unfavorable.

3.1.2 Chemical Vapor Transport

CVT growth of semiconducting silicides is performed only in closed systems. The crystal growth is simply realized in a sealed quartz ampoule with typical dimensions of 20–30   mm diameter and from 90 to 200 mm length. The ampoule must be cleaned carefully, filled with the source material consisting of either pure elements, or the compound synthesized before, or an appropriate precursor mixture. A transport agent is added in an appropriate amount. The ampoule is evacuated, sealed and placed into a two-zone furnace to maintain a temperature gradient with well defined and stabilized temperatures at the two ends of the ampoule, as schematically shown in Fig. 3.1.

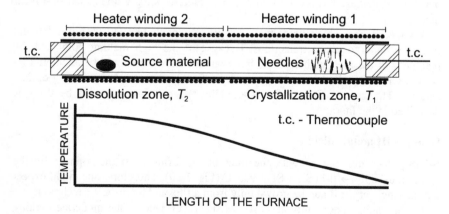

Fig. 3.1. Chemical vapor transport in a closed system

The transport agents used for semiconducting silicides are halogens I_2, Br_2, and Cl_2 [e.g. 3.5, 3.6] or $CuCl_2$, $MnCl_2$ [3.7].

The basic steps of CVT are: vaporization of the source compound, its reaction with the transport medium at the temperature T_2, transport of the gaseous products by diffusion into the zone with the temperature T_1, solidification of the compound at this temperature, and return of the transport medium into the T_2-zone. When halogens are used as the transport medium, the process can be described by the reaction

$$\text{MeSi} + 2X_{2/g} \underset{T_2}{\overset{T_1}{\rightleftharpoons}} \text{MeX}_{2,g} + \text{SiX}_{2,g}, \tag{3.1}$$

where Me represents metal atoms and X corresponds to a halogen.

In closed ampoules this process mostly takes place near a thermodynamic equilibrium. In a multiphase system like Me-Si, the transport medium X_2 forms the gas phase in an incongruent reaction with the solid phase. In particular, the gas phase normally contains metal and silicon in a ratio different from that of the solid after the equilibrium has been reached [3.5, 3.8]. This incongruent solution can be represented by the combination of the reactions:

$$\text{MeSi}_x, s + m\,(x - y)/2X_2, g = \text{MeSi}_y, s + (x - y)\text{SiX}_m, g \tag{3.2}$$

$$\text{MeSi}_x, s + (2m + n)/2X_2, g = \text{MeX}_n, g + x\text{SiX}_m, g, \tag{3.3}$$

where x, y are stoichiometric coefficients of coexisting metal silicides and m, n are the stoichiometric coefficients of the metal and silicon halides.

A sufficient concentration of all components to be transported in the gas phase is necessary to make the transport possible. The balance pressure p_A^* of the component A (say metal or silicon) is used as a measure of this concentration N_A [3.9]:

$$p_A^* = \sum_i \alpha_{i,A} \cdot p_i = N_A R_0 T \tag{3.4}$$

with $\alpha_{i,A}$ as the stoichiometric index of A in the species i with the partial pressure p_i, R_0 is the universal gas constant.

The transport rate (mole per hour) for any component A is proportional to the difference in p_A^* and p_B^* between the solution and crystallization zones at T_2 and T_1. Therefore, as a simple approach, the ratio of the balance pressure p_{Me}^*/p_{Si}^* should be in the range between 10^{-4} and 10^4 to achieve sufficient transport rates of the semiconducting silicides, according to Krausze et al. [3.10]. Transport of the highest silicides in the silicide forming systems takes place from a higher to a lower temperature or vice versa, depending on the transport agent used.

Details about thermodynamic and kinetic evaluation of the growth process can be found for instance in [3.5, 3.6, 3.10]. The advantage of the method is that the crystals grown usually have a low concentration of point defects and lower density of dislocations as compared with melt grown crystals. Furthermore, the method is relatively easy to be realized with a low cost. The most important advantage, however, is that it can be used for the crystal growth of materials unstable at the melting point (e.g. β-FeSi$_2$). For some of the semiconducting silicides (CrSi$_2$ and β-FeSi$_2$) the small crystal size is compensated by the needle-like shape, which is easy to handle for electrical and thermoelectric measurements.

3.1.3 Flux Growth

This method involves crystal growth from a melt solution of the material to be grown in a solvent. Its most important advantage is the reduced process temperature to below the melting point of the material. As a result, appropriate crucible materials can by easily found and, as for CVT, high temperature phase transitions or incongruent melting can be avoided. Therefore, technical efforts can be reduced compared with the common melt growth methods. The main disadvantages of the method are limited purity of the crystal grown due to incorporation of the flux material and small crystal size. The main problem, which has to be solved in this method is the choice of the solvent material. The properties desirable for the solvent are:
1. The compound to be grown is the only stable solid phase under the growth conditions (main parameters are temperature and concentration).
2. The solubility of the grown compound (crystal composition) in the solvent is not too low and must decrease with decreasing temperature.
3. The melting point of the flux material must be much lower than the stability limit of the phase to be grown.
4. The components of the solvent exhibit a very low solubility in the grown crystal or at least the incorporated components are electrically inactive.
5. A choice of a crucible material, which does not react with the melt solution and is not wet by the solution, must be possible.
6. The viscosity of the solution should be low, in the range 10^{-3}–10^{-2} Pa s.
7. The residual melt should be easily separated from the grown crystals.

For semiconducting silicide growth from the flux tin, zinc, Cu$_3$Si, or PdSi have been used preferentially as a solvent material. The solubility of different components in the flux can be estimated from binary phase diagrams (see e.g. [3.11, 3.12]). A more detailed analysis needs ternary phase diagrams, especially in the section MeSi$_x$ - solvent. Alumina and quartz are good candidates for crucibles.

The growth process must be performed in an oxygen-free atmosphere in a furnace with a good thermal stability and well operated temperature. Vertical tube furnaces with a flowing argon atmosphere usually meet the requirements. The weight ratio of a solvent to the growing material is in the range from 10:1 to 30:1. The precursor can be introduced in the form of elemental components or as a presynthesized compound. The crystal growth takes place with slow cooling of the furnace. After heating to a maximum temperature and waiting for an equilibration time the temperature is slowly decreased with a rate of some degrees per hour.

Spontaneous nucleation occurs first and one has to control the number of nuclei to obtain only a few large single crystals for subsequent growth. In order to minimize the number of growing nuclei an oscillation technique in the starting phase of the cooling can be used [3.13]. A typical programmable temperature–time diagram for the flux growth of silicides is shown in Fig. 3.2. Depending on the crystal/solvent system different cooling rates with a constant linear growth rate, a constant cooling rate or a maximum stable growth rate can be applied during the crystallization. For a detailed description see Scheel and Elwell [3.13]. The residual tin or zinc solvent can be removed from the crystals by treating in diluted KOH, while PdSi is dissolved with aqua regia.

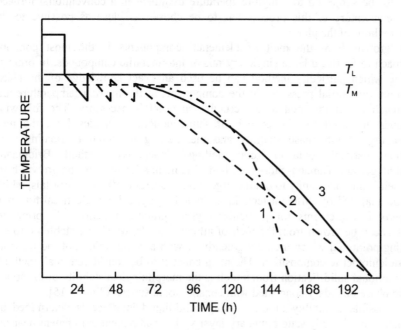

Fig. 3.2. Typical temperature program for the flux growth with a temperature oscillation in the metastable region below the liquidus temperature T_L around the melting point T_M: 1 - constant linear growth, 2 - constant cooling rate, 3 - maximum stable growth rate [3.13]

A second method to produce single crystals from a melt solution of the components is the travelling solvent method (TSM). The method is similar to the vertical zone melting in a crucible described in Sect. 3.1.4. An alloy consisting of the components of the compound to be grown and the solvent is used as a liquid zone. This zone is moved through the material with a very low pulling rate. The use of a seed for a growth of oriented crystals is possible. Other details related to the crystal growth of refractory compounds from flux can be found in the comprehensive review of Gurin [3.14] who analyzed the complete process.

3.1.4   Melt Growth

Samples needed for crystal structure investigations can usually be very small with a size of some tens of micrometer. Therefore, small single crystal fragments of about $40 \times 45 \times 15 \ \mu m^3$ can be prepared from an arc-melted and heat-treated mixture of elemental components if the phase diagram allows this method of synthesis of the compound. The elements are melted in a water-cooled copper hearth in purified argon at a pressure of about 6×10^4 Pa. The melting procedure has to be repeated several times in order to improve the alloy homogeneity. For further improvement of the homogeneity and enlargement of crystal grains the samples can also be subjected to a high-temperature treatment in a conventional furnace. The temperature of this process has to be chosen as close as possible to the stability limit of the phase.

A growth from the melt of elemental components is the most common approach to fabricate large single crystals of intermetallic compounds. In order to decide which particular method can be used an exact knowledge of the phase diagram, the partial pressures of the components at the melting temperature and the reactivity of the melt to the crucible material is necessary. The following methods are used for the growth of semiconducting silicides from the melt depending on the phase stability and the melting temperature: floating zone melting, vertical container zone melting, Czochralski method, Bridgman–Stockbarger and Tamann–Stöber method. The main advantage of the growth from the melt is an opportunity to obtain large (up to several cm^3) single crystals, which can be analyzed by different techniques including optical and electrical transport property measurements for particular crystallographic directions. In principle, crystals can be grown from the melt of all compounds, which are stable up to the melting point and which melt congruently or with a composition not too far from the stoichiometric composition. The main concept to be considered in all methods of directional solidification from a melt is the morphological stability of the solid–liquid phase boundary connected with constitutional supercooling [3.15].

A possible instability of the moving solid/liquid interface is determined by formation of a steady state boundary layer with a component enrichment near the interface caused by distribution coefficients $k < 1$ for all deviations from the stoichiometric compositions in the metal-silicon phase diagrams shown in Sect. 1.2. Constitutional supercooling occurs if $G_L < mG_C$, where G_L is the temperature gradient in the liquid near the solid/liquid interface, G_C is the gradient of solute concentration in the liquid phase in the steady-state diffusion boundary layer, and m is the liquidus slope in the equilibrium phase diagram. In the steady state conditions of unidirectional solidification with a planar solid/liquid interface this gradient is [3.16]:

$$G_C = (1 - k)/k \cdot N_\infty \cdot V/D , \tag{3.5}$$

where k is the distribution coefficient taken from the phase diagram, N_∞ is the solute concentration in the melt far from the solid/liquid interface, V is the growth rate, and D is the diffusion coefficient of the solute in the melt. Because of the lack of exact experimental data for silicide forming systems the accuracy of (3.5) seems to be sufficient to describe the phenomenon. Any unstable solid/liquid interface will lead to cellular growth with undesired second phase formation for the silicides. While the distribution coefficient and the diffusion coefficient in the melt are fixed by the system a high temperature gradient in the liquid near the solid/liquid interface must be realized by the design of the crystal growth facility. Moreover, the pulling rate must be chosen to be as low as necessary to guarantee a stable growing interface.

The different demands determined, e.g. by the melting temperature, the reactivity, the partial pressure of the components at the melting point of the compounds to be grown can be fulfilled using a wide variety of growth methods. The Tamann–Stöber method, the Bridgman and Bridgman–Stockbarger methods, the Czochralski method, zone melting and zone floating methods are among them.

The Tamann–Stöber method is characterized by melting the material in a vertical crucible followed by slow cooling. The crucible is located in a temperature zone with a gradient so that the melt at the bottom of crucible crystallizes first. For seed selection the bottom of the crucible is usually rounded or formed as a tip. This method is also referred to as vertical gradient freezing (VGF) method. In the literature dealing with the single crystal growth of semiconducting silicides this method is sometimes called Bridgman-like. The Bridgman technique uses a relative translation of the crucible containing a completely molten material to an axial temperature gradient in a furnace. In the crystal growth of semiconducting silicides only vertical configurations, shown in Fig. 3.3, are used.

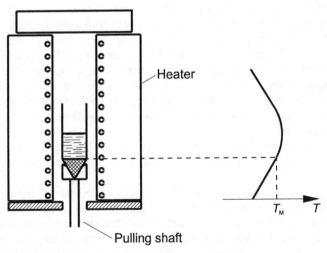

Fig. 3.3. Vertical Bridgman method and related temperature profile

The temperature gradient near the solid/liquid interface can be made more stable and sharper by applying the vertical Stockbarger configuration consisting of two furnaces with different temperature levels divided by an adiabatic loss zone, as illustrated in Fig. 3.4.

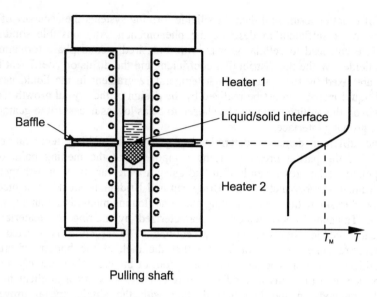

Fig. 3.4. Vertical Stockbarger configuration and related temperature profile

The crucible is moved into the temperature gradient zone with a velocity which must be very well controlled. For a convenient seed selection the bottom of the crucible usually ends in a rounded tip with a small radius of curvature.

A problem of general interest in all crucible methods is the choice of the crucible material. It must be chemically inert with respect to the melt, should not be wetted by the melt, and should have a thermal expansion coefficient less than the crystal itself. Depending on the silicide to be grown graphite, graphite covered with boron nitride, alumina, boron nitride, quartz and Al_2O_3-supported silica and welded molybdenum crucibles are used. Melts of semiconducting silicides react with oxygen and form SiO or SiO_2, depending on the temperature. Therefore, an inert atmosphere of purified argon or helium and in some cases high vacuum conditions must be used.

Czochralski growth by crystal pulling from a melt contained in a crucible is a widely used method in research and industry. It is schematically illustrated in Fig. 3.5.

The apparatus consists of the crucible, which can be rotated, the upper pulling shaft, which enables a rotation and the velocity of which must be stabilized, the heater and a vacuum containment. Seed crystals are used if available. The choice of the crucible material and the atmosphere have to meet the same requirements as discussed for the Bridgman method. The Czochralski growth of semiconducting silicides is carried out using silica crucibles in an atmosphere of Ar containing 10% H_2 by using an RF induction furnace for $MnSi_{2-x}$ and alumina crucibles for $FeSi_2$. To avoid any contact with the crucible for higher melting silicides a Hukin-type cold copper crucible is also used.

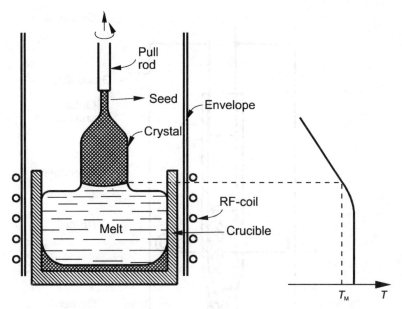

Fig. 3.5. Czochralski method of the crystal growth and related temperature profile

A common procedure to allow the growth of only one crystallite in all pulling methods is the bottle-neck technique. Starting from a polycrystalline seed the diameter of the growing crystal is reduced to a minimum in the growth process followed by enlarging the diameter to the normal rod size. In the narrow part of the crystal only a few or sometimes only one crystallite can propagate. Otherwise, a growth process without a seed can be performed from the polycrystalline material for formation of small single-crystalline seeds for further growth experiments.

Zone melting in which only a relatively small zone of a cylindrical rod is molten and the zone travels from one end of the rod to the other one can be realized in containers or as crucible-free zone melting. The first method schematically shown in Fig. 3.6, is applied with silica containers to grow $CrSi_2$ single crystals [3.17]. During the preparation of $CrSi_2$ single crystals one starting material CrSi is purified by this method using the redistribution of impurities according to their distribution coefficients $k < 1$.

The second method, schematically presented in Fig. 3.7, is used for the growth of $CrSi_2$ single crystals. The melting may be accomplished by RF heating. In order to avoid a reaction with oxygen either vacuum or an inert gas ambient is used. Special pulling heads with stabilized drivers are designed for the motion and rotation of the upper and the lower pulling shafts.

The floating zone method is normally preferred for refractory materials for which no crucible material is available. In [3.3] the growth of high temperature melting $MoSi_2$ is described using a floating zone technique with an optical double-ellipsoidal mirror furnace.

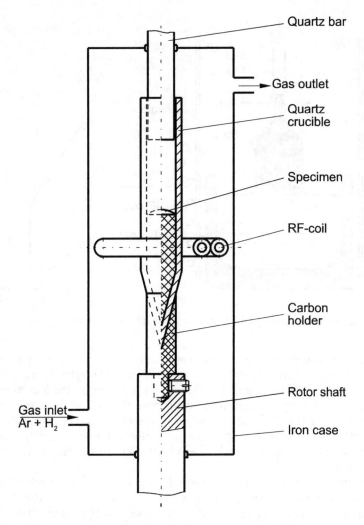

Quartz bar

Gas outlet

Quartz
crucible

Specimen

RF-coil

Carbon
holder

Rotor shaft

Iron case

Gas inlet
Ar + H₂

Fig. 3.6. Vertical container zone melting [3.17]

The apparatus is schematically shown in Fig. 3.8. The shape and stability of the molten zone play an important role in floating zone melting. The molten zone is hanging free between the two rods, and its length is limited to about the rod diameter for diameters up to 10 mm used in the silicide crystal growth.

A special method, which combines the melt levitation in a cold copper crucible with the zone melting where only a small part of the material is molten, was described by Thomas [3.18] and used by Gottlieb and co-workers [3.19, 3.20] to grow single crystals of Ru_2Si_3 and $ReSi_{1.75}$. The scheme of this method standing between floating zone and Czochralski method is shown in Fig. 3.9.

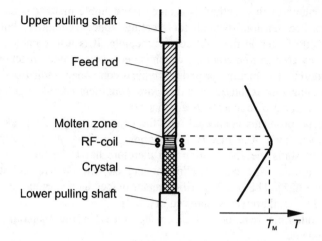

Fig. 3.7. Floating zone method and related temperature profile

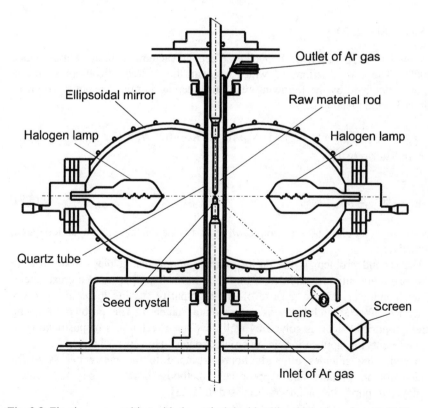

Fig. 3.8. Floating zone melting with the optical double ellipsoidal mirror furnace [3.3]

The advantage of this method is a much more stable melt. It is achieved not only by surface tension, typical for floating zone techniques, but also by electromagnetic forces in the cold copper crucible. It is also combined with the absence of any contact to a crucible, which is necessary to avoid reaction with the crucible material. The method provides a typical component redistribution known from the floating zone technique and therefore, congruent and incongruent melting compounds can be grown in a stationary manner.

Two special problems connected with the crystal growth of silicide materials should be mentioned. The first is silicothermic reactions of oxygen traces with silicon, which shifts the composition and therefore has to be taken into account during preparation. The consideration of the oxygen content in materials is described in [3.21]. The use of purified metal oxides as a starting material for a silicide single crystal growth is analyzed in [3.22].

The silicothermic reactions of all oxide traces in the materials have the following scheme

$$MeO_x + (2+x/2)Si = MeSi_2 + x/2SiO_2, \tag{3.6}$$

$$SiO_2 + Si = 2SiO. \tag{3.7}$$

Complete evaporation of SiO takes place in vacuum at temperatures above 1000°C. The mass fractions M, which result in the desired silicon concentration N_{Si} are calculated by the following relations, keeping in mind weight fractions of silicon K_{Si} and metal K_{Me}:

$$M_{Me} = M_{total}(1+(N_{Si}(1-K_{Me})+(1-N_{Si})A_{Me}/A_O\,K_{Me})/((1-N_{Si})(A_{Me}/A_{Si}(1-K_{Si})-A_{Me}/A_O\,K_{Si}))^{-1}, \tag{3.8}$$

$$M_{Si} = M_{total} - M_{Me}, \tag{3.9}$$

where A_{Me}, A_{Si} and A_O are atomic weights of the metal, silicon and oxygen, respectively.

The second problem is the purity of the components. It plays a key role in semiconducting silicides and determines their intrinsic properties to a great extent. In semiconductor science the use of ultrahigh-purity starting materials with 6N purity (99.9999%) and better is an absolute standard. The problem of using ultrahigh-purity silicon is solved by well developed techniques of purification of this element for application in microelectronics. However, the metals in the semiconducting silicides often do not correspond to the purity necessary for semiconductor preparation. Appropriate methods have to be additionally employed to purify the metals as discussed in [3.23].

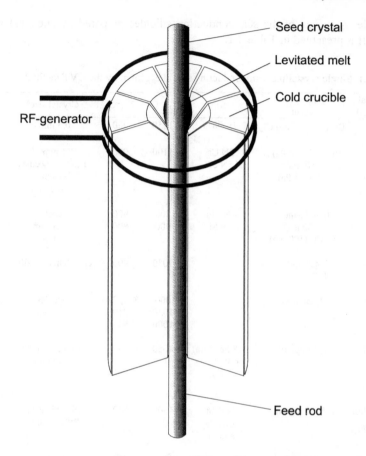

Fig. 3.9. Floating zone method with an additional levitation of the melt in the cold copper crucible

3.2 Crystal Growth by Chemical Vapor Transport Reactions

Chemical transport reactions in closed systems have been used for the growth of single crystals of $CrSi_2$, $MnSi_{2-x}$, $ReSi_{1.75}$, and β-$FeSi_2$. Thermodynamical analysis of the processes involved [3.6] shows that the content of silicon and metal in the gas phase is essential for the prediction of the transport behavior. Results of the theoretical estimation in comparison with experimental observations are discussed in this section. Furthermore, a summary of related thermochemical data is available. A survey of the semiconducting silicides prepared by chemical vapor transport is presented in Table 3.1.

Table 3.1. Single crystalline semiconducting silicides grown by the CVT method

Material (dopants)	Transport agent	Purity	T_1 (°C)	T_2 (°C)	Crystal size[a]	Refs
CrSi$_2$	Br$_2$ (1.3×10^4 Pa) Cl$_2$	95–98%	1050	900	0.5 mm^3	3.24
	Br$_2$ (1.3×10^4 Pa) I$_2$ (10^5 Pa) Cl$_2$ (10^5 Pa)	2N Cr 3N Si	1100	900	1 mm^3 1 mm^3 + needles l=13 mm d=0.2 mm	3.10
MnSi$_2$	I$_2$ and MnCl$_2$ (100 mg) CuCl$_2$ (100 mg)	4N Mn 5N Si	800 900	900 800	Needles l<0.1 mm 3 mm^3	3.7 3.25
ReSi$_{1.75}$	Cl$_2$ and I$_2$ 5 mg/cm^3	-	1050	900	Very small crystallites	3.26
β-FeSi$_2$	I$_2$, 50 kPa	-	1000– 1030 1000	830, 870 910 830	Needles l≤3 mm	3.27, 3.28
(Cr, Al)	I$_2$ 5 mg/cm^3	4N8 Fe (Alfa) or 3N Fe + 6N Si	1050	750	Twinned needles l=10 mm	3.29
(Cr, Co, Mn, Ni)	I$_2$ 5–10 mg/cm^3	2N FeSi$_2$ ≥3N Fe and Si	1000	680	Twinned needles l=10–15 mm	3.30, 3.31
(Cr, Co)	I$_2$ (100 mg)	4N8 Fe (ZFW) >5N Si	1050	750	Needles (5–15)×2×0.5 mm^3	3.23
	I$_2$ (100 mg)	^{57}Fe 4N8 Fe >5N Si	1050	750	Needles (5–15)×2×0.5 mm^3	3.32, 3.23

[a] l – length, d – diameter.

3.2.1 Silicides of Group VI Metals

CrSi$_2$. Single crystals of CrSi$_2$ were grown by the CVT method in closed silica ampoules by Nickl and Koukoussas [3.24] and by Krausze *et al.* [3.10]. A powder of 1–10 µm particles of the silicide with a purity of 95–98% was used in [3.24] as a starting material. It was heat treated at 500–1000°C in vacuum of about 10^{-2} Pa. Analytical grade Br$_2$ or purified Cl$_2$ was employed as a transport agent. The latter was used in preliminary experiments only because its transport rate observed was always smaller than that for Br$_2$. The transport proceeded from the higher temperature T_2 (1100–900°C) to the lower one T_1 (1000–700°C) with a temperature difference of 120–200 degrees. With the silicides as starting materials a congruent transport was observed for $T_2 \rightarrow T_1$ without enrichment or depletion of one of the components. The best crystals were grown at $T_1 = 900°C$ and $T_2 = 1050°C$ during 216 h at the Br_2 pressure of 1.3×10^4 Pa. They were 0.5 mm^3 in size with well developed growth facets.

Larger crystals with a higher purity were grown by Krausze *et al.* [3.10]. Polycrystalline $CrSi_2$ synthesized from a mixture of 2N chromium and 3N silicon powders at 1000°C was used. The transport agent (Cl_2, Br_2 or I_2) was introduced into the silica ampoules before sealing in an amount ensuring a pressure of 10^5 Pa at the process temperature. Chemical vapor transport experiments were performed in the temperature range between 900 and 1100°C. In the cases of Cl_2 and Br_2 the transport direction was found to be from the hot to the cold side of the ampoule. There was a reverse transport direction, i.e. from the cold to the hot side, observed for I_2. The maximum transport rate of 1.8 mg/h was realized with Br_2 in a temperature gradient from 1100 to 900°C. When Cl_2 was used as a transport agent, silicon also crystallized in the cold zone of the ampoule. For Br_2 and I_2 only crystals of $CrSi_2$ were found. The habit and size of the crystals fabricated are nearly independent of the transport medium. Typical isometric crystals grown are demonstrated in Fig. 3.10a. They are about $1 \times 1 \times 1$ mm^3 in size. Additionally, the hexagonal needles with a diameter of 0.2 mm and a length of 13 mm shown in Fig. 3.10b were observed when Cl_2 was used as a transport agent. Despite the success in single crystal growth of $CrSi_2$ by the CVT technique, the single crystals obtained were not used for electrical measurements.

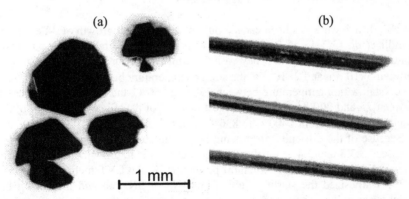

(a) (b)

1 mm

Fig. 3.10. Isometric (a) and needle-like (b) $CrSi_2$ crystals grown by the CVT method [3.10]

3.2.2 Silicides of Group VII Metals

MnSi$_{2-x}$. Single crystals of $Mn_{15}Si_{26}$ were grown by Kojima and Nishida [3.7, 3.25]. The starting material was prepared from a mixture of manganese (4N) and silicon (5N) by arc melting in a purified argon atmosphere. The ingot was identified to be $MnSi_{1.73}$ by XRD. The compound was crushed in a mortar and then sealed with a transport agent into a quartz tube. Three different transport agents, i.e. I_2, $CuCl_2$ and $MnCl_2$, were used. Thermal processing was performed in a furnace with two temperature zones kept at 900 and 800°C for a few weeks. With I_2 and $MnCl_2$ transport agents the crystals grew in the high temperature zone. The growth rate was low and the crystals were small (~0.1 mm). In the case of $CuCl_2$ the crystals grew in the zone of lower temperature. The crystals were identified by XRD to be $Mn_{15}Si_{26}$. The average crystal size was 1.5 mm. In contrast to all single crystals grown from the melt, no striations of MnSi were observed along the $Mn_{15}Si_{26}$ crystals but, unfortunately, the size of the grown crystals was too small to perform electrical transport measurements.

$ReSi_{1.75}$ ($ReSi_2$). There has been only one attempt, by Khristov *et al.* [3.26], to grow crystals of this semiconductor by the CVT method. A detailed thermodynamic analysis of the process is not possible because there are insufficient data. The transport proceeded with a measurable rate in the presence of Cl_2 and I_2 as a transport media. The experiments were performed in sealed silica ampoules (20×150 mm) for 96 h with a concentration of the transport agents of 5 mg/cm³. The transport proceeded between 1050°C and 900°C from the hot to the cold side with a rate lower than 0.035 mg/h. Very small crystallites (< 0.1 mm) identified by XRD as $ReSi_2$ were grown. They were too small to perform electrical transport measurements.

3.2.3 Silicides of Group VIII Metals

β-FeSi$_2$. Much work has been carried out on single crystal growth of β-FeSi$_2$. First Wandji *et al.* [3.27] and later Ouvrard *et al.* [3.28] reported the crystal growth using I_2 as a transport agent. They varied the initial pressure of the transport agent between 5×10^3 Pa and 2×10^5 Pa, the source temperature between 750 and 1050°C, the crystallization temperature between 750 and 960°C, the temperature difference between 75 and 200 degrees. The largest crystals of β-FeSi$_2$ with a length of 2 to 3 mm were grown from 1000°C to 830°C with I_2 pressure of 5×10^4 Pa for 196 h.

Because of the difficulty of growing large single crystals from the flux or melt (see Sect. 3.3.3 and 3.4.4) much effort has been made to grow high-purity β-FeSi$_2$ single crystals with a good structural perfection by the CVT method. Kloc *et al.* [3.29] synthesized the starting FeSi$_2$ from 4N8 or 3N iron and 6N silicon in a Czochralski puller. Then, about 5 g of this material and the I_2 carrier (5 mg/cm³) were placed inside a clean quartz ampoule. The temperature gradient was stabilized from 1050°C to 750°C. The formation of the β-FeSi$_2$ phase was not affected by the excess silicon in the ampoule which was added during the previous melting of silicon with iron. Silicon condensed at the cold part of the ampoule and β-FeSi$_2$ crystals grew on this layer in the form of thin, frequently twinned needles, each about 10 mm long. Chromium and aluminum doped crystals were also grown by the same method. The crystals were used for electrical measurements reported in [3.33] and [3.34]. A thermal heat treatment in vacuum of 1.3×10^{-2} Pa at 600°C for 5 h did not change the n-type conductivity, which was attributed to possible iodine doping.

Pure and chromium, cobalt, manganese and nickel doped β-FeSi$_2$ crystals were grown by Tomm *et al.* [3.30] using I$_2$ as a transport agent (5–10 mg/cm^3). FeSi$_2$ powder (2N) was used as a starting material. The crystal growth was performed in a temperature gradient from 1000°C to 680°C for 200 h. The crystals obtained were needle-like up to 20 mm in length and 1–2 mm in diameter. The habit was not really prismatic in all cases. The measured XRD pattern correlates only with the orthorhombic β-FeSi$_2$ phase, no α-FeSi$_2$ was found. Dopant elements (chromium, cobalt, manganese, and nickel) were added to the source material at concentrations up to 2 wt %. The concentration of dopants incorporated in the single crystals was about one order of magnitude lower, as estimated from EDX and EPR measurements. Doping was found to influence the growth process. The doped crystals were always larger and the transport rates were higher. Identical results were obtained in the experiments of Brehme *et al.* [3.31] when the concentration of the same dopants in the source material was increased up to 5 wt %.

It is well established that doping with transition metals locating at the right side of iron in the Periodic Table, like cobalt and nickel, yields β-FeSi$_2$ with n-type conductivity. Transition metal impurities from the left side, like chromium and manganese, result in p-type material. However, the electrical effect of the neighboring elements of silicon — aluminum, boron, gallium, phosphorus, nitrogen and arsenic — and of interstitial elements mainly boron, carbon and nitrogen must be taken into account as well. On the other hand, it seems that the intrinsic properties of undoped β-FeSi$_2$ in the first experiments were always measured on the crystals doped with uncontrolled impurities. In contrast to the doped single crystals the behavior of the undoped samples was irregular due to nonintentional doping which changed from sample to sample [3.35]. In spite of the use of higher-purity iron as a starting material, the purity of the grown crystals was not reported [3.29, 3.33]. Furthermore, there was insufficient information about the intrinsic properties of undoped β-FeSi$_2$. Beyond this, main component solubility limits in β-FeSi$_2$ are not well known. They are assumed to be close to the stoichiometric composition. But, in semiconductors even small deviations can influence electrical properties significantly. Therefore, high-purity single crystals of β-FeSi$_2$ are necessary to elucidate the properties of undoped samples as a function of the deviation from the stoichiometric composition. To achieve high-purity semiconductor-grade β-FeSi$_2$ single crystals it is necessary to use high-purity starting materials and to optimize the entire preparation process to maintain the purity. Ultrahigh-purity silicon can be provided from the well developed technique of purification of this element for application in microelectronics. The preparation of high-purity iron with a very low content not only of metallic but also of nonmetallic impurities, which might substitute for silicon in the crystal lattice, and of interstitials is a large-scale process. To our knowledge the highest purity of commercially available iron is 4N8 metal base (Alfa - Puratronic$^{®}$) but without any specification of the nonmetallic impurities.

Behr *et al.* [3.23] prepared β-FeSi$_2$ single crystals starting from high-purity elements. Silicon with a purity exceeding 5N and high-purity iron (>4N8), the chemical composition of which is shown in Table 3.2, was used as source materials. The preparation of pure iron was reported by Weise and Owsian [3.36] and in [3.23], too. The purification process started from the technologically available iron powder. A vapor deposition technology with several steps including an exchange distillation in order to separate nickel from iron was used. The purity of the iron is characterized by the ratio of residual resistivities at 293 K and 4.2 K, which is as large as about 3000.

Table 3.2. Impurities in Fe and β-FeSi$_2$ measured by spark source mass spectrography (MS), gas hot extraction and carrier gas fusion methods

Element	Fe-wire (wt ppm)	FeSi$_2$ (wt ppm)
B	< 1	< 1
S	< 1	< 0.1
P	< 1	< 1
Na	< 1	2
Ba	< 1	< 0.1
Cl	2	< 1
Br	< 1	< 0.1
K	< 1	< 1
Ca	< 0.1	2
Ti	< 1	< 1
Cr	< 1	7
Mn	< 1	<0.1
Co	1	1
Pb	< 1	< 0.1
W	< 1	< 0.1
Ni	2	11
Cu	5	2
Zn	< 1	< 1
Ga	1	1
Zr	1	1
Nb	< 0.1	< 1
Sb	< 1	< 0.1
Sn	< 1	< 0.1
As	< 1	< 0.1
Mo	< 2	< 0.1
Ag	< 0.1	1
I	< 0.1	19
Ta[a]	13	19
C[b]	10	not determined
O[b]	9	not determined
N[b]	6	not determined
H	< 6	not determined

[a] The MS-sample holder is made of tantalum, [b] 70% of the impurity is located at the surface [3.37].

The chemical transport reaction was performed in closed silica ampoules filled with I$_2$. In most cases the composition of the starting material was equal to the atomic ratio Fe/Si = 2, but the ratios 2.5 and 1.5 were also chosen to ensure crystal growth at upper and lower boundaries of the homogeneity range of β-FeSi$_2$. The ampoules cooled by liquid nitrogen were evacuated to about 0.1 Pa and then sealed off. The chemical vapor transport was performed in a horizontal configuration and proceeded from the source at the temperature of 1100°C (T_2) to the crystallization zone kept at a lower temperatures T_1. Several runs with temperature differences between T_2 and T_1 of 100 to 300°C were performed. Usually, after 24 h preheating in a reversal temperature gradient $T_2 < T_1$, the ampoule was kept for about 240 h at the process temperatures T_1 and T_2. Special attention must be paid to the maintenance of the purity of the starting agents. This demands high-purity iodine.

The single crystals grown are demonstrated in Fig. 3.11. They have a needle-like shape with dimensions of (5–10)×2×0.5 mm^3. The measured XRD peaks correlate exclusively to the orthorhombic β-FeSi$_2$ phase. No trace of α-FeSi$_2$ is found. Only at the crystallization temperature of 1000°C were plate-like crystals of the α-FeSi$_2$ phase obtained. The habit of the crystals does not change if different source compositions and temperatures in the crystallization zone of the transport ampoule are used. In contrast to the crystals reported before [3.29], selected crystals have a flat surface indicating that the usual twinning can be avoided in the material of higher purity. Further efforts are necessary to find optimum conditions for this type of crystal growth.

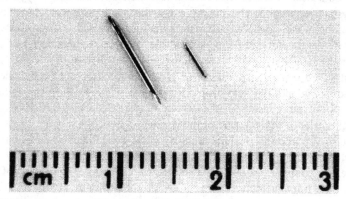

Fig. 3.11. Single crystals of high-purity β-FeSi$_2$ [3.23]

The analyses of the single crystals display no significant increase in impurities compared to the starting materials (see Table 3.2) but a decrease in the concentration of elements which are not transported by iodine. Only the concentration of the transport agent (I$_2$) is increased in the single crystals to 19 wt ppm. The slight increase in nickel and chromium content could be attributed to some contamination from stainless steel, the source of which is not yet identified.

Room temperature electrical measurements on single crystals grown from different compositions of the source material and at different crystallization temperatures are summarized in Table 3.3. In contrast to literature data all crystals show only n-type conductivity. Significant variation is evident for the undoped crystals, which can be correlated with a deviation from the stoichiometric composition. From the data it can be concluded that not only the source composition but also the crystallization temperature T_1 yields different compositions within the homogeneity range of β-$FeSi_2$, the extension of which has not been determined yet. The different behavior of the crystals grown at the upper boundary of the homogeneity range is also believed to be an effect of intrinsic defects.

Table 3.3. Room temperature electrical parameters of β-$FeSi_2$ single crystals grown at $T_1 = 750°C$ and $T_2 = 1050°C$

Source composition	Resistance (ohm cm)	Thermopower ($\mu V/K$)	Activation energy (meV)
$FeSi_{1.5}$	18	−650	23
$FeSi_{2.0}$	20	−600	45
$FeSi_{2.0}$	230	−750	50
$FeSi_{2.5}$	8	−300	~40

Starting from a certain purity, only n-type single crystals grew even in equilibria both with FeSi and with silicon. It is concluded that the p-type undoped single crystals reported in the literature are the result of a very high impurity concentration in the crystals. Differences in the electrical transport properties between crystals prepared at the upper and the lower phase boundaries of β-$FeSi_2$ were found, indicating that properties of these high-purity crystals were significantly influenced by intrinsic (deviation from the stoichiometric composition) but not by extrinsic (impurities) factors. Further studies are necessary for a more detailed elucidation of the tendencies observed and an improved knowledge of the nature of the intrinsic donors.

As electrical properties depend strongly on the crystallization temperature one can conclude a strong dependence of the lower and upper boundary of the homogeneity region of β-$FeSi_2$ on temperature. Therefore, annealing of single crystals grown from the source materials with compositions $FeSi_{1.5}$ or $FeSi_{2.5}$ to fix the lower and upper boundary of the homogeneity region of β-$FeSi_2$ and applying I_2 as a transport agent to establish the thermodynamical coexistence of the source and crystals at a well defined temperature, becomes very important. The homogeneity region of β-$FeSi_2$ was found to be only at the Fe-rich side from the 1:2 stoichiometry. This is characterized by n-type conductivity. The silicon content for both the lower and the upper phase boundary of the homogeneity region increases with the temperature rise. Thus, the crystals annealed at higher temperatures have lower carrier concentrations.

Single crystals including up to 40% of the isotope [57]Fe were grown by the method described above [3.32]. This was done for investigations of particular positions of iron atoms in β-$FeSi_2$ by Mössbauer spectroscopy. The samples were needle-like oriented in $\langle 110 \rangle$ direction.

Furthermore, β-FeSi$_2$ single crystals doped with chromium and cobalt were grown using the CVT method described in [3.23]. For doping, 4N chromium and 4N cobalt were used. The relation between the impurity concentration in the source and grown crystals is illustrated by the data in Table 3.4.

Table 3.4. Impurity concentration in the source and grown β-FeSi$_2$ single crystals [3.23]

Source composition	Cobalt concentration (wt ppm)		Chromium concentration (wt ppm)	
	in the source	in the crystal	in the source	in the crystal
Fe(ZFW)Si$_{2.5}$	0	< 8.6	0	< 8
Fe(Alfa)Si$_2$	3000	240	3000	230
Fe(Alfa)Si$_2$	10 000	210	10 000	3100
Fe(Alfa)Si$_2$	30 000	780	30 000	36 000
Fe(ZFW)Si$_2$	10 000	340	50 000	19 000

In the cobalt-doped samples a strong depletion of this impurity in the crystals was found. Only 2.6–8% of the source cobalt was in the single crystals. At the beginning of the experiments it seemed that the purity of the starting material does not play an important role in the doped crystals and therefore, 4N8 iron (Alfa) was used for the crystal growth. But the depletion in cobalt makes other impurities, first of all nickel, important in the crystal properties. Therefore, the problem of purity is much more serious than originally believed. Electrical properties of lightly doped materials can be significantly influenced by accompanying impurities. Quite a different picture was found for chromium doping. The amount of chromium built into the lattice of β-FeSi$_2$ is much higher than that of cobalt (see Table 3.4). It practically reaches the source concentration if one takes into account the uncertainty of the concentration measurements. In this case the purity of the starting elements does not play such a role as for cobalt doping.

The growth of β-FeSi$_2$ single crystals by the CVT method has appeared to be the best developed among the semiconducting silicides. The crystals have a considerable size, high lattice perfection and excellent purity. They enable measurements of intrinsic electronic properties as a function of deviation from the stoichiometry and impurity doping.

3.3   Crystal Growth from the Flux

Crystal growth from high-temperature solutions, usually referred to as growth from the flux, has been used to obtain crystals of CrSi$_2$, ReSi$_{1.75}$(ReSi$_2$) β-FeSi$_2$, and OsSi$_2$. A comprehensive analysis of the problems connected with the flux growth of refractory compounds was given by Gurin [3.14]. In this section we focus on the above mentioned silicides. An overview for the semiconducting silicides grown by this method is presented in Table 3.5.

Table 3.5. Single crystalline semiconducting silicides grown from the flux

Material	Solvent	Purity	Max. temperature (°C)	Cooling rate	Crystal size[a]	Refs
CrSi$_2$	Zn	≥2N	900, 2 h	18–180 K/h	Needles l=5–7 mm	3.38 3.39 3.40 3.41
				9 K/h	Small isometric crystals	
	Sn	≤4N	1400, 10 h	50 K/h	Several tenth of mm need-like l≤10mm	3.42
	Sn	3N Sn 2N Cr 3N Si	1400	20–60 K/h	Needles l<10 mm	3.43
ReSi$_{1.75}$	Sn and Zn	-	1000	slow cooling	Very small crystals	3.44
	Sn	3N	1400, 10 h	20–50 K/h	Needles l=3 mm Polyhedra 0.5×0.5×0.5 mm^3	3.26
β-FeSi$_2$ (TSM)	Cu$_3$Si	4N Fe, 4N6 Cu >5N Si	950	Travelling rate 0.7 mm/h	Grains of β-FeSi$_2$	3.45
OsSi$_2$	PdSi	~2N	1227	6 K/h	Small fragments	3.4

[a] l – length, d – diameter.

3.3.1 Silicides of Group VI Metals

CrSi$_2$. Single crystals of this compound were grown from the flux by Gurin [3.14, 3.38–3.41], Okada and Atoda [3.42] and Peshev *et al.* [3.43]. Gurin *et al.* [3.38] reported the growth process with the use of zinc as a solvent metal. The starting materials were silicon and chromium of 2N or higher purity. There was an excess of silicon in relation to the stoichiometric disilicide composition. The synthesis of the compound took place at the boiling temperature of zinc (907°C) during 1–2 h. The melt was cooled at a rate of 0.3–3 K/min to the solidification temperature of zinc. The single crystals obtained were needle-like of 5–7 mm length. The composition was determined to be CrSi$_{1.93}$, the lattice parameters were $a = 0.4427$ nm, $c = 0.6391$ nm and $c/a = 1.444$.

In later works of Gurin *et al.* [3.40, 3.41, 3.14] crystal growth from a zinc flux was extensively investigated in a wide range of regimes including heating for high temperature homogenization and crystal growth during subsequent cooling.

Heating was performed with high (1.7–3.3 K/s), medium (0.083–0.16 K/s) and low (3.3–6.7 mK/s) rates up to 600, 700, 800, and 900°C. Then rapid quenching, annealing up to 24 h, and cooling with 50–83 mK/s and 1.7–5 mK/s were performed. In the heating stage at the lowest heating rate up to 900°C needle-like crystals up to 150 µm were grown. They were enlarged to 250 µm after holding the temperature at 900°C for 24 h. During the cooling process the largest needle-like crystals obtained were grown with a cooling rate of 2.5 mK/s. Under the same conditions isometric crystals with a size up to 360 µm were also grown.

Okada and Atoda [3.42] prepared $CrSi_2$ single crystals from a tin flux. The elemental starting materials were mixed with atomic ratios $Si:Cr = 2.0$ and $Sn:Cr = 6.4$. The mixture was heated in an argon atmosphere at the rate of 300 K/h to 1400°C, held at that temperature for 10 h to ensure uniform solution of the components, and then cooled at the rate of 50 K/h to 500°C. Single crystals of hexagonal prismatic shape extending along ⟨0001⟩ direction were obtained. The measured lattice constants were $a = 0.4426 \pm 0.0002$ nm and $c = 0.6375 \pm 0.0002$ nm. The composition was identified by EPMA to be $CrSi_{1.96}$. There were 10–100 wt ppm tin and less than 10 wt ppm aluminum, calcium, iron, copper, and magnesium in the crystals.

Peshev et al. [3.43] discussed a choice of solvent metals. Copper, silver, zinc, aluminum, tin, lead, and bismuth were concluded to be promising solvents for $CrSi_2$ taking into consideration their partial pressures, the toxicity of the vapors and costs. After estimation of solubility and phase formation in the silicon-metal and chromium-metal phase diagrams, only tin was chosen for experimental evaluation. An ingot of tin of 3N purity, chromium powder of 2N purity and silicon powder of 3N purity were used as starting materials. Chromium and silicon, taken at the ratio 1:2, were initially mixed for 30 min in a Specamill homogenizer. The mixture was then placed into a corundum crucible containing tin shavings with a weight of 5–20 times higher than that of chromium. The parameters of the growth process were the same as described by Okada and Atoda [3.42], only the cooling rate was chosen between 20 and 60 K/h. The crystals were separated by dissolving the tin-flux in dilute HCl. In all solutions only $CrSi_2$ was crystallized.

Experimental conditions affected the crystal size and shape. Needle-like crystals with a maximum length of 10 mm were predominant. For the Sn:Cr ratio of 15, small polyhedra appeared. Experiments demonstrated the possibility to use $CrSi_2$ as a starting material, but it resulted in smaller and usually coalesced crystal formation. Chemical analysis of the crystals obtained showed their composition to be $CrSi_{1.98}$.

Despite the success in $CrSi_2$ crystal growth from the flux the single crystals obtained were not electrically characterized.

3.3.2 Silicides of Group VII Metals

$ReSi_{1.75}$. Zinc and tin fluxes were used by Siegrist et al. [3.44] to grow crystals of this semiconductor. At the time it was believed to have produced the stoichiometric disilicide composition $ReSi_2$. Cooling from 1000°C was reported to be employed but without further specification . The crystals were very small and displayed the same structure as those obtained by arc melting (see Sect. 3.4.3).

Analysis of the available data proves that the most suitable solvent for the preparation of $ReSi_{1.75}$ crystals is tin. Parallel attempts to grow single crystals from aluminum flux were not successful. Powdered rhenium and silicon and a tin ingot were used by Khristov et al. [3.26]. All reagents were of 3N purity. Rhenium and silicon in the atomic ratio of 1:2 or 1:1.8 were preliminarily mixed for 30 min in a Specamill homogenizer. Then the mixture was placed into a corundum crucible containing tin shavings with a weight 5–20 times that of the rhenium in the powder mixture. The crucible was placed in a furnace with a purified argon atmosphere, heated to 1400°C, held 10 h for homogenization and cooled at the rate of 20–50 K/h. The crystals were separated from the flux material by dissolving the tin in diluted HCl. Needles up to 3 mm in length and polyhedra up to 500 μm were found. The best crystals were obtained using the atomic ratio Si:Re = 2.0, the weight ratio Sn:Re = 20, and the cooling rate of 20 K/h. Starting from the atomic ratio Si:Re = 1.8 only mixtures of $ReSi_{1.75}$ and Re_5Si_3 could be found. Electrical measurements were not performed on these crystals.

3.3.3 Silicides of Group VIII Metals

β-$FeSi_2$. Because of the peritectoid formation of β-$FeSi_2$ at temperatures far below the liquidus temperature Abrikosov and Petrova [3.45] crystallized this phase from a flux at temperatures below the peritectoid reaction. The solvent was Cu_3Si. First the ternary phase diagram Fe-Cu-Si was investigated by means of microstructural, thermal and XRD analysis. The initial materials were 4N pure iron carbonyl, 4N6 pure copper, and zone-refined polycrystalline p-type silicon. The region of the primary crystallization of the β-$FeSi_2$ phase ranging from 50 to 82 mol % of Cu_3Si was preselected from the phase diagram. Then, a possibility to obtain single crystals of β-$FeSi_2$ by a zone melting-like method with Cu_3Si as a solvent (travelling solvent method) was investigated using a zone width of 10 mm and a temperature of 945–950°C. A graphitized quartz boat containing the materials was sealed in the evacuated quartz capsule. A seed crystal of β-$FeSi_2$ grown from the $FeSi_2$ alloy annealed at 900°C for 300 h was also placed there. The zone part consisted of the alloy containing 40 mol % of $FeSi_2$ and 60 mol % of Cu_3Si. The composition of the feed rod material was $FeSi_2$. The pulling rate was about 0.7 mm/h. Considerable enlargement of the β-$FeSi_2$ grains in the bar in comparison with the cast alloy and only a small amount of Cu_3Si were observed.

$OsSi_2$. Single crystals of this semiconductor were grown from the flux of PdSi by Mason and Müller-Vogt [3.4]. The solvent compound has a melting point of 1087°C, which allowed the growth temperature to be lowered significantly and silica glass to be used. Mixtures of PdSi with $OsSi_2$ with concentrations up to 30 mol % were prepared. It was found that those with contents less than 15 mol % were completely liquid at 1223°C. Thus, the concentration of 10 mol % was chosen for crystal growth experiments. Cooling rates were varied from 1 to

300 K/h. In the range of 1–10 K/h there was no significant effect on the crystal perfection. Cooling of PdSi 10 mol % $OsSi_2$ melts from 1223°C to room temperature at 6 K/h yielded crystals with a well-defined morphology. The fragmented crystals were separated from the flux material by dissolving PdSi in aqua regia. They contained 0.15 at % of palladium, 1.5 at % of aluminum, and 0.2 at % of iron impurities, which came from the precursor or crucible materials. The authors pointed out that this level of contamination is considered to be too small to have a significant effect on the electrical properties of the semiconductor. From the experience with high-purity β-$FeSi_2$ single crystals [3.23] this opinion cannot be confirmed.

3.4   Crystal Growth from the Melt

The growth from the melt using Bridgman-like methods, Czochralski method, vertical container and floating zone melting is mostly applied in the preparation of semiconducting silicide single crystals. Attempts have been made to grow crystals of Mg_2Si, $BaSi_2$, $CrSi_2$, $MnSi_{2-x}$, $ReSi_{1.75}$, β-$FeSi_2$, Ru_2Si_3, $OsSi_2$. These are overviewed in Table 3.6.

3.4.1   Silicides of Group II Metals

Mg_2Si. In 1955 Whitsett and Danielson [3.69] first reported on electrical resistivity and Hall effect measurements on Mg_2Si single crystals grown from the melt. Unfortunately, they did not give any details about the growth procedure. Later, Morris *et al.* [3.46] described the growth of n- and p-type Mg_2Si single crystals. They started from sublimed magnesium with a purity of 4N or higher and semiconductor-grade silicon. Stoichiometric portions of magnesium and silicon were melted together in a graphite crucible with a spectrographically-pure graphite liner, which had an inner diameter of 6.4 cm and a length of 6.4 cm. Thermocouples were placed in the crucible wall near the top and bottom of the melt. Melting was carried out under an argon pressure of 2–3×10^5 Pa in order to reduce the loss of magnesium. The partial pressure of pure magnesium is 10^5 Pa at 1090°C, but it is reduced over melted Mg_2Si. A temperature gradient of 25 K/cm was established in the melt with the top at the higher temperature. The melt was then solidified within one more hour and the ingot was cooled to room temperature at 50 K/h. The crystals grown were always n-type. P-type samples were obtained when the melt was doped with 0.2–0.02 wt % silver or copper. Doping with aluminum or iron did not result in p-type conductivity.

LaBotz and Mason [3.47] chose another preparation route for Mg_2Si single crystals. Polycrystalline samples of various compositions were prepared by fusion of pure elements. Semiconductor-grade silicon and sublimated magnesium with 4N or better purity were used.

Table 3.6. Single crystalline semiconducting silicides grown from the melt

Material (dopants)	Method	Purity	Crucible material	Pulling rate, rotation	Crystal size[a]	Refs
Mg_2Si (Al, Fe)	Bridgman-like (Tamann–Stöber)	≥4N Mg >5N Si	Graphite	25 K/cm	$1\times1\times4$ mm^3 $2\times2\times10$ mm^3	3.46
	Bridgman-like (Tamann–Stöber)	≥4N Mg >5N Si	Graphite in stainless steel	15 K/h, T_M–30 K 10 h	d=13 mm l=30 mm	3.47
$BaSi_2$	Bridgman-like (Tamann–Stöber)	~4N	Welded Mo crucibles	10 K/h	Polycrystals	3.48
$CrSi_2$ (Mn)	Czochralski	4N Cr >5N Si	No details	No details	>15×5×1 mm^3	3.49, 3.50
	Bridgman-like (Tamann–Stöber)	4N Cr >5N Si	Quartz	1600°C, 50 K/h	>15×5×1 mm^3	3.49
	Vertical container zone melting	3N Cr 3N +>5N Si 3N3 Cr 4N+>5N Si	Quartz	0.17– 0.3 mm/min 0.17 mm/min 30 rpm	d=8 mm l=35 mm d=8 mm l=35 mm	3.17 3.51
	Floating zone (RF)	4N Cr, >5N Si		24 mm/h 30 rpm	d=4 mm l=40 mm Mosaic substructure	3.21
	Floating zone (optical)			6–10 mm/h 15 rpm	d=10 mm l=50 mm	3.52 3.3
$MnSi_2$	Arc melting	2N5 Mn 3N8 Si		Homogenization 750°C, 168 h	$41\times14\times500$ μm^3	3.53
	Bridgman	4N Mn >5N Si	Quartz	2 mm/h	d=10 mm l=40 mm	3.54 3.55
		4N Mn 5N Si	Quartz	1170°C 0.33–6 mm/h	d=15 mm l=35 mm	3.56
		4N Mn 5N Si	Quartz	1200°C 0.33–8 mm/h	d=18 mm l=50 mm	3.25
		5N Mn >5N Si	Quartz	1200°C, 8 mm/h	d=15 mm l=35 mm	3.57
	Czochralski	4N Mn >5N Si	Quartz	5–10 mm/h 6–8 rpm	d≈10 mm l=30 mm	3.58
		4N5 Mn 5N Si	Quartz	0.6–0.9 mm/min	d≈10 mm l=40 mm	3.59
		4N5 Mn 5N Si		15–18 mm/h 10 rpm 27 mm/h	d=20 mm l=50 mm d=10 mm l=40 mm	3.60 3.61
		~4N5 Mn 5N Si		18–180 mm/h 10–20 rpm	d=20 mm l=60 mm	3.62
(Cr, Ge) (Fe)		<4N5 Mn <3N5 Fe 5N Si		6 mm/h ±20 rpm	d=20 mm l=60 mm	3.63

Material (dopants)	Method	Purity	Crucible material	Pulling rate, rotation	Crystal size[a]	Refs
ReSi$_{1.75}$	Arc melting	4N Re 4N Si		Slow cooling	Well crystallized facets	3.44
	Zone floating with levitation	4N Re 6N Si	-	10 mm/h	d=9 mm l=20–30 mm	3.20 3.18
β-FeSi$_2$	Czochralski + annealing	4N8 Fe or 3N Fe 6N Si	Alumina	800°C, 240 h	Precipitation of FeSi	3.29
	Horizontal gradient freezing	4N Fe 9N Si or 3N FeSi$_2$	Graphite covered with BN powder	1450°C, 1 h 60–200 K/h to 800 or 900°C, 100 h or 900°C, 300–1000 h	Polycrystalline 2–3 mm grains	3.64
Ru$_2$Si$_3$	Bridgman-like (Tamann–Stöber)	4N7 Re >5N Si	Thin-walled pyrolytic BN	1 K/h	Cracks and grain boundaries	3.37
	Bridgman-like (Tamann–Stöber)	4N7 Re >5N Si	Thin-walled pyrolytic BN	1770°C hold, 4 K/h⇒ 1.2 mm/h	2×10×10 mm^3	3.65
	Zone floating with levitation	3N Ru 6N Si	-	10 mm/h	d=8 mm l=50 mm	3.19 3.18
OsSi$_2$	Bridgman-like (Tamann–Stöber)	~2N	Quartz	1–300 K/h	Some mm fragments	3.4
Ir$_3$Si$_5$	Arc melting + annealing	Ir and 3N Si 4N Ir+3N Si	Evacuated quartz tubes	-	40×50×15 μm^3	3.66 3.67
(Pt, Os)	Bridgman-like (Tamann–Stöber)	3N5 Ir 5N5 Si 5N Pt 2N8 Os	-	-	no details about perfection	3.68
Mg$_2$Si$_{1-x}$Ge$_x$	Bridgman-like (Tamann–Stöber)	≥4N Mg, > 5N Si + Ge	Graphite in stainless steal	15 K/h	-	3.47 3.46

[a] l – length, d – diameter.

The magnesium was cut into a few pieces totaling 20–25 g, etched until bright in dilute HNO$_3$, rinsed in deionized water, dried, and weighed accurately. Silicon was added in the amount required to produce a stoichiometrically correct final composition. Because of the high reactivity and high vapor pressure of magnesium at elevated temperature and the tendency of silicon to wet and react with various materials, considerable experimentation was required before a satisfactory combination of crucible material and fusion cycle was obtained. After working with alumina, thoria-lined, boron nitride, and graphite crucibles the latter was finally preferred. It provides easy machining and sample removal, along with low cost, and availability in a high-purity form. Although an amount of silicon soaked into the graphite, it was possible to compensate it by adding an excess of silicon, and minimizing the time for which the sample was at a maximum temperature. Since the vapor pressure of magnesium is equal to 10^5 Pa at 1090°C, it was necessary to enclose the graphite crucible in a closed container made of stainless steel.

The whole system is shown in Fig. 3.12. The graphite crucible is placed inside the stainless steel fusion container, to the closed end of which is attached a stainless steel pipe.

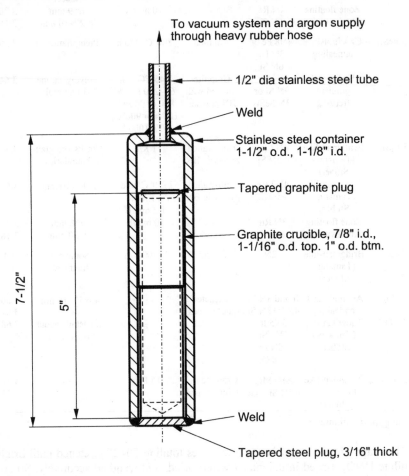

To vacuum system and argon supply through heavy rubber hose

1/2" dia stainless steel tube

Weld

Stainless steel container 1-1/2" o.d., 1-1/8" i.d.

Tapered graphite plug

Graphite crucible, 7/8" i.d., 1-1/16" o.d. top. 1" o.d. btm.

7-1/2"

5"

Weld

Tapered steel plug, 3/16" thick

Fig. 3.12. Bridgman-like (Tamann) crystal growth attachment [3.47]

The system can be evacuated and then pressurized up to 1.4×10^5 Pa of argon. For the crystal growth the container was heated as rapidly as possible by the induction heater to a temperature 30–40°C above the melting point of the compound. Once this temperature was attained the container was transferred into a resistance furnace, which had been preheated to the same temperature. After soaking within this furnace for about 5 min the sample was inverted a few times to ensure good mixing of the molten ingredients. The sample was placed near the end of the furnace so that it was in the zone with a temperature gradient. The furnace was then cooled at 15 K/h to a temperature 15°C below the freezing temperature. After the crystallization process was completed the sample was held 30°C below the freezing temperature for approximately 12 h. In order to promote grain growth, it was cooled to room temperature at about 100 K/h.

In order to compensate the loss of silicon, which penetrated into the graphite, it was necessary to add about 1% silicon, when a fusion cycle with maximum temperature approximately 35°C above the liquidus temperature for about 5 min was used. The amount of silicon penetrated into graphite depends very strongly on the maximum temperature and time of the fusion cycle.

The authors pointed out that it was necessary to make over 90 samples to improve the method to a good success rate of the growth process. The specimen prepared from the crystals to perform measurements of the thermal conductivity had a cross-section of $1.57\,cm^2$ and a length between the differential thermocouples of 1.61 cm. It was coarse-grained and contained no visible cracks. At a large magnification only traces of magnesium-rich eutectic could be observed.

$Mg_2Si_{1-x}Ge_x$. Single crystal growth of this ternary semiconductor was reported by LaBotz et al. [3.70]. The equipment and the process used is the same as for Mg_2Si [3.47]. Specimens of Mg_2Si, $Mg_2Si_{0.8}Ge_{0.2}$, $Mg_2Si_{0.6}Ge_{0.4}$, $Mg_2Si_{0.4}Ge_{0.6}$, $Mg_2Si_{0.2}Ge_{0.8}$, and Mg_2Ge were obtained. The latter three compounds tended to decompose in water or humid air and so were stored in acetone.

$BaSi_2$. Evers and Weiss [3.48] attempted to use the Bridgman-like (Tamann–Stöber) technique to grow single crystals of this material, as well as other silicides and germanides. The preparation route involves melting of stoichiometric mixtures of UHV-"reactively"-distilled alkali earth-metals [3.71, 3.72] with semiconductor-grade silicon in evacuated electron-beam-welded molybdenum crucibles [3.73]. To avoid stoichiometry changes of the charge due to partial condensation of one of the components on the lid of the crucible, the lid temperature was kept 40°C above that of the charge. The melt was heated to 1200–1400°C and slowly cooled down at 10 K/h. Among the materials tested only $CaSi_2$, $CaGe_2$ and $SrSi_2$, were obtained as single crystals up to $8\times8\,mm^2$. Despite considerable efforts they did not obtain single phase $SrGe_2$, $BaSi_2$ and $BaGe_2$. Polycrystalline $BaSi_2$ was only prepared by inductive heating of stoichiometric mixtures of the components in a cold-boat under argon atmosphere. No evaporation losses were detected. Spark source mass spectrographic analyses of all impurities displayed a total purity of the compounds of 3N5 to 4N. Copper was not observed at least above the detection limit (0.1 ppm).

3.4.2 Silicides of Group VI Metals

$CrSi_2$. This compound melts congruently at a moderate temperature of 1490°C. Therefore, it is possible to grow single crystals from the melt by different methods. Czochralski growth [3.49, 3.50], melting or zone melting in quartz crucibles [3.17, 3.49, 3.51] and floating zone melting [3.21, 3.52] were reported.

Shinoda [3.49] prepared crystals of undoped and silicon doped $CrSi_{2+x}$ $(0 \leq x \leq 0.02)$ from mixtures of proper proportions of chromium ($\geq 4N$) and silicon (n-type, 100 ohm cm) RF melted in a quartz crucible in argon. The cast materials were porous and very fragile. The crystals were grown by the

Czochralski method from the melts of as-prepared $CrSi_{2+x}$, corresponding to the atomic proportion of chromium and silicon in the melt. Unfortunately, no details of the pulling process were described.

The balanced samples ($x = 0$) showed regular Laue patterns typical for the $CrSi_2$ single crystal, while silicon-enriched samples were polycrystalline. The observed lattice constants were $a = 0.4431$ nm, $c = 0.6364$ nm and $c/a = 1.44$, which agree well with those reported previously [3.74]. It was found by XRD examination that the silicon-rich crystals were single phase for $x \leq 0.02$ and had almost the same lattice constants as those of the stoichiometric crystals. If x exceeded 0.02 the crystals contained a small amount of free silicon.

When crystals were grown from melts of $CrSi_2$ and $MnSi_2$ (2–12 mol % in the mixtures) by the Czochralski method, the quantity of manganese dissolved in the as-grown crystals was found to be relatively small. The electrical properties of these crystals were almost the same as those of the silicon-rich crystals. No further modifications could be obtained. Conversion of the conductivity type was achieved in crystals prepared by melting of the mixtures of $CrSi_2$ and $MnSi_2$ (10–12 mol % in the mixture). This was done in a quartz crucible at 1600°C cooled to 1300°C at a rate of 50 K/h by using a Bridgman-type furnace. The XRD patterns showed that the crystals were single phase and that the lattice constants did not vary with the amount of $MnSi_2$ added.

Voronov et al. [3.50] reported the growth of $CrSi_2$ single crystals by the Czochralski method. The crystals had a size greater than $10 \times 10 \times 10$ mm^3, but no details about the starting materials and the growth conditions were given.

Zone melting in quartz crucibles was used to grow single crystals of $CrSi_2$ by Nishida [3.17]. He started from a CrSi powder prepared from 3N chromium and 3N silicon by a solid state reaction in a sealed quartz tube at temperatures of 800–1000°C. The powder was pressed at 1250°C into a rod (6 mm square, 80 mm length) and then refined by floating zone melting at the speed of 3 mm/min in an argon atmosphere containing 10 vol % of hydrogen. Repeating the zone refining process provided formation of small CrSi crystals (9 mm^3 on average) in the rod. These individual crystals in the coarse-grained rod were used as a matrix for the $CrSi_2$ single crystal growth. The apparatus for this is schematically shown in Fig. 3.6. It is a unit commonly utilized for refining of materials, but with various improvements. The carbon holder had a conical hole, which could be set up on the rotor shaft as the after heater. Consequently, the specimen contained in the conical quartz crucible was induction-heated and rotated with the carbon holder and the rotor shaft.

A $CrSi_2$ ingot was prepared by melting of the mixture of CrSi crystals adding the high purity single crystalline silicon (purity 5N with n-type conductivity) in an argon atmosphere. The ingot was so porous and brittle that it was easily crushed to a size of about 3 mm^3. The individual crystals were transferred to a transparent conical quartz crucible. They were repeatedly induction-melted to obtain a large single crystal. The refining speed ranged from 0.17 to 3 mm/min. The process was performed at about 1550°C in an argon atmosphere containing 10 vol % of hydrogen. The single crystal of $CrSi_2$ grown was 8 mm in diameter and 35 mm long.

The orientation and the lattice parameters of the crystal showed that the growth direction was approximately parallel to the c axis. The measured lattice constants $a = 0.4424$ nm, $c = 0.6347$ nm and $c/a = 1.44$ were in good agreement with those reported by other authors. The chemical composition was determined to be 33.58 at % chromium and 66.42 at % silicon. From the as-grown crystals a rectangular $\langle 1000 \rangle$ oriented parallelepiped was cut with $0.1 \times 4 \times 8$ mm^3 size and annealed in a sealed quartz tube at 1100°C for 200 h for further analysis.

Oshugi et al. [3.51] used a method very similar to those reported previously [3.17], but they started from 3N3 pure chromium powder and 4N silicon powder. Chromium and silicon in the composition ratio 1:1 were mixed and encapsulated in a quartz tube. Then CrSi was synthesized from the mixture by a solid-state reaction at 997°C within 24 h. The monosilicide was ground into a fine powder with particles of a few micrometers using an agate mortar. The powder was pressed at 1197°C into a bar of 6 mm$^2 \times 80$ mm size. The sintered material was refined by a few repetitions of zone melting at 1597°C in an argon atmosphere containing 10 vol % of hydrogen at the refining rate of 3 mm/min and rotation speed of 40 rpm. The square specimen completely transformed into a rod consisting of many tiny crystals with average size of 9 mm. An individual crystal in the coarse-grained rod was ground into powder and identified to be in the cubic system T^4_1 with a lattice constant of $a = 0.4607 \pm 0.0004$ nm by the XRD technique. Qualitative analysis with an x-ray microanalyser confirmed no segregation of silicon and no chemical element, other than chromium and silicon in the crystal.

The CrSi crystal and a silicon single crystal (n-type, 100 ohm cm) were weighed in a chemical composition ratio of 1:1 and then arc-melted in an argon atmosphere. The CrSi$_2$ compound obtained was brittle and readily broken into granular particles of about 5 mm in size. The particles of CrSi$_2$ were put into a quartz tube with a conical end and a diameter of 8 mm, and then refined in a high frequency floating-zone apparatus with a carbon holder for preheating. The first cycle of refining was performed at a rate of 10 mm/h and rotation speed of 30 rpm with a zone temperature of 1597°C. The CrSi$_2$ boule grew to dimensions of about 8 mm diameter and 35 mm length. A portion of the boule was ground into powder and identified by XRD to be CrSi$_2$. Moreover, the boule was confirmed by Laue and rotational crystal methods to be single crystal with lattice constants $a = 0.4424 \pm 0.0004$ nm and $c = 0.6342 \pm 0.0002$ nm, which were in agreement with those of polycrystalline specimens already published. The growth direction was along the c axis with a tilting angle less than 9°.

In contrast to the growth methods already discussed, Behr [3.21] and Hirano and Kaise [3.52] used a containerless floating zone method to prepare CrSi$_2$ single crystals. The main advantage of this method is the absence of any crucible as a source of impurities. Therefore, it allows the growth of high-purity single crystals.

As starting material for the crystal growth CrSi$_2$ Behr [3.21] used powder compacts with the diameter ranged from 6 to 90 mm depending on the required crystal size. They were prepared from Cr/Si powder mixtures (particle size 45–80 μm) without precompaction in molybdenum containers at 1200°C for 5 h at an argon pressure of 5×10^4 Pa. Electrolytic chromium (4N) and semiconductor-grade silicon (>5N) were used. After milling in a planetary ball mill with agate as a material for the vessel and balls the purity of the silicon for components other than chromium, oxygen and nitrogen proved to be better than 4N. However, chromium powder could be produced only in a mill made of stainless steel or hard alloy. Therefore, the purity of the starting chromium was deteriorated during powder production. Depending on the vessel material iron or tungsten, molybdenum and cobalt were found as the main impurities [3.75]. The oxygen content of chromium and silicon powders was 1 at % and did not change after production of the powders and was also found in the compacts. Balancing the concentration of the powder compacts, silicon and chromium losses were accounted for by evaporation of the elements and of SiO, which forms by reduction of the corresponding oxides during heat treatment at 1000°C.

The crystal growth from the powder compacts was carried out by crucibleless RF floating zone (4 kW, 3.4 MHz) under a static atmosphere of argon (7×10^4 Pa) and hydrogen (10^4 Pa), for avoiding electric discharges. The inert gas was purified by metallic titanium kept at 800°C and controlled during melting by a quadrupole-mass detector to contain < 50 vol ppm nitrogen (detection limit), < 5 vol ppm oxygen, and < 10 vol ppm hydrocarbons (see also [3.76]). A two-step process was used to grow oriented single crystals. In the first step the powder compact was melted and transformed into a feed rod of nearly constant diameter. In the second step the single crystal was grown in the end of the rod. For this it was melted once again, connected with a seed crystal and drawn into the opposite direction to the alloying process. The pulling speed was 0.4 mm/min with a seed rotation of 30 rpm. By changing the speed of the feed rod, the diameter fluctuations of the feed rod could be reduced. The grown single crystals were 4 mm in diameter and about 50 mm in length. They exhibited a mosaic substructure with small inclusions of CrSi or silicon. Impurities were detected in the samples: 18 at ppm of tungsten, 10 at ppm of copper, 23 at ppm of iron, 22 at ppm of nickel, 66 at ppm of vanadium and 32 at ppm of titanium, and 3000 at ppm of oxygen. For the crystal growth, a method avoiding oxygen contamination is preferred.

Single crystals of $CrSi_2$ of about 10 mm diameter and 50 mm length were grown by Hirano and Kaise [3.52]. The raw material was prepared by arc melting, in argon, of 6N silicon and 3N chromium. The crystal growth was performed in an optical furnace for floating zone melting with two halogen lamps of 3.5 kW power and a double-ellipsoidal reflector [3.3] under a flowing argon atmosphere (5N) with a flow rate of 24 l/h. The pulling rate varied between 6 and 10 mm/h, the crystal was rotated at 15 rpm. The chemical composition of the as-grown crystal analyzed by an x-ray microanalyser with an energy-dispersive detector showed it to be stoichiometric with 33.3 ±0.5 at % of chromium and 66.7 ±0.4 at % silicon. Unidentified inclusions with a very small volume fraction were observed in the crystals.

The $CrSi_2$ single crystals obtained from the melt were grown mostly for use in electrical measurements. The perfection of these crystals seems to be lower compared with those grown by the CVT method, the purity, however, seems to be at the same level or higher. Further improvements would be needed to determine how the intrinsic properties of this semiconductor depended on the deviation from the exact stoichiometry.

3.4.3 Silicides of Group VII Metals

MnSi$_{2-x}$. Single crystals of MnSi$_{2-x}$ with a composition of about MnSi$_{1.73}$ were grown from the melt by different authors using arc melting followed by heat treatment, Bridgman and Czochralski methods. All crystals grown from the melt exhibited a two-phase structure with plate-like inclusions, "striations" of MnSi which disappeared during annealing only in the CrSi$_2$ doped crystals. The reasons for the occurrence of these inclusions are not well understood yet.

In 1967 Knott et al. [3.53] reported the preparation of single crystals to determine the exact crystal structure. Mn-Si alloys of various compositions were made from electrolytic manganese of 2N5 purity and 3N8 silicon in an arc melting furnace. They were annealed for homogenization for one week at 750°C and water quenched. Since these alloys were very brittle the cast buttons were crushed and a small portion ground giving a suitable material for XRD pattern determination. The grown single crystals had dimensions of 0.042×0.014×0.5 mm^3 and a composition of MnSi$_{1.74}$.

The Bridgman method was successfully applied to grow single crystals of MnSi$_{2-x}$ by Levinson [3.54, 3.55], Kawasumi et al. [3.56, 3.57] and Kojima et al. [3.25].

Levinson [3.54, 3.55] reported the growth of MnSi$_{1.73}$ single crystals with MnSi striations. The starting material was 4N manganese flake and semiconductor-grade silicon. Melt compositions ranged from MnSi$_{1.69}$ to MnSi$_{2.00}$. The elemental manganese and silicon were prereacted by induction heating in a water-cooled copper boat in dry purified argon. The resulting ingot was placed in a thin-walled pointed quartz crucible, which was in turn placed in the close-fitting supporting mullite sheath. The crystal growth followed the usual Bridgman technique. The weight loss after premelting and crystal growth was less than 0.01%. Growth rates of 2 mm/h were typical, and crack-free boules of about 10 mm diameter and 40 mm length were obtained. Metallographic and XRD analysis of the boule verified the presence of only the MnSi$_{2-x}$ phase. Laue back-reflection photographs indicated the material to be single crystalline. The exact type of superstructure of the crystals was not verified. From the magnetic measurements about 2 wt % of MnSi precipitate was assumed. The starting and the end parts of the boule contained phase distributions in agreement with the peritectic formation similar to those reported by Morokhovets et al. [3.59]. The central part examined in polarized light showed the presence of fine striations of MnSi as found by Morokhovets et al. [3.59] and Ivanova et al. [3.60]. The composition was 49 ±1 at % manganese, 51 ±1 at % silicon in the MnSi phase and 37 ±1 at % manganese, 63 ±1 at % silicon in the MnSi$_{2-x}$ matrix. The MnSi striation as a metallic inclusion affects electrical properties of the semiconducting phase.

To examine the existence of MnSi striations in the crystals and the orientation relationships between the striations and the matrix, MnSi$_{2-x}$ crystals were grown using the Bridgman method by Kawasumi et al. [3.56]. Starting Ingots were prepared by arc melting in an argon atmosphere using manganese flakes with 4N purity and silicon single crystals with 5N purity weighted in desired atomic ratios from 1.715 to 1.750. Since each ingot obtained was slightly different in the composition, a correction was made by using a method similar to the Coble analysis [3.77] from optical micrographs. It was confirmed by the XRD measurements that the main diffraction peaks of the ingots coincided with those of $Mn_{15}Si_{26}$. The silicon-rich and silicon-deficient ingots yielded additional peaks corresponding to silicon and MnSi. Metallographic examination showed that the MnSi striations exist in the grains in all the samples and even appear in some grains containing free silicon. The ingot was placed in a quartz crucible protected by a recrystallized alumina tube, and $MnSi_{2-x}$ crystals were grown by the Bridgman method in an argon atmosphere with growth rates ranging from 0.33 to 6 mm/h and a starting temperature of 1170°C. In both silicon-rich and silicon-deficient boules $Mn_{15}Si_{26}$ existed along with silicon and a small amount of MnSi in the upper end crystallized last and in the lower end crystallized first. The MnSi phase in $Mn_{15}Si_{26}$ always exists in the form of striations. The ratio of the striation to the striation spacing is nearly constant and is about 0.02 for all crystals. This value is equal to those estimated magnetically by Levinson [3.54]. The striations are perpendicular to the c axis of $Mn_{15}Si_{26}$ with a correlation $[1\bar{1}2]MnSi//(001)matrix$.

The same behavior was found by Kojima et al. [3.25] who started from 4N manganese flakes and 5N silicon single crystal. The materials were weighed in atomic ratios from 1.00 to 1.73 and arc melted in an argon atmosphere. Coarse grains were charged into a quartz tube with a pointed base 18 mm inner diameter and 250 mm long. It was placed in a Bridgman furnace and subjected to unidirectional solidification in argon. The temperature gradient in the vicinity of the liquid/solid interface was about 70 K/m and the starting temperature 1200°C. The pulling rates could be varied between 0.33 and 8 mm/h. The size of the as-grown boule was 18 mm in diameter and 90 mm in length. No influence of the pulling rate on the geometry of the striations was observed.

Later, Kawasumi and Nishida [3.57] reexamined their own investigations starting from 5N manganese and n-type single crystalline silicon (100 ohm cm). These materials were weighed in atomic ratios Si/Mn from 1.70 to 1.75 and then arc melted. The alloys obtained were enriched with silicon, because of the manganese vaporization. Therefore, MnSi single crystals were first grown by a Bridgman method and then used as a starting material to grow $MnSi_{2-x}$ boules. The MnSi crystals were grown in a quartz tube with pointed base from the MnSi alloy arc melted and crushed in an agate mortar. The starting temperature was 1300°C, the temperature gradient near the liquid/solid interface was 2000 K/m and the pulling rate 8 mm/h. The size of the boule obtained was about 15 mm in diameter and 150 mm in length. It had a composition of $MnSi_{1.04}$. It was used for synthesis of $MnSi_{2-x}$ boules (silicon was also added) with $x = 0.30, 0.28$ and 0.25 in the Bridgman furnace. The furnace was first kept at 1300°C for 1 h in order to melt the mixed powder completely in the quartz tube. Then the temperature was reduced to 1200°C and the $MnSi_2$ boules were grown at a rate of 8 mm/h. Each boule, about 15 mm in diameter and 35 mm in length, was then subjected to annealing at 1100°C for 200 h in an evacuated quartz tube. In the single crystals with the main phase $Mn_{15}Si_{26}$ the same striations of MnSi were found as previously with the direction perpendicular to the c axis. The average dimension of striations was 0.6 μm in width and 175 μm in length, and the mean interparticle distance was 30 μm in the (100) plane of the boule.

The single crystal growth of $MnSi_{2-x}$ by the Czochralski method was reported by Fujino et al. [3.58], Morochowez et al. [3.59], Ivanova et al. [3.60], Abrikosov et al. [3.62], Abrikosov and Ivanova [3.63] and Zwilling and Nowotny [3.61].

The Czochralski method was applied first to $MnSi_{2-x}$ by Fujino et al. [3.58]. The master alloys were prepared from mixtures of 4N manganese and n-type (100 ohm cm) silicon in silica crucibles in an argon atmosphere containing 10% hydrogen by using a RF-induction furnace. It was found that molten manganese reacted with the silica crucible resulting in an appreciable decrease in weight of the ingot. Therefore, silicon blocks were charged in a silica crucible and broken pieces of manganese were charged in the middle portion of a pile of silicon blocks. This method seems to be effective in preventing direct contact of molten manganese with a silica crucible. The observed weight loss of the ingot was < 0.2%. The crystal growth was performed with pulling rates of 5 to 10 mm/min and rotation of the crystal from 6 to 8 rpm. When a crystal is pulled from the melt containing 46.80 wt % silicon, almost all of the melt can be used by the growing single crystal. On the other hand, crystal pulling from the melts containing from 47.20 to 50.56 wt % silicon results in single crystal growth only in the early stage. The lower portions become polycrystalline with precipitated silicon. The volume of the single crystal decreases with increasing silicon content. No single crystals could be grown from the melt containing 46.40 wt % of silicon or less. Chemical analysis of the grown crystals yielded their composition to be $MnSi_{1.72}$. They had tetragonal symmetry and pronounced cleavage perpendicular to the c axis. The lattice parameters are $a = 0.5526$ nm, $c = 1.7455$ nm and $c/a = 3.159$. These values are in good agreement with those reported previously for $MnSi_2$ by Borén [3.78]. Thus, it is concluded that $MnSi_{1.72}$ is the only intermetallic compound in the neighborhood of $MnSi_{2-x}$.

Morochovez et al. [3.59] reported the single crystal growth by the Czochralski method using starting components of higher purity. The starting materials were 5N silicon and 4N5 manganese purified in vacuum. The syntheses of compounds with compositions in the range from 46.3 to 47.0 wt % of silicon was performed in evacuated quartz ampoules by RF heating. Crystal growth was performed with a pulling rate of 0.6–0.9 mm/min. The growth process was stoped when about 10% of the melt was left in the crucible. In the part of the boule crystallized first precipitates of MnSi were detected. They indicated the peritectic formation of the $MnSi_{1.72}$ phase. The middle part of the boule was found to be single crystalline or to have single crystalline blocks. The best results were obtained at the silicon content of 46.8–47.0 wt %. In the end part all the boules were polycrystalline with a silicon eutectic.

Ivanova et al. [3.60] grew $MnSi_{2-x}$ using 5N silicon and manganese, which contained 10 ppm of magnesium and less than 10 ppm of copper. The synthesis was performed in quartz ampoules at the pressure of 0.13 Pa with RF heating. To remove the oxide layer from the surface the synthesized material was etched in a solution of 50% HF in H_2O and then boiled in distilled water. Crystal growth was carried out in a helium atmosphere at 1.2×10^5 Pa. The pulling speed was chosen between 0.5 and 1 mm/min. The crystal and the crucible were counter-rotated at 10 rpm. At the end of the growth process the crystal was cooled at 70 K/h to 400°C. The crystals grown in this regime exhibited a lot of cracks and therefore, the pulling rate was lowered to 0.25–0.3 mm/min and the crystal was cooled more slowly using an after heater. The crystals were 15–20 mm in diameter and 40–50 mm in length. They contained much fewer cracks, but there were striations in all crystals with composition between $MnSi_{1.67}$ and $MnSi_{1.75}$. Heat treatment at 1000°C for 1000 h did not change the microstructure and the properties significantly.

Single crystals of $MnSi_{2-x}$ were fabricated by Zwilling and Novotny [3.61] using a Czochralski growth facility described by Zwilling [3.79]. Starting from the elements with a purity of 2N5 and using a pulling rate of 0.4 mm/min single crystals 40 mm length and from 8 to 10 mm diameter were grown. They were $Mn_{27}Si_{47}$ ($MnSi_{1.74}$) with a mosaic microstructure.

Single crystals of the solid solution of germanium and $CrSi_2$ in $MnSi_2$ were grown by Abrikosov et al. [3.62] by the Czochralski method as described by Ivanova et al. [3.60]. The rotation rates of the crucible and the crystal were 10 to 20 rpm, the pulling rate was varied between 1.8 and 18 mm/h. The orientation of the seed crystals was [001] and [100]. In contrast to the undoped $MnSi_{2-x}$ single crystals only the [001] orientation showed a stable crystal growth. The crystals obtained were 15 to 20 mm in diameter and 50 to 60 mm in length. In the single crystals the same striations were observed as in the undoped samples but, after annealing at 1000°C for 400 h the samples containing 6 mol % $CrSi_2$ became single phase in contrast to the crystal with 4 mol % of $CrSi_2$ which showed MnSi as a second phase. In this case, MnSi did not exist as striations but as small inclusions. Crystals doped with germanium differed significantly from the undoped ones. Striations were found in the (001) plane in a grid-like structure consisting of small inclusions of MnSi or germanium. After a heat treatment at 1000°C for 400 h samples with a composition $Mn(Si_{0.99}Ge_{0.01})_{1.72}$ and $Mn(Si_{0.98}Ge_{0.02})_{1.72}$ became single phase.

Single crystals of the solid solution of $FeSi_2$ in $MnSi_{2-x}$ were grown by Abrikosov and Ivanova [3.63] by the Czochralski method. They started with 5N silicon, manganese containing less than 50 ppm of aluminum and copper, iron containing less than 500 ppm of aluminum, copper and nickel were used. The synthesis of the compound was performed in quartz ampoules evacuated to 0.013 Pa using an RF furnace. Crucible and crystal were rotated 20 rpm, the pulling rate was 6 mm/h. The grown crystal was cooled to 400°C at 40 K/h and then to room temperature during 1.5–2 h. The crystals obtained were 15–20 mm in diameter and 50–60 mm in length. The single crystals with 3 and 5 mol % of $FeSi_2$

were grown in the [001] and [100] directions and with 8 mol % in [001] only. In the crystal with 8 mol % of FeSi$_2$ twins were detected. The microstructure was similar to that of undoped crystals [3.60]. The striations were thought to be small inclusions of MnSi, the content of which decreases with an increase of FeSi$_2$ concentration.

Quite similar behavior was found for CrSi$_2$ doped MnSi$_2$ crystals [3.62]. The content of MnSi phase could be reduced by annealing at 900°C and the MnSi phase vanished after 370 h in crystals with 5 and 8 mol % of FeSi$_2$. In crystals containing 3 mol % of FeSi$_2$ small inclusions were found even after 370 h annealing. Therefore, the increase of the homogeneity region with increasing FeSi$_2$ concentration was inferred.

Much effort has been given to grow large, perfect and highly-pure single crystals of MnSi$_{2-x}$ from the melt. Unfortunately, all crystals grown exhibited a two-phase structure with plate-like inclusions of MnSi which fix the stoichiometry of the main phase to the lower phase boundary. For an evaluation of the electrical properties dependence on the composition of the MnSi$_{2-x}$ phase it would be desirable to determine them on single crystals grown by CVT using different source compositions.

ReSi$_{1.75}$(ReSi$_2$). Siegrist *et al.* [3.44] prepared small single crystals of ReSi$_2$ by arc melting. Rhenium (4N) and silicon (4N) in powder form were used as starting materials. Stoichiometric amounts of the elements were melted several times under purified argon in a water-cooled copper hearth. On slowly cooling the ReSi$_2$ buttons, well crystallized facets were formed on their surface. The crystals did not agree with the reported tetragonal MoSi$_2$ structure but displayed an orthorhombic distortion of this structure type.

Large single crystals of the semiconducting rhenium silicide were grown by Gottlieb *et al.* [3.20] using a zone melting technique with electromagnetic levitation of the melt. They started from 3N rhenium and 6N silicon, which were melted together in the argon atmosphere in a Hukin-type cold crucible (see Fig. 3.9). The samples prepared had compositions between ReSi$_{1.75}$ and ReSi$_2$. They were subsequently annealed for a few hours in the same crucible at 800–850°C. The single crystal was produced by zone melting in the same crucible with the electromagnetic levitation of the molten zone. During the pulling, a polycrystalline rod of the compound was pushed inside the cold crucible keeping the melt volume constant and stabilizing the liquid/solid interface. The pulling rate was about 2.8 μm/s and the common dimensions of the crystal were from 8 to 10 mm in diameter and from 2 to 30 mm in length. The composition of the grown single crystals was determined by XRD to be ReSi$_{1.75}$ with the lattice constants $a = 0.3139(2)$ nm, $b = 0.312(2)$ nm, $c = 0.7670(3)$ nm, and $\alpha = 89°87(5)'$.

We tried to grow ReSi$_2$ single crystals by RF floating zone in an argon atmosphere but, due to the large decrease of the electrical resistivity during melting of the semiconducting phase, it was impossible to establish a stable molten zone. A further possibility to grow high quality single crystals from the melt would be the use of optical heating.

3.4.4   Silicides of Group VIII Metals

β-FeSi₂. This phase does not coexist with the melt. Therefore, there is no opportunity to grow single crystals of β-FeSi$_2$ directly from the melt. Kloc *et al.* [3.29] and Shibata *et al.* [3.64] tried to fabricate crystals of this semiconductor via the peritectic reaction by annealing of α-FeSi$_2$ in the temperature range where the β-phase is the stable one.

For the growth of α-FeSi$_2$ Kloc *et al.* [3.29] used 4N8 or 3N iron and 6N silicon. The apparatus used to grow α-FeSi$_2$ consisted of a 400 mm high and 250 mm diameter stainless steel water cooled chamber. Two sealed shafts for rotation and pulling from the top and bottom, respectively, allowed independent positioning of the crucible and the seed. The melt (100–200 g) was contained in a 50 mm high and 50 mm diameter alumina crucible. It was RF inductively heated. The synthesis and pulling of α-FeSi$_2$ was done within one run. Before melting the charge material, the apparatus was evacuated to 10^{-3} Pa, back-filled with 5N argon, and again evacuated. Growth runs were carried out in a dynamic vacuum of 10^{-3} Pa. The pure elements reacted inside the Al$_2$O$_3$ crucible. An excess 5% silicon was added to the stoichiometric amount to compensate its evaporation. After about 30 min, which was needed for reaction and homogenization, the seed crystal was partly immersed into the melt and the Czochralski process was started. The α-FeSi$_2$ single crystals remained metastable during cooling because of the very sluggish transformation [3.80] to the β-phase. To transform the α-FeSi$_2$ phase into β-FeSi$_2$ the single crystals were annealed at 800°C for 240 h. The annealed samples contained FeSi precipitates.

Commercially available chunk and powdered FeSi$_2$ as well as a mixture of 4N iron and 9N silicon, were used to grow β-FeSi$_2$ crystals by Shibita *et al.* [3.64]. For solidification of starting compositions FeSi$_2$, Fe$_2$Si$_5$ and FeSi$_3$ they used the horizontal gradient freeze method. Crucibles were made from high purity graphite and covered with a boron nitride powder. The material was melted at 1450°C for 1 h in pure argon for homogenization and then cooled at rates between 60 and 200 K/h down to the in-situ annealing temperature T_1, varying between 800 and 900°C. At that temperature the peritectoid reaction took place. The duration of this annealing process was 100 h. Ex-situ annealing at 900°C for 200–1000 h was also applied. Only very small crystals were obtained using the method described.

Ru₂Si₃. Single crystals of Ru$_2$Si$_3$ were grown from the melt by Vining and Allevato [3.65] using the Bridgman-like method. Samples were prepared from the 4N7 ruthenium powder and 1–2 ohm cm, n-type silicon. Stoichiometric quantities of the components were placed in a thin-walled, tapered pyrolytic boron nitride crucible. The bottom of the crucible ended in a rounded tip with approximately 2 mm radius of curvature. The crucible with the charge was placed in a vertical two-zone furnace in an atmosphere of flowing helium. The temperature profile along the furnace axis was recorded in a separate experiment with an empty crucible. The temperature gradient was estimated to be 32 K/cm in the operating region. After raising the temperature to about 1765°C at the bottom of the crucible, the setpoint temperature for both furnace heating elements was simultaneously lowered at a rate of 4 K/h. The estimated growth rate was about 1.2 mm/h. After cooling below the melting point, the furnace power was cut off and the sample was quenched. The samples prepared at a slower cooling rate following the crystal growth procedure resulted only in polycrystalline ingots. The shape of the base of the ingot did not conform with the base of the crucible, indicating the presence of gas trapped between the melt and the crucible during the crystal growth. Similar indications of trapped gas were observed in several other trials [3.37].

The grown single crystals had lattice constants consistent with Poutcharovsky and Parté's [3.81], i.e. $a = 1.1057$ nm, $b = 0.8934$ nm and $c = 0.5533$ nm. The growth direction was about 8.8° off the [010] direction. There were plate-like inclusions with a submicrometer width parallel to the (010) and (100) planes. The inclusions comprised about 1% of the area of the sample. They were rich in boron.

The same method was described by Ohta *et al.* [3.37], but they used graphite crucibles. Furthermore, the cooling rate was 1–2 K/h, which was lower than that reported in [3.65]. All the samples prepared were constitutionally supercooled displaying a cellular substructure and either RuSi or silicon at the cell boundaries. The compound was found to be single-crystalline in regions up to 2–3 mm each.

Large single crystals of Ru_2Si_3 were grown by Gottlieb *et al.* [3.19] (see also [3.82]) by the zone melting technique with an electromagnetic levitation of the melt. They started from 3N ruthenium and 6N silicon, which were melted at the stoichiometric quantities in pure argon in the Hukin-type cold crucible. As the resistivity of Ru_2Si_3 is too high, it cannot be melted by RF heating directly. Hence, the zone pulling could not start with a solid bar of an alloy with the composition of the desired crystal. Preheating of the bar was achieved by placing a small amount of pure ruthenium and silicon (about 10% of the total mass of the final crystal was of correct stoichiometric for Ru_2Si_3) in the crucible above the solid bar. The prepared single crystals were of cylindrical shape with 8 mm diameter and 50 mm length. It had the orthorhombic structure.

$OsSi_2$. Mason and Müller-Vogt [3.4] found that the alloy consisting of 15 at % osmium and 85 at % silicon was completely liquid at 1457°C. During its slow cooling, $OsSi_2$ crystallized until the eutectic was reached at 1357°C. The growth experiments were carried out in Al_2O_3-supported silica crucibles in a high-purity argon atmosphere at cooling rates of 1 to 300 K/h. The growth runs yielded crystals as irregular fragments, some with edges several millimeters long, but without a well defined morphology. The poor crystal growth indicates disruptions in the process, affected by evaporation of volatile OsO_4 produced in reactions of SiO or SiO_2 with osmium. This process could cause spurious nucleation, leading to disruption of the single-crystal growth. Improved crystals were grown at lower temperatures from the PdSi flux (see Sect. 3.3.4).

Ir_3Si_5. Ir_3Si_5 crystals doped with platinum or osmium were grown from the melt by Allevato and Vining [3.68]. They used 3N5 iridium powder, 5N5 silicon lump, 5N platinum and 2N8 osmium. The iridium powder was first arc-melted into beads, in order to avoid weight losses, and then arc-melted together with silicon. Afterwards, crystal growth was performed in a Bridgman-like furnace. Ir_3Si_5 was formed peritectically and the sample preparation was accomplished with a silicon-rich melt containing 65 at % of silicon. Density and XRD pattern were in good agreement with previous results [3.83]. Details on growth conditions, microstructure and crystallinity were not provided, but one suspects them to be identical to the above discussed growth of Ru_2Si_3 single crystals [3.65], performed within the same method by the same authors.

3.5 Summary

State-of-the-art in single crystal growth of semiconducting silicides has been shown in this chapter. It is characterized by a widespread activity ranging from early studies, mainly focused on chemical transport processes, to the high-level growth of single crystals for the determination of intrinsic properties of the semiconducting phases. The problem of structural perfection and purity was not so vital for the crystal property investigations, but it became extremely important for those crystals grown for electrical and optical measurements. A sufficient purity allowing investigation of electrical and optical parameters of these semiconductors as a function of their chemical composition has been achieved only for the β-$FeSi_2$ phase. The knowledge about the homogeneity range of other semiconducting silicides and the composition dependence of the properties within this range is still limited.

The growth of Mg_2Si has been accomplished exclusively from the melt. The crystals display a quality corresponding to the 4N purity of the starting materials. They contain small inclusions of a magnesium-rich eutectic. Thus, the component composition can be fixed at the lower limit of the homogeneity region, only. Crystal growth of $BaSi_2$ has been unsuccessful up to now.

Single crystals of $CrSi_2$ can be grown practically by all the methods discussed. The crystals are large enough to perform electrical and optical measurements. The purity and perfection of the crystals remain satisfactory. An impurity concentration less than 100 ppm can be achieved in the best crystals.

Much effort has been directed to grow large and perfect high-purity single crystals of $MnSi_{2-x}$ from the melt. Unfortunately, all crystals grown exhibited a two-phase composition with plate-like inclusions of MnSi which fix the stoichiometry of the principal phase to the lower phase boundary. For an evaluation of the electrical properties as a function of the composition single crystals grown by the CVT method with different source compositions are desired. Unfortunately, the crystals grown by CVT are relatively small, thus restricting electrical measurements.

$ReSi_{1.75}$ ($ReSi_2$) single crystals are grown by different methods. The crystals obtained by CVT and from the flux are usually very small and poor in purity (3N or worse). These are not convenient for the measurements. In contrast, the crystals grown from the melt are rather large and their purity (4N or better) corresponds to the starting materials.

Among the group VIII metal silicides crystals of β-FeSi$_2$, Ru$_2$Si$_3$, OsSi$_2$ and Ir$_3$Si$_5$ have been grown successfully. The OsSi$_2$ crystals grown from a flux or melt and the Ir$_3$Si$_5$ crystals grown from the melt are small, of low purity and undefined microstructure. Large Ru$_2$Si$_3$ single crystals are grown from the melt only. Unfortunately, the starting ruthenium for the crystals grown by a containerless method was only of 3N purity while crystals grown from high-purity materials were grown in BN crucibles and exhibited boron-rich precipitates. Successful growth of β-FeSi$_2$ crystals is performed only by the CVT method. The grown crystals have a considerable size, high crystal perfection and excellent purity, which permit their electrical and optical characterization.

In conclusion, further developments improving crystal perfection and regulating the impurity content are important in order to obtain single crystals of semiconducting silicides for precise investigations of their fundamental properties.

4 Fundamental Electronic and Optical Properties

Victor L. Shaposhnikov and Victor E. Borisenko

Belarusian State University of Informatics and Radioelectronics
Minsk, Belarus

Horst Lange

Hahn-Meitner-Institute
Berlin, Germany

CONTENTS

The behavior of electrons in a solid determines its main properties. It is most conveniently specified on the basis of quantum mechanical models describing the electronic band structure of the solid and related optical and transport properties. Computer simulation is employed more and more to validate phenomenological models, to provide complementary information to experimental measurements, and to understand physical phenomena at the atomic scale. In this chapter we focus on the fundamental electronic properties of semiconducting silicides starting from their electronic band structures and extending the analysis to the nature of the orbital band composition, density of states and effective masses of the charge carriers. Related optical properties represented by interband optical spectra, photo- and electroluminescence, infrared optical response are then shown and discussed.

4.1 Basic Relationships

The short introduction to the quantum mechanics of electrons in solids presented below aims at a better understanding of theoretical and experimental results discussed in the following sections of this chapter. More details can be found in the standard textbooks [4.1–4.5].

The time-independent Schrödinger equation

$$H\psi_n(r) = E_n\psi_n(r) \qquad (4.1)$$

with the one-electron Hamiltonian operator including kinetic energy term $\dfrac{1}{2m}p^2$ and potential energy term $V(r)$

$$H = \frac{1}{2m}p^2 + V(r) \qquad (4.2)$$

is usually applied to describe the behavior of electrons. Here ψ_n and E_n are, respectively, the wave function and energy eigenvalue representing an electron in an eigenstate labeled by n. One should remember that according to Pauli's principle each eigenstate can only accommodate up to two electrons of the opposite spin.

The energy eigenvalues determined as a function of the wave vector $k = p/\hbar$, where p is the electron momentum, are the energy bands. The information contained in the functions $E_n(k)$ is referred to as the electronic band structure of the crystal. For a given band n, $E_n(k)$ usually does not have a simple analytical form.

A theoretical simulation of the band structure involves two major steps. In the first step the one-electron potential $V(r)$ is determined. This can be performed with the first principle, so-called *ab initio*, calculations, in which the number of atoms and their positions are the only input parameters. Simpler empirical approaches using parameters which fit experimental results can also be applied to calculate this potential.

The tight-binding model and local-density approximation are often employed for simulation of the one-electron potential. In the tight-binding model the atomic potential is assumed to be so strong that the electron is essentially localized at a single atom and the wave functions for neighboring sites have a little overlap. Consequently, there is basically no overlap between electron wave functions that are separated by two or more lattice sites. This is reasonable for the narrow, inner bands in solids. Within the local-density approximation (LDA) many-body effects are accounted for in the one-electron potential by the exchange and correlation term, which is assumed to be a function of the local charge density only. It gives good results for ground state properties such as the cohesive energies and charge densities of the valence electrons. However, the calculated fundamental energy gaps in semiconductors appear to be underestimated due to neglect of nonlocality in the many-body self-energy operator.

For the known potential $V(r)$ the Schrödinger equation (4.1) is solved. In order to simplify the calculations, translation symmetry of the crystal properties is used. It remains to find the solution in the form of Bloch functions

$$\psi_{nk}(r) = u_{nk}(r)\exp(ikr) \quad \text{with} \quad u_{nk}(r+R) = u_{nk}(r), \qquad (4.3)$$

where R is the translation vector. Thus, the one-electron wave functions can be indexed by constants k, which are the wave vectors of the plane waves forming the "backbone" of the Bloch function. The k-vector is situated in the region which corresponds to the first Brillouin zone or can be defined by the primitive parallelepiped in reciprocal space: $-\pi < ka_i \le \pi$, $i = 1, 2, 3$, where a_i is the translation vector. The Brillouin zones of the crystal lattices typical for semiconducting silicides are shown in Fig. 4.1. Special points in the Brillouin zones, which are points of high symmetry, are indicated. Band structure calculations are usually performed in the vicinity of these points and along the directions between them. The calculation methods and their abbreviations are listed in the Nomenclature section in the front of this book.

Extremum points of $E_n(k)$ forming the energy gap in a semiconductor are important for understanding its electronic properties. A Taylor expansion of $E_n(k)$ around the extremum at $k = k_0$ and ignoring the high-order terms gives

$$E_n(k) = E_{n0} + \frac{\hbar^2(k - k_0)^2}{2m*} \text{ with } m* = \hbar^2 \left(\frac{d^2 E}{dk^2} \bigg|_{k=k_0} \right)^{-1}. \qquad (4.4)$$

The equation describes a parabolic energy band with a constant electron mass $m*$, referred to as the effective mass. The effective mass is clearly inversely proportional to the curvature of $E_n(k)$ at $k = k_0$. Unlike the gravitational mass, the effective mass can be positive, negative, zero, or even infinite. An infinite effective mass corresponds to an electron localized at a lattice site. In this case, the electron bound to the nucleus assumes the total mass of the crystal. A negative $m*$ corresponds to a concave downward band and means that the electron moves in the opposite direction to the applied force. If the band structure of a solid is anisotropic, the effective mass becomes a tensor, characterized by $m*_x$, $m*_y$, $m*_z$ components. Four types of critical points, namely M_0, M_1, M_2, and M_3, are distinguished as a function of the sign of these components in a particular point of the Brillouin zones with a zero energy gradient on k. In the M_0 critical point all three components are positive. In the M_1 and M_2 one and two components, respectively, are negative. In the M_3 point all three components are negative. An average effective mass can be calculated as $m* = (m*_x m*_y m*_z)^{1/3}$. This is used for simplified calculations of optical and transport properties of solids.

Another important characteristic of the electronic system of a solid is the density of states (DOS). It is a number of electron states in the energy interval between E and $E+dE$ per spin orientation per unit volume of real space. The total DOS may be calculated as

$$N(E) = \frac{2}{V_{BZ}} \sum_n \int_{V_{BZ}} \delta(E - E_n(k)) dk, \qquad (4.5)$$

where V_{BZ} is the Brillouin zone volume.

The integral is taken over the surface of constant energy E and n is the band index. The projected DOS of the atom type t is

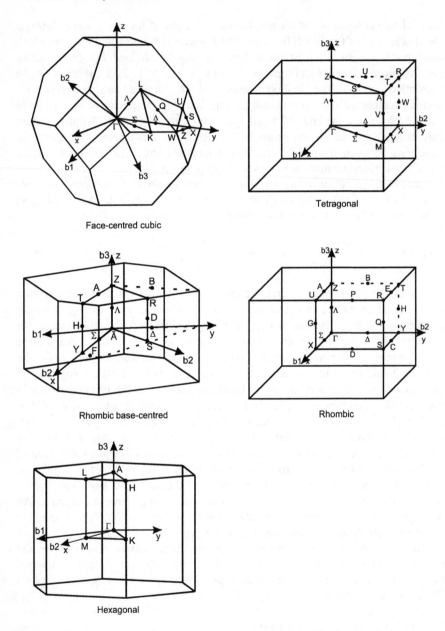

Fig. 4.1. Brillouin zones and high symmetry points of different crystal lattices

$$N'_l(E) = \frac{2}{V_{BZ}} \sum_n \int_{V_{BZ}} Q'_l(k) \delta(E - E_n(k)) dk, \tag{4.6}$$

where Q'_l is the partial charge and l is the orbital quantum number.

A calculated band structure creates a background for theoretical prediction of optical properties of the solid. This can be done with the use of the dielectric function $\varepsilon(\omega)$ relating to the applied electromagnetic field E and the produced displacement of charges D in the system via $D = \varepsilon(\omega)E$. The dielectric function is a complex quantity, so $\varepsilon(\omega) = \varepsilon_1(\omega) + i\varepsilon_2(\omega)$ with the corresponding real $\varepsilon_1(\omega)$ and imaginary $\varepsilon_2(\omega)$ parts.

The imaginary part of the dielectric function is related to the quantum-mechanical transition rate W_{ij} in the random phase approximation for transitions from the state i to j via

$$\varepsilon_2(\omega) = (n^2 / \omega) \sum W_{ij}(\omega). \tag{4.7}$$

The summation runs over all occupied and unoccupied states. For transitions between one-electron Bloch states from the filled valence into empty conduction band states the transition rate within the first-order perturbation theory is

$$W_{ij}(\omega) = (2\pi/h) |V_{ij}(k)|^2 \delta(E_{ij}(k) - \hbar\omega), \tag{4.8}$$

where the interband matrix element $|V_{ij}|$ contains the Bloch factors of the valence and conduction band wave functions and the dipole operator of the incident light wave, which is considered as a perturbation. The interband matrix element depends on the symmetry of the states participating in the optical transitions and the light polarization. The dipole operator has odd parity, therefore V_{ij} vanishes unless the Bloch factors have odd parity. Transitions are termed allowed if V_{ij} is nonzero, otherwise they are called forbidden. Generally V_{ij} is dependent on the electron wave vector k, however, near critical points it is generally independent of k.

The real and imaginary parts of the dielectric function are connected by the Kramers–Kronig relation:

$$\varepsilon_1(\omega) = 1 + (2/\pi)P \int_0^\infty \frac{\varepsilon_2(\omega')\omega' d\omega'}{\omega'^2 - \omega^2}, \tag{4.9}$$

where P means principal value of the integral.

The real and imaginary parts of the dielectric function are used to calculate the macroscopic optical properties of solids represented by the refractive index n^*, extinction index k^*, and absorption coefficient α^*

$$n^* = \{(1/2)[\varepsilon_1 + (\varepsilon_1^2 + \varepsilon_2^2)^{1/2}]\}^{1/2}, \tag{4.10}$$

$$k^* = \{(1/2)[-\varepsilon_1 + (\varepsilon_1^2 + \varepsilon_2^2)^{1/2}]\}^{1/2}, \tag{4.11}$$

$$\alpha^* = 4\pi k^*/\lambda \quad (\lambda \text{ is the wavelength of the light in vacuum}). \tag{4.12}$$

In standard optical spectroscopy experiments the intensity of reflected and/or transmitted light is measured, which is reflected at an air/semiconductor interface or transmitted through an absorbing slab. In the case of normal light incidence on the surface of the sample, the reflectance is defined as

$$R = [(n^* - 1)^2 + k^{*2}]/[(n^* + 1)^2 + k^{*2}].$$ (4.13)

The transmittance T^* of an absorbing slab of thickness d is

$$T^* = [(1 - R)^2 \exp(-\alpha^* d)]/[1 - R^2 \exp(-2\alpha^* d)].$$ (4.14)

The intrinsic optical absorption and reflectance spectra of a semiconducting material provide fundamental insight into its electronic band structure. One method of experimental determination of the absorption coefficient for a thin semiconducting film is to measure its normal incident reflectance R and transmittance T^*. The photon energy dependence of the absorption coefficient may then be derived from these data, the film thickness, and an optical model of the sample. In the case of allowed direct (momentum conserving) electron transitions across the band gap E_g

$$\alpha^* \sim (E_g - \hbar\omega)^{1/2}.$$ (4.15)

If the minimum of the conduction band and the maximum of the valence band lie at different points in the k-space the participation of phonons is necessary in order to provide the momentum balance (indirect transitions). In this case the optical transitions can proceed with phonon absorption or emission. Second-order perturbation theory is required for determining the optical transition probability. Therefore much weaker absorption is expected than for direct absorption. For symmetry-allowed optical transitions around an indirect gap E_g one obtains

$$\alpha^* \sim (\hbar\omega + E_{ph} - E_g)^2/[\exp(E_{ph}/k_B T) - 1]$$ (4.16)

for transitions with photon absorption and

$$\alpha^* \sim \exp(E_{ph}/k_B T)\,(\hbar\omega - E_{ph} - E_g)^2/[\exp(E_{ph}/k_B T) - 1]$$ (4.17)

for transitions with phonon emission, where E_{ph} is the energy of the phonon involved.

Thus, optical interband investigations of semiconductors provide a wealth of information on the electronic structure of these materials. The optical parameters are directly related to the joint density of interband states. Critical points in the joint density of interband states are reflected as distinct line shapes in optical spectra and photon energy dependencies of optical constants.

Theoretical simulation of the electronic band structure and related optical properties of a semiconductor often show more details than an appropriate experimental technique. Nevertheless, experimental analysis of the samples including XPS, UPS, EELS, AES, BIS, SXES and others (see the Nomenclature section for decoding the abbreviations) as well as optical, electrical resistivity, and thermoelectric property measurements have been long used to test band models.

Both theoretical and experimental results characterizing band structures and related electronic and optical properties of semiconducting silicides are presented and discussed in the subsequent sections.

4.2 Electronic Band Structure

Theoretically simulated band structures and their orbital compositions, densities of states, effective masses of charge carriers in semiconducting silicides in comparison with the available experimental data are discussed in this section. Except for Mg_2Si and $BaSi_2$, all other semiconducting silicides belong to the family of transition metal silicides, characterized by participation of valence d electrons from the metal in the bond formation. This is used for a comparison and searching for general features of their fundamental electronic properties.

4.2.1 Silicides of Group II Metals

The semiconducting silicides of this group, which are Mg_2Si and $BaSi_2$, are quite different in stoichiometric proportions of the components and crystalline structure (for structure details see Table 1.1). This completely eliminates the possibility of looking for any identity in their electronic band structures and related properties. Moreover, they have been studied at different theoretical and experimental levels: Mg_2Si is much better investigated than $BaSi_2$, which is seen from the data presented below.

Mg_2Si. This compound belongs to so-called electron-deficient semiconductors [4.6]. The other members of the family are Mg_2Ge and Mg_2Sn. They crystallize in the fcc antifluorite structure. The number of valence electrons per unit cell in this family of semiconductors is less than that necessary to form all the single bonds, which are compatible with the crystallographic structure. While each magnesium atom is tetrahedrally coordinated to the silicon anions, the anion has 8 equivalent nearest neighbors with which, according to crystallographic data, it could form 8 single bonds. This, however, can not occur since each unit cell contains 8 electrons only. Under these conditions, one expects that in the ground state, each anion will form only 4 bonds and that these bonds will undergo pivoting resonance. The nature of the empty conduction states in the material is not immediately evident. Explanation of the electronic ground state requires the concept of an electron transfer in addition to that of pivoting resonance.

Fundamental electronic properties of Mg_2Si have been the object of extensive theoretical [4.6–4.12] and experimental [4.13–4.26] studies. The theoretical simulation was performed within PP and HF models. Bulk single crystals, rather than thin films, were examined in the experimental studies. Thin film formation is difficult because of the low condensation coefficient and high vapor pressure of magnesium.

One of the most representative calculated band structures of Mg_2Si is shown in Fig. 4.2. The material is definitely an indirect gap semiconductor. The top of the valence band is at the center of the Brillouin zone (Γ point) with symmetry Γ_{15}. The conduction band minimum is at the X point. It is either X_1 or X_3, but these two levels are close together. The precise assignment of the lowest state is questionable. From the symmetry properties and the mobility data, the minimum of the conduction band is concluded to have X_3 symmetry [4.10]. This is also consistent with measurements of the piezoelectric effect, which show the conduction band minimum to occur in the [100] crystal direction coinciding with the Γ-X direction in the Brillouin zone [4.8]. Thus, the direct gap corresponds to the $\Gamma_{15} \rightarrow \Gamma_1$ transition and the indirect gap is formed by the $\Gamma_{15} \rightarrow X_3$ transition.

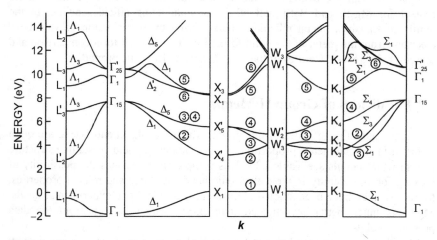

Fig. 4.2. Band structure of Mg_2Si calculated by the PP method [4.12]

Theoretically calculated values of the indirect and direct gaps are collected in Table 4.1 in comparison with experimental data obtained in optical and temperature-dependent resistivity measurements. In their critical analysis one should account for the well known underestimation of the gaps calculated within the local-density formalism relative to the experiment. Although there is scatter in the values characterizing the gaps, it is evident that the indirect gap of 0.74–0.78 eV and the direct gap of about 2 eV are most believable. The temperature-dependent shift of the indirect energy gap derived from infrared absorption measurements in the range 85–370 K is estimated to be -5.0×10^{-4} eV/K for Mg_2Si (-6.5×10^{-4} eV/K for Mg_2Ge) [4.24]. While no absolute value of the gap was determined, it was assumed to be 0.78 eV at 0 K, as had been found by electrical measurements [4.15].

The overall nature of the valence band of Mg_2Si is very similar to other semiconductors with the fcc translation symmetry and eight electrons per unit cell to go into this band. This is found not to change very much both in the group IV semiconductors and the III-V compounds.

Table 4.1. Band gaps in Mg_2Si (d – direct, in – indirect)

Theoretical calculations			Optical measurements – temperature		Resistivity measurements (temperature dependence)	
Method[a]	E_g (eV)	Refs	E_g (eV)	Refs	E_g (eV)	Refs
PP	0.37 (in)	4.7	0.60 (in) – 90 K	4.23	0.65–0.70	4.18
PP	0.49 (in)	4.10	0.66 (in) – 90 K	4.22	0.77	4.14
PP	0.5 (in)	4.6	0.74 (in) – RT	4.17	0.78	4.15
PP	0.53 (in)	4.12			0.78	4.16
HF	0.6 (in)	4.8				
PP	0.7 (in)	4.11				
PP	1.3 (in)	4.9				
PP	1.8 (d)	4.10	0.83 (d) – RT	4.17		
PP	1.90 (d)	4.7	0.99 (d) – RT	4.17		
PP	2.06 (d)	4.12	2.17 (d) – RT	4.20		
HF	2.11 (d)	4.8				
PP	2.84 (d)	4.9				

[a] Abbreviations of the methods are given in the Nomenclature section in the front of this book.

The lowest valence band has been supposed to originate from the $3s$ silicon orbitals [4.6]. The magnesium $3p$ electrons participate in the states of the upper three valence bands, as experimentally confirmed by AES of Mg_2Si [4.13]. Meanwhile, the $3p$ silicon orbitals dominate there. The compound is therefore quite ionic but Mg $\rightarrow$ Si valence-electron transfer is not complete, in agreement with charge distribution measurements by x-rays. In agreement with the XPS data [4.19] a total valence band width of 9.9 eV was calculated [4.6]. The lowest valence band is theoretically predicted to be separated from the higher bands by an intervalence gap of 3.5 eV, but it is at variance with the interpretation of the XPS spectra.

From the fact that most of the valence-charge density is concentrated near silicon sites, it appears likely that the core levels which are most affected by the valence exchange are the silicon core levels. Silicon core shifts are much less critical than magnesium core shifts. The silicon sites form a tetrahedron centered about a magnesium site. Thus, as in the case of silicon, the valence charge density about a magnesium site may be represented by an sp^3 hybrid, but with a much weaker s contribution than in the case of silicon.

The lowest conduction band states are expected to correspond to nonbonding magnesium orbitals, i.e. Γ_1(Mg $3s$), Γ'_2(Mg $3s$) and Γ'_{25}(Mg $3p$) [4.6]. The Γ_1(Mg $3s$) antibonding combination of $3p$ silicon and magnesium orbitals occurs at higher energies. The lowest three conduction levels, in order of increasing energy, are Γ_1, Γ'_{25}, and Γ'_2, in agreement with LCAO predictions. However, their wave functions do not correspond to nonbonding magnesium orbitals, but rather to Mg-Si bonding states where silicon necessarily contributes its excited atomic orbitals. The Mg-Si bonding character of the lowest conduction bands is valid over most of the Brillouin zone. Therefore, the energy gap in Mg_2Si separates bonding states from each other and nonbonding from antibonding states as in diamond and zincblende compounds. This is probably why the compound has a small energy gap.

The two electrons of magnesium are essentially expelled from the magnesium spheres leaving the system with ionic character [4.6, 4.7]. At the silicon site, the net potential is more attractive resulting in a significant transfer of charge from magnesium to silicon. This disrupts the charge anisotropy around the silicon, and leads to the ionic-like structure in Mg_2Si.

The electron-density distribution calculated from x-ray data [4.25] indicates spherical magnesium ions but shows that the outer electron shells of the silicon ions are distorted towards the nearest magnesium neighbors, thus forming an electron bridge of 2×10^{-4} e/nm^3. The average charge density between the ions is 5×10^{-5} e/nm^3, which means that 8 electrons are distributed within the unit cell in regions between ions. The ionic charges have been found to be approximately $Mg^{+1.5}$ and Si^{-1}. This is also confirmed by PP calculations of the valence-band density of states [4.11]. According to the calculations the upper valence bands correspond to an incomplete shell of the s and p electrons of silicon with some admixture of the s and p electrons of magnesium. So, the valence electrons of the magnesium atoms are incompletely transferred to their nearest neighbors. The estimated radii of magnesium and silicon in Mg_2Si are 0.100 nm and 0.175 nm, respectively [4.25]. They lie between corresponding covalent and ionic radii.

The upper limit for the magnesium ionicity has been estimated to be 0.66 electrons from the magnesium binding energy in Mg_2Si [4.8]. One can obtain this result by assuming that 2 electrons in the Si–Mg bond are shared on a 1/3 and 2/3 basis. Although, the extent of the ionic character of the magnesium atom is not determined, it is clear that a self-consistent calculation of the electronic energy levels must be sufficiently flexible to allow for the charge transfer from a magnesium to a silicon atom and to include the effects of the unscreened magnesium nucleus.

The effective mass for electrons derived from the experimental mobility data under the assumption of electron scattering by polarization waves has been found to be $0.46m_0$ [4.15]. This correlates with the theoretical calculations predicting $m^*{}_{//} = 0.69m_0$ and $m^*{}_\perp = 0.25m_0$ [4.9].

BaSi$_2$. There are no published data concerning details of the electron energy bands in this compound. In a simple valence bond picture of the semiconducting orthorhombic phase [4.27] the valence band "bonding $Si_4\sigma$" (formed by electrons in sigma-bonded Si_4 clusters) is filled by 80 electrons, which correspond to eight formula units in a unit cell with ten valence electrons in each formula unit. The energetically higher empty conduction band should be formed by $5d$-$6s$ states of barium (lower lying) and antibonding $Si_4\sigma^*$ (higher lying). Valence electrons are more or less localized at the Si_4 clusters, which are separated from each other by at least 0.40 nm. This distance has to be compared with the average Si–Si bonds in the clusters, which is 0.24 nm. Indeed, the localization of the valence electrons corresponds to insulating or semiconducting behavior.

The band gaps extracted from the temperature dependence of the material resistivity are quite scattered. The gap is estimated to be 1.3 eV from the measurements on high purity polycrystalline samples prepared by melting of the stoichiometric mixture of the components [4.28]. However, the gap of 0.48 eV is mentioned in [4.29] for the disilicide synthesized in the reaction $BaO + 3Si = BaSi_2 + SiO$. The first value for the gap looks more reliable, particularly as the correct test measurements were performed in this research for Ge and InSb. More theoretical and experimental work has evidently to be done for better characterization of semiconducting $BaSi_2$.

4.2.2 Silicides of Group VI Metals

This group of semiconducting silicides is represented by $CrSi_2$, $MoSi_2$, and WSi_2. They are compounds with the same hexagonal crystal structure and constituent atoms, which vary within a given column of the Periodic Table. Since the outer electronic configurations of the atoms in this family are the same, one expects close similarities in their energy bands, and hence in their optical properties. The band gaps of the compounds have been found to be sensitive to both structural and chemical changes. In this family of semiconducting disilicides $CrSi_2$ is the most widely studied compound. This should allow a reasonable prediction of electronic properties of the two others.

$CrSi_2$ (3d). Among the semiconducting silicides, $CrSi_2$ is one of the best-characterized materials. It is an indirect-narrow-gap semiconductor, which has been confirmed by numerous theoretical simulations and experiments. Its calculated band structure is presented in Fig. 4.3.

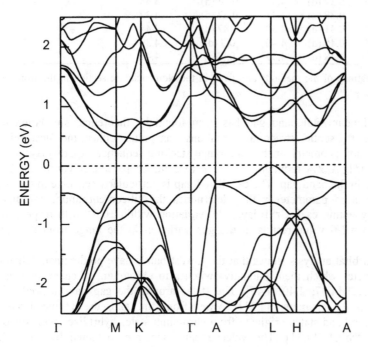

Fig. 4.3. Band structure of $CrSi_2$ calculated by the LMTO method [4.30]

The indirect gap occurs between the highest filled valence-band state at L ($n = 21$) and the lowest unoccupied conduction-band state at M ($n = 22$). Calculated and experimentally measured gap values are summarized in Table 4.2. The discrepancy in the calculated gaps may either be due to the spherical averaging of the potentials in LMTO and ASW methods, which is not performed in the LAPW method, or to the different exchange correlation potentials used in these three simulation procedures. Accounting for the typical theoretical underestimation, the indirect gap of 0.35 eV and direct gap of about 0.9 eV are evident to characterize this semiconductor, while the direct gap value has to be determined more carefully.

Table 4.2. Band gaps in $CrSi_2$ (d – direct, in – indirect)

Theoretical calculations			Optical measurements – temperature		Resistivity measurements (temperature dependence)	
Method[a]	E_g (eV)	Refs	E_g (eV)	Refs	E_g (eV)	Refs
ASW	0.21 (in)	4.31	0.35 (in)	4.32	0.27	4.33
LMTO	0.25 (in)	4.34	0.35 (in)	4.35	0.30	4.36
LMTO	0.25 (in)	4.37	0.35 (in)	4.34	0.30	4.38
LMTO	0.29 (in)	4.30	0.5 (in)	4.39	0.35	4.32
LAPW	0.30 (in)	4.40	0.55 (in)	4.41	0.35	4.42
LMTO	0.38 (in)	4.43			0.84	4.44
LMTO	0.37 (d)	4.37	0.34 (d)	4.44		
ASW	0.38 (d)	4.31	0.67 (d)	4.32		
LMTO	0.41 (d)	4.30	0.87 (d)	4.44		
LAPW	0.45 (d)	4.40	0.9 (d)	4.39		
LMTO	0.48 (d)	4.43	0.9 (d)	4.41		

[a] Abbreviations of the methods are given in the Nomenclature section in the front of this book.

The fundamental band gap has been found by theoretical simulation to be sensitive to subtle structural variations that alter either the near-neighbor coordination geometry or the long-range stacking sequence of $CrSi_2$ layers [4.37, 4.40, 4.45]. A 2% reduction in the c axis lattice parameter causes an 8 meV decrease in the band gap. The calculated gap is particularly sensitive to the silicon atom position parameter x that determines the silicon-chromium coordination geometry within each $CrSi_2$ layer. Its variation within ±0.01 with respect to the "ideal" $x = 1/6$ value results in the gap variation in the range from 0.28 eV to 0.39 eV.

The orbital analysis shows that the lowest bands have predominant silicon $3s$, $3p$ character which then gradually switches to chromium $3d$ character near the Fermi level [4.37, 4.40, 4.45]. The states around the gap have a predominant chromium $3d$ character [4.31]. Away from the gap, the valence-band and conduction-band states originate from chromium states hybridized with silicon $3s$ and $3p$ orbitals [4.31]. The valence band spectrum obtained by XPS [4.46] enhances the importance of multiplet coupling between the $3p$ and $3d$ holes in the band structure of $CrSi_2$.

A comprehensive overview of the CrSi$_2$ electronic properties is ensured by DOS results. Fig. 4.4 demonstrates one of the most representative calculated DOS curves. There are two main peaks at binding energies of 0.6 eV and 2.3 eV. These features are also seen in the single-integrated photoemission spectra [4.47, 4.48]. The valence-band data emphasizing the chromium 3d component demonstrate two dominant binding energies of 0.6 eV and 1.7 eV. While being in agreement with the theoretically simulated density of states, the experimental XPS spectra contain little in the way of discernible structure at higher binding energies.

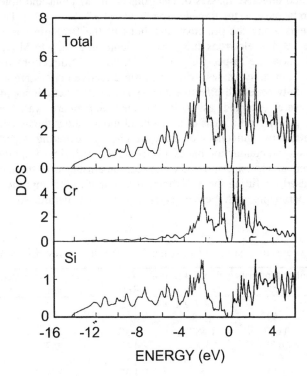

Fig. 4.4. Total and projected density of states (states/eV/cell) in CrSi$_2$ calculated by the LAPW method [4.45]

The XPS and BIS spectra recorded at binding energies less than 7 eV have provided details about the silicon p states, which are centered 3 eV below the Fermi level with an additional structure around 6 eV and a broad distribution in the unoccupied states above the Fermi level [4.49, 4.50]. The relative weight of the silicon s electrons is enhanced with respect to the general intensity.

Silicon s states are not restricted to the bottom of the valence band, but also show strong admixture to states at higher energies. Examples include s character associated with those 4p states around 6 eV below the Fermi level and even with the unoccupied bands between 1 and 3 eV above it. Moreover, both silicon s and d state densities contribute significantly in the low binding energy region. These silicon d states are in some way linked to the chromium d wave function. The silicon configuration is approximately $s^{1.6}p^{2.4}$, hence silicon s states contribute ~0.4 electrons to the bond order. Both silicon s and p states have a low DOS at the Fermi level due to the quasigap.

Photoabsorption measurements at the chromium L_2 and L_3 edges in $CrSi_2$, analyzed in terms of the final d-DOS, allow us to identify the structure of Si–Cr bonds at about 6.5 eV above the Fermi level to be the d-symmetry part of the bands with dominant pd antibonding character [4.51, 4.52].

The calculated effective masses of electrons at the M point and holes at the L point appear to be close to the free-electron values [4.37, 4.40, 4.45]. The effective mass components along the principal directions of the Brillouin zone edges are listed in Table 4.3. For electrons at M x corresponds to Γ-M, y to M-K, z to M-L. For holes at L these are respectively A-L, L-H, and L-M directions. Holes have nearly isotropic effective masses, while electron effective masses are anisotropic. The experimentally observed effective masses are 3–20 times higher [4.36, 4.38, 4.42]. They were obtained as a fit to the temperature variation of the thermoelectric power on the basis of a two-band model under the assumption that $CrSi_2$ is a degenerate semiconductor, acoustic phonon scattering is predominant, and both valence and conduction bands are spherical. Such significant discrepancy between theoretical and experimental values can be attributed to the imperfection of the model used for fitting the experimental data and to the impurity influence. Additional experiments with perfect crystals would be helpful to resolve the problem.

Table 4.3. Carrier effective masses in $CrSi_2$ (in units of the electron rest mass m_0)

Calculated for principal-axis directions				Experimental	
m^*_x	m^*_y	m^*_z	Refs	m^*	Refs
		e l e c t r o n s			
0.66	0.59	1.35	4.37	7	4.42
0.685	0.657	1.487	4.30	12	4.36
0.7	0.7	1.4	4.40	20.2	4.38
		h o l e s			
0.94	1.14	0.81	4.37	3.2	4.36
1.103	1.195	0.819	4.30	3.2	4.38
1.2	1.3	0.9	4.40	5	4.42

$MoSi_2$ ($4d$) and WSi_2 ($5d$). The question as to whether hexagonal $MoSi_2$ and WSi_2 are semiconductors or semimetals remained unanswered for a long time. Molybdenum and tungsten as well as chromium belong to the same group VI of the Periodic Table. They have identical configurations of the valence electron shells. Thus, for the same hexagonal lattice structure of the disilicides produced by chromium, molybdenum, and tungsten one can expect their electronic properties to be closed. This was confirmed by theoretical calculations in the early 1990s [4.30, 4.45]. As usual in the families of semiconductors composed of atoms of the same column of the Periodic Table, the gap has a tendency to decrease in the sequence $CrSi_2$-$MoSi_2$-WSi_2.

According to the theoretical simulation $MoSi_2$ is a narrow-gap semiconductor, while WSi_2 has been calculated to be a gapless semiconductor in [4.45], but it is still thought to be a narrow-gap semiconductor in the later calculations performed for the recently obtained lattice parameters [4.30]. The calculated band structure of MoSi₂ is demonstrated in Fig. 4.5.

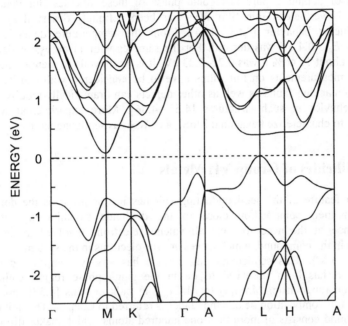

Fig. 4.5. Band structure of MoSi₂ calculated by the LMTO method [4.30]

It has the same specific features as those of the above discussed CrSi₂, namely the direct transition at the L point and the L-M indirect transition. The band structure of narrow-gap WSi₂ is identical. The calculated band gap parameters are summarized in Table 4.4.

Table 4.4. Calculated band gaps (d – direct, in – indirect) and carrier effective masses (in units of the electron rest mass m_0) in hexagonal MoSi₂ and WSi₂

Silicide		Band gaps		Effective mass components [4.30]			
	Method[a]	E_g (eV)	Refs	carriers	m^*_x	m^*_y	m^*_z
MoSi₂	LAPW	0.02 (in)	4.45	electrons	0.375	0.325	0.729
	LMTO	0.07 (in)	4.30	holes	0.433	0.549	0.499
	LMTO	0.36 (d)	4.30				
WSi₂	LAPW	–0.03 (in)	4.45	electrons	0.328	0.294	0.654
	LMTO	0.07 (in)	4.30	holes	0.390	0.572	0.451
	LMTO	0.44 (d)	4.30				

[a] Abbreviations of the methods are given in the Nomenclature section in the front of this book.

Effective masses of the carriers in $MoSi_2$ and WSi_2 are almost half those in $CrSi_2$, but their anisotropy for electrons and isotropy for holes are preserved.

Experimental information about fundamental electronic properties of hexagonal $MoSi_2$ and WSi_2 remains poor. This is due to the thermodynamic and kinetic problems of obtaining pure hexagonal phase of these silicides (for details see Chap. 1). Nevertheless, thin film WSi_2 has been definitely observed to exhibit semiconductor behavior [4.53, 4.54]. From the temperature dependence of resistivity $E_g = 0.4$ eV was derived. For the samples with an excess or deficit of silicon at about 10 at % it was 0.22–0.23 eV. It is worthwhile to mention here that electrical measurements on thin films have to be considered with caution. High density of stacking faults as well as other defects can dramatically modify charge carrier behavior in such structures [4.55, 4.56]. More experimental work is necessary to characterize hexagonal $MoSi_2$ and WSi_2 as semiconductors.

4.2.3 Silicides of Group VII Metals

The main feature of the semiconducting silicides in this group is the deficit of silicon in their composition close to the disilicide stoichiometry. It is a consequence of the need to have all valence electrons involved in the bond formation in the compound, which leaves no free electrons in the system.

$MnSi_{2-x}$ (3d). This semiconducting silicide has a very complicated crystal structure. It has been observed to be in the single-phase region within the composition range from $MnSi_{1.72}$ to $MnSi_{1.75}$ in bulk crystals [4.57]. For other structural and composition details the reader is referred to Chap. 1. The unit cell of the compound consists of more than one hundred atoms, which makes difficult to study its band structure and related electronic properties by a computer simulation. Thus, the material has been investigated exclusively by experimental techniques.

The data characterizing the band gap of this semiconductor are collected in Table 4.5. Optical measurements performed on polycrystalline films exhibit a minimum direct gap of 0.68–0.70 eV [4.58]. Another direct transition at 0.82 eV is also mentioned. Both are in general agreement with the observation of two maxima in the refractive index curve. It should, however, be remembered that this interpretation is based on measurements on polycrystalline films. Meanwhile, the gap of about 0.70 eV is mainly confirmed by resistivity measurements.

Table 4.5. Experimental band gaps in $MnSi_{2-x}$

Optical measurements at room temperature			Resistivity measurements (temperature dependence)		
composition	E_g (eV)	Refs	composition	E_g (eV)	Refs
$MnSi_{1.7}$	0.7[a]	4.58	$MnSi_{1.7}$	0.5	4.36
$MnSi_{1.73}$	0.78–0.83	4.59	$MnSi_{1.7}$	0.68	4.58
			$MnSi_{1.72-1.75}$	0.4	4.57
			$MnSi_{1.73}$	0.42–0.45	4.60
			$MnSi_{1.73}$	0.70	4.61

[a] Direct gap is referred to in the reference.

Considerable anisotropy of the carrier effective masses can be expected as the hole mobility measured in the [100] direction is almost twice that in the [001] direction [4.61].

$ReSi_{1.75}$ (5d). Among rhenium silicides the stoichiometric phase $ReSi_2$ was believed to be a semiconductor for a long time. It was observed in experiments with bulk [4.62, 4.63] and thin film [4.64–4.66] samples. However, in the first theoretical calculations of the band structure [4.67, 4.68] $ReSi_2$ was concluded to have metallic properties. The discrepancy between experimental observations and theoretical predictions arose from the fact that a tetragonal ($C11b$, $MoSi_2$-type) structure of the stoichiometric rhenium disilicide ($ReSi_2$) was used for calculations. The total number of valence electrons in this unit cell is odd (7 valence electrons for rhenium plus $4 \times 2 = 8$ for silicon) resulting in metallic properties of the disilicide. In fact, $ReSi_{1.75}$ or equivalently $Re_2Si_{3.5}$ seems to follow the empirical rule of "14" according to which silicides with 14 electrons per d-metal atom are semiconductors [4.69]. In the case of $Re_2Si_{3.5}$ one actually has $7 \times 2 = 14$ electrons given by rhenium and $4 \times 3.5 = 14$ electrons given by silicon, thus there are $28/2 = 14$ valence electrons per rhenium atom in the compound. The $ReSi_{1.75}$ phase has the alternative triclinic ($P1$ space group) structure. These stoichiometry and structure type have been experimentally confirmed [4.69, 4.70].

Theoretical simulation of the band structure of $ReSi_{2-x}$ as a function of the silicon content has definitely shown dramatic changes of the electronic properties from metallic to semiconducting in the order $ReSi_2$-$ReSi_{1.875}$-$ReSi_{1.75}$ [4.71]. For simplicity, accounting for the monoclinic distortion, i.e. the deviation from 90° of the c axis angle with respect to the ab plane (which is 89.90° [4.69]), to be negligible, the lattice structure was approximated to have an orthorhombic symmetry. In $ReSi_{1.75}$ the occupation factor of 0.75 for Si3 and Si4 sites is used. In the case of $ReSi_2$ both empty spheres were replaced by silicon atoms, while for $ReSi_{1.875}$ only one of them was replaced.

The calculated band structures along the most essential high-symmetry directions are shown in Fig. 4.6. The bands from 52nd to 61st are plotted. For $ReSi_2$ complete degeneracy of some bands occurs as it should be for a simple multiplication of the unit cell. The compounds $ReSi_2$ and $ReSi_{1.875}$ turn out to be essentially metallic. However, it is interesting to note that, besides some lowering of the Fermi level, in $ReSi_{1.875}$ one can already trace the beginning of the band gap formation in the region about 1.0 eV below the Fermi level just between the 56th and the 57th bands.

The band structure of $ReSi_{1.75}$ is typical for indirect semiconductors. There is an energy gap between the 56th and the 57th bands. The valence band maximum is at the Γ point and the conduction band minimum lies in the S point. An indirect band gap of 0.16 eV has been calculated between these points. It correlates well with the experimental gap values 0.12–0.19 eV obtained in optical and resistivity measurements as is evident from the data summarized in Table 4.6. The first well pronounced direct transition is located at the S point and has the value of 0.30 eV.

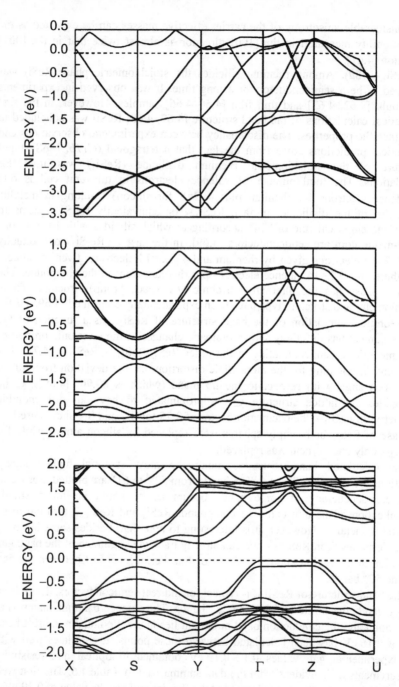

Fig. 4.6. Band structure of ReSi$_2$, ReSi$_{1.875}$, and ReSi$_{1.75}$ calculated by the LMTO method [4.71]. Zero of the energy scale corresponds to the Fermi energy

Table 4.6. Band gaps in ReSi$_{1.75}$ (d – direct, in – indirect)

LMTO calculations [4.71] E_g (eV)	Optical measurements at room temperature E_g (eV)	Refs	Resistivity measurements (temperature dependence) E_g (eV)	Refs
0.16 (in)	0.12 (in)	4.64	0.125	4.62
	0.13 (in)	4.63	0.14	4.64
	0.15 (in)	4.72	0.17–0.19	4.65
	0.15 (in)	4.66	0.21–0.22	4.63
0.30 (d)	0.36 (d)	4.64	0.16 and 0.30	4.69

The energy dependence of the total and projected densities of states for ReSi$_{1.75}$ is presented in Fig. 4.7.

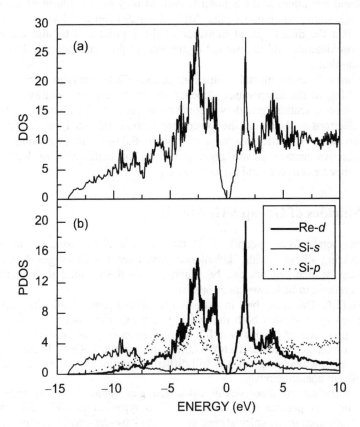

Fig. 4.7. Total (a) and projected (b) density of states (states/eV/cell) in ReSi$_{1.75}$ calculated by the LMTO method [4.71]. PDOSs are summed over the atomic types

The valence band extends down to about 14.0 eV below the Fermi energy. There are three main regions. The first is from the bottom up to about −7.5 eV and corresponds mainly to pure silicon s-like states. The second wide region is from

−7.5 eV up to about −1.5 eV with the main features at around −4.1 and −2.6 eV. Bonding silicon p and rhenium d states play a dominant role in this. The third region starts at −1.5 eV. It may be characterized as a region of nonbonding rhenium d states. The main peak is located at −0.9 eV. The results are in good agreement with the experimental data [4.66] where the dominant d-band structure in the photoemission spectra occurred around −3.0 eV with a broad shoulder down to −1.5 eV. The dominant rhenium d-derived structures in the photoemission spectra are shifted to higher binding energies compared to the $3d$ band silicides.

The eigenstate of the conduction band minimum at the S point is mainly defined by d states (~60%) of rhenium atoms with the admixture of p (~27%) and d (~11%) states of silicon atoms. The valence band maximum at the Γ point consists of almost pure d-like states of rhenium (~90%). Contrary to this, the valence band maximum at the S point is defined only by one fourth of the metal d states with strong admixture of the rhenium p states (up to 14%). So, one may conclude that the direct optical transition at the S point will be allowed in the dipole approximation and the oscillator strength for this transition should be of a sizable magnitude.

There are only experimental estimations of the effective masses of the carriers [4.64]. Fitting of the temperature dependence of the carrier mobility assuming acoustic phonon scattering and two conduction band model gives the effective mass of electrons as $0.01m_0$ in the lower conduction band and $1.8m_0$ in the upper conduction band. The effective mass of holes is $0.04m_0$. The low values of the carrier effective masses ensure good prospects for application of $ReSi_{1.75}$ in ultrahigh-speed electronics and thermoelectrics.

4.2.4   Silicides of Group VIII Metals

Transition metals from group VIII of the Periodic Table form a variety of semiconducting silicides with fundamental gaps over a wide range. Both direct and indirect semiconductors can be found among them, while their electronic properties remain to be known more precisely.

β-FeSi₂ (3d). This is the best investigated semiconducting silicide. Its electronic structure has been studied both theoretically and experimentally. This is mainly due to the prospects of this material for an application in optoelectronics assuming it to be a direct gap semiconductor with a gap of about 0.8 eV. There is a rich collection of data available, but it is still difficult to define the nature and value of the energy gap unambiguously.

The comparison of theoretically calculated gap values with experimentally obtained ones is presented in Table 4.7. Two types of orthorhombic crystal structure were used in the calculations: simple and base-centered, but if everything is correct, there should be no difference in the resulting band spectra. One of the most representative band structures is shown in Fig.4.8. This was calculated by the LMTO method [4.73] and later confirmed by the plane-wave density-functional calculations [4.82]. A base-centered orthorhombic unit cell *Pbcn* was used in [4.73]. The total number of valence electrons is equal to 128 per unit cell and the gap appears between the 64th and 65th bands. The discrepancies in the theoretical data may be due to neglecting combined correction terms in LMTO calculations. The difference between calculated and experimentally obtained gap values may occur due to underestimation of all methods within the LDA theory and the influence of correlation effects. Nevertheless, the LDA scheme, as well as the more precise generalized gradient approximation, reliably predicts trends in the gap character.

Table 4.7. Band gaps in β-FeSi$_2$ (d – direct, in – indirect)

Method[a]	E_g (eV)	Refs	E_g (eV)	Refs
	Theoretical calculations		Optical measurements – temperature	
LMTO	0.44 (in)	4.74	0.53–0.69 (in)-RT	4.75
LMTO	0.44 (in)	4.76	0.68 (in)-RT	4.77
ASW	0.44 (in)	4.78	0.76 (in)-RT	4.79
	0.52 (in)	4.80	0.77 (in)-RT	4.73, 4.81
FLAPW	0.73 (in)	4.82	0.78 (in)-RT	4.83, 4.84
LMTO	0.74 (in)	4.73	0.79 (in)-5 K	4.85
LMTO	0.80 (in)	4.86	0.79 (in)-RT	4.87, 4.88
			0.80 (in)-RT	4.89
			0.81 (in)-RT	4.90
			0.86 (in)-77 K	4.83
ASW	0.46 (d)	4.78	0.78–0.83 (d)-RT	4.91
LMTO	0.51 (d)	4.76	0.8 ±0.2 (d)-RT	4.92
LMTO	0.52 (d)	4.74	0.80 (d)-RT	4.93
LMTO	0.53 (d)	4.94	0.80–0.87-(d) RT	4.95
FLAPW	0.78 (d)	4.92	0.80–0.88-(d) RT	4.96
LMTO	0.78 (d)	4.73	0.82–0.85-(d) RT	4.97
LMTO	0.82 (d)	4.86	0.83 (d)-RT	4.84
FLAPW	0.82 (d)	4.82	0.84–0.88 (d)-RT	4.75
			0.84 (d)-RT	4.77, 4.81, 4.83, 4.98, 4.99
			0.85 (d)-RT	4.87–4.89, 4.100–4.103
			0.85–0.95 (d)-RT	4.104
			0.87 (d)-77K	4.83
			0.87 (d)-RT	4.79, 4.105–4.109
			0.87–0.89 (d)-RT	4.110
			0.88 (d)-RT	4.111, 4.112
			0.89 (d)-85 K	4.113
			0.89 (d)-RT	4.74, 4.114
			0.94 (d)-5 K	4.85
			0.96 (d)-RT	4.90
			1.0 ±0.2 (d)-RT	4.115

[a] Abbreviations of the methods are given in the Nomenclature section in the front of this book.

The energy spectrum is characterized by a direct band gap of 0.742 eV at the Λ point located at half of the distance between the Γ and Z points. The conduction band minimum at the Y point is only 8 meV higher than the one at the Λ point. The second slightly larger direct gap of 0.82 eV can be seen at the Y point. Because of such a small difference between the energy values of conduction band minimum at the Y and L points, it is difficult to define the nature of the gap unambiguously, so one can speak in terms of a quasi-direct gap.

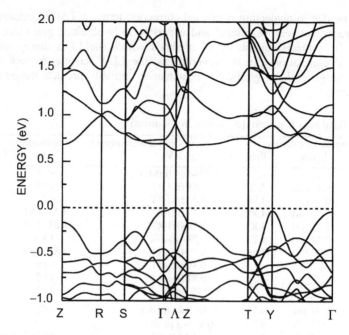

Fig. 4.8. Band structure of β-FeSi$_2$ calculated by the LMTO method [4.73]

Besides the results obtained in the LMTO [4.73] and plane-wave [4.82] calculations, practically the same band structures were obtained by the FLAPW method [4.92]. The difference is that in [4.92] the eigenvalues at the Y point are shifted slightly up on the energy scale, and instead of two approximately similar conduction band minima at Λ and Y points there are valence band maxima.

In the case of the simple orthorhombic unit cell the electronic structure calculations characterize β-FeSi$_2$ as an indirect gap semiconductor [4.74, 4.78, 4.86]. The direct gap transition is located in the Γ point, while the indirect one is located along Γ-R [4.86] or Γ-Z [4.74, 4.78] directions. The difference between direct and indirect gaps is small, while the gaps obtained differ from each other. The across-gap oscillator strength at the Γ point between the top of the valence and bottom of the conduction bands was found to be extremely small ($<10^{-5}$) [4.78].

Such a value can be explained by small contributions of iron and silicon p states. It is necessary to note that calculations performed in [4.86] by the LMTO method without the so-called "combined correction term" are not sufficiently precise, because the correction term might influence the band spectra and energy values.

In the case of the base-centered orthorhombic unit cell the main difference in the results is connected with the relative positions of the Y and Λ points. For example, in [4.94] the upper filled valence band at the Y point lies higher in comparison with the Λ point. Opposite results have been obtained in [4.76], where a valence band maximum is located at the Γ point and a conduction band minimum is located at the Λ point.

Some difference in the band gap values is seen from experimental optical and electrical data. Optical absorption measurements revealed the direct transition in the range 0.7–1.0 eV. Experimental evidence for the existence of an indirect absorption edge at energies of 0.53–0.81 eV has also been found. The energy of the participating phonon is estimated to be 35 meV [4.73]. Extrapolation of the optical data to zero temperature showed direct energy gaps with the following values: 0.89 eV [4.116], 0.90 eV [4.117], 0.92 eV [4.93]. In all the optical investigations on β-FeSi$_2$ the direct interband edge is clearly identified irrespective of the difference in the structural quality of the films which seems to have only a small effect on the direct gap value. Resistivity measurements reported the values of the energy gap as 0.98 eV [4.118], 0.88 ±0.04 eV [4.119], 0.95 eV [4.120]. According to most of the theoretical and experimental data the value of the direct energy gap of 0.87 eV is most probable to characterize β-FeSi$_2$.

The energy dependence of the total and projected DOS presented in Fig. 4.9 shows details of the orbital composition around the gap.

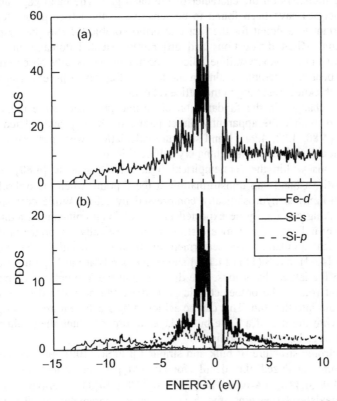

Fig. 4.9. Total (a) and projected (b) density of states in β-FeSi$_2$ calculated by the LMTO method [4.73]

The valence band from the bottom up to about −7.0 eV corresponds to pure silicon s-like states. The second wide region is from −7.0 eV up to about −2.0 eV with main features at −5.7 and around −3.0 eV. Bonding silicon p and iron d states play a dominant role in this. The third part starts at −2.0 eV. It may be characterized as the region of nonbonding iron d states. The main peaks are located at −1.4, −1.2, and −0.4 eV. These results are in good agreement with both theoretical [4.92] and experimental [4.115] data. Theory predicts indirect transitions between 0.738 and 0.750 eV starting from the valence band maximum at the Λ point. Besides, theory also yields direct transitions with an energy of 0.742 eV occurring at the Λ point having a much lower oscillator strength than those at the Y point. Thus, some deviations from the straight-line approximation in the energy region below 0.8 eV might be due to a superposition of both indirect and direct low-oscillator strength transitions. Good agreement is observed in the comparison of XPS and BIS spectra with theoretically calculated DOSs [4.115].

The estimation of the carrier effective masses at Γ shows $m^*_h = 0.85 m_0$ and $m^*_e = 0.80 m_0$ [4.86].

The search for conditions in which β-FeSi$_2$ would behave as a direct gap semiconductor has stimulated theoretical estimation of the influence of changes in its lattice parameters on the character of the band gap. The band gap nature and the gap energies have been found to be sensitive to the lattice parameters. It is very important to account for the fact that orthorhombic β-FeSi$_2$ poorly matches cubic silicon. Misfit dislocations and differently oriented domains in epitaxial films grown onto monocrystalline silicon substrates are usually observed. Thus, one can expect considerable strains in the lattice of epitaxial β-FeSi$_2$ films below the critical thickness resulting in the lattice relaxation.

Relevant changes in the fundamental electronic properties were theoretically estimated to modify the apparently direct nature of the gap and related optical transitions [4.80, 4.82, 4.94]. Band structure calculations were performed for the β-FeSi$_2$ unit cell parameters matching crystalline silicon.

Calculations within the semiempirical tight-binding scheme [4.80] show an indirect-to-direct band gap transformation at the Γ point, when the orthorhombic lattice of β-FeSi$_2$ is hydrostatically compressed by 3.8%. While corresponding lattice shrinking can hardly be expected in conventional epitaxial structures, this result clearly indicates that lattice distortions are critically important in β-FeSi$_2$. The theoretical simulation later performed for strained epitaxial β-FeSi$_2$ thin films on Si(100) [4.94] and Si(111) [4.121] shows the bands at the Y point to be very sensitive to the lattice distortions. The direct gap at the Y point, which may play an important role in the optoelectronic properties, can be tailored by a suitably induced strain into the film. This can be achieved by a coherent growth of β-FeSi$_2$ on silicon with the use of a SiGe buffer layer in order to tune the strain and get sufficiently large high-quality grains.

The electronic structure of bulk and strained β-FeSi$_2$ (100) epitaxial layers on Si(001) were analyzed in detail for two types of orientations [4.82]: β-FeSi$_2$ [100]//Si⟨110⟩ (A-type) and β-FeSi$_2$ [100]//Si⟨001⟩ (B-type). For the A-type epitaxial relationships, the b and c lattice parameters of β-FeSi$_2$ were reduced by 1.4% and 1.8%, respectively, to fit the silicon substrate. This corresponds to the structure being slightly compressed in comparison with the bulk lattice. For the B-type relationships, the unit cell was expanded along b and c directions by approximately 4% with respect to the equilibrium bulk case. The material with the undisturbed lattice was predicted to be an indirect-gap semiconductor with the energy gap of 0.73 eV between the Y and the Λ points and the first direct gap of 0.82 eV located at the Y point, which acceptably correlated to the results of previous band structure calculations. Another direct gap of 0.90 eV was found at the Λ point. For the B-type interface the energy gap is reduced. The indirect gap has become 0.53 eV while two direct gaps are 0.62 and 0.83 eV at the Y and L points, respectively. For the A-type interface two different band spectra were obtained. The first, which corresponds to β-$FeSi_2$ constrained to the experimental lattice parameter of silicon, is characterized by an indirect gap of 0.82 eV and two direct gaps of 0.85 and 0.92 eV. The second corresponds to β-$FeSi_2$ constrained to *ab initio* lattice parameters of silicon and the resulting band structure has been found to be direct with the energy gap of 0.93 eV at the Λ point. The second direct gap at the Y point is 1.04 eV. Thus, the A-type epitaxial relationships have been concluded to favor direct gap transitions in β-$FeSi_2$. Substitution of one of the host atoms in β-$FeSi_2$ by an appropriate impurity also provides possibilities for modification of the energy gap. Thin film β-$FeSi_{1.92}Ge_{0.08}$ fabricated by reactive atomic deposition of iron onto strain relaxed $Si_{0.93}Ge_{0.07}/Si$ substrates showed to have a direct band gap of 0.83 eV [4.122]. The reference β-$FeSi_2$ film formed at the same technological conditions was characterized by the direct gap of 0.87 eV. Although details of the electronic structure of the ternary silicide remain unknown, the observed red shift of the band gap indicates practical prospects of energy gap engineering in ternaries.

Ru_2Si_3 (4d). Orthorhombic ruthenium silicide Ru_2Si_3 and isostructural germanide Ru_2Ge_3 belong to a large family of transition-metal compounds with the atomic arrangements known as Nowotny "chimney-ladder" structures. A striking feature of these compounds is that their composition is controlled by a valence-electron concentration (VEC) in such a way that the number of valence electrons per transition-metal site practically never exceeds the "magic" number of 14. This appears to be a good criterion for the occurrence of a band gap, which should be formed between bonding and antibonding states. It has been suggested that VEC = 14 might correspond to a complete filling of the 10 transition-metal d states and 4 sp silicon (germanium) valence bands [4.123–4.125].

According to the experimental data [4.126] these materials exist in two modifications: low-temperature orthorhombic $Pbcn$ phase and high-temperature tetragonal $P\bar{4}c2$ one. Because of the prospects for practical applications, the orthorhombic phase has appeared to be mostly investigated.

Theoretical study of the electronic structure of orthorhombic Ru_2Si_3 within the LDA scheme has found it to be a direct-gap semiconductor [4.127–4.129]. One of its representative calculated band structures is shown in Fig. 4.10a.

The gap is formed between the 112th and 113th bands, and is characterized by a direct transition at the Γ point. It is worthwhile to note that there is only one distinct extremum in the band structure obtained. The orbital composition of the extreme states in the gap region shows that the eigenstates of the conduction band minimum are mainly defined by ruthenium d with small admixture of silicon p electrons. The contribution of ruthenium $3d$ states at the Γ point is much larger (about 60%) than the shares of ruthenium $1d$ and $2d$ states [4.127, 4.129]. The valence band maximum is characterized by large admixture of the other electron states to the eigenfunctions, which are silicon p and ruthenium d with a smaller contribution of silicon s and ruthenium p states. The interesting feature here is the second direct transition in the Γ point between the 111th and 113th bands. The corresponding energy gap is about 0.9 eV.

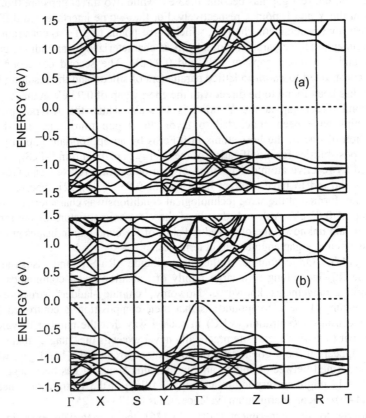

Fig. 4.10. Band structures of orthorhombic Ru_2Si_3 (a) and Ru_2Ge_3 (b) calculated by the LMTO method [4.129]

As for orbital composition, it is formed mainly by ruthenium d states and one can expect that its oscillator strength should be small.

Theoretically calculated and experimentally obtained gap values in orthorhombic Ru_2Si_3 and Ru_2Ge_3 are compared in Table 4.8. For reference, the energy gaps in tetragonal Ru_2Si_3 and Ru_2Ge_3 (high-temperature phases) derived from the resistivity measurements are 0.44 and 0.34 eV, respectively [4.126].

Electronic property calculations for the orthorhombic ruthenium germanide Ru_2Ge_3 have revealed this compound as a quasi-direct-gap semiconductor, as far as the energy difference between direct and indirect transitions lies within the error of the calculation procedure employed [4.129]. Its band structure is shown in Fig. 4.10b. The indirect gap is located between the Γ and Y points of the Brillouin zone. The lowest conduction band near the Y point is almost flat in the Γ-Y and Γ-X directions. The direct gap is located in the Γ point. The orbital composition of the extreme states in the gap region for ruthenium germanide is similar to that observed for the silicide.

The total and projected DOSs as a function of energy are shown for Ru_2Si_3 in Fig. 4.11. The projected DOSs of chemically inequivalent types of atoms are practically the same, so they are plotted for the first types of the atoms only. The valence band extends down to about −13 eV. The energy range from the bottom up to −7.0 eV is attributed to pure silicon s-like states. Bonding silicon p and ruthenium d electron states occupy the wide region up to the top of the valence band. These results are in good agreement with those obtained in [4.127]. The main feature is a small distinct gap of about 0.1 eV within the valence band, which splits off the silicon s states region from the bonding pd hybridization region [4.127], while in [4.129] only a considerable drop of the DOSs is observed in this energy region. This could be ascribed to the fact that the dispersion of some bands is somewhat larger compared to the one presented in [4.127]. The DOS spectra obtained in [4.123, 4.133] also have a broad sp band with a high-density region due to ruthenium d orbitals. The deep minimum in the high-density region at a VEC of 14, which might correspond to the experimental gap, is observed.

The character of DOSs in Ru_2Si_3 and Ru_2Ge_3 is quite similar. The main feature is that in the case of Ru_2Ge_3 there is a distinct gap within the valence band near −7.0 eV just between the region of the germanium s states and the one composed by bonding germanium p and ruthenium d states, that was also observed for Ru_2Si_3 in [4.127].

Table 4.8. Band gaps in orthorhombic Me_2Si_3 silicides and related germanides (d – direct, in – indirect)

	Theoretical calculations		Resistivity measurements (temperature dependence)	
Method[a]	E_g (eV)	Refs	E_g (eV)	Refs
		Ru_2Si_3		
LMTO	0.41 (d)	4.129	0.70	4.126
LMTO	0.42 (d)	4.128	1.08	4.130
FLAPW	0.42 (d)	4.127	1.19	4.131
		Os_2Si_3		
LMTO	0.95 (d)	4.129	2.3 ±0.2	4.132
		Ru_2Ge_3		
LMTO	0.31 (in)	4.129	0.52	4.126
LMTO	0.33 (d)	4.129		
		Os_2Ge_3		
LMTO	0.87(in)	4.129		
LMTO	0.91(d)	4.129		

[a] Abbreviations of the methods are given in the Nomenclature section in the front of this book.

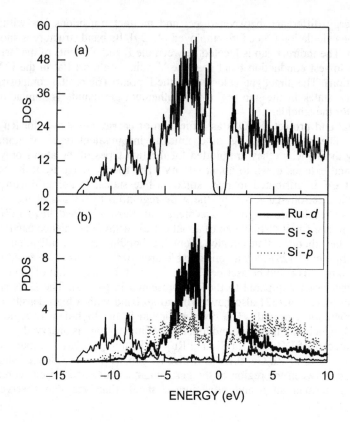

Fig. 4.11. Total (a) and projected (b) density of states (states/eV/cell) in orthorhombic Ru$_2$Si$_3$ calculated by the LMTO method [4.129]

Effective masses of the carriers in Ru$_2$Si$_3$ were experimentally estimated from the temperature dependence of Hall mobility and Seebeck coefficient. They are listed in Table 4.9 as well as theoretical data obtained by FLAPW calculations in the Γ point. The principal axes were chosen to be along the axis of the reciprocal-lattice vectors, so that x corresponds to the Γ-X direction, y to Γ-Y and z to Γ-Z.

Table 4.9. Effective masses of electrons and holes in Ru$_2$Si$_3$ (in units of the electron rest mass m_0)

Experimental [4.131] m^*	Calculated for principal-axis directions [4.127]		
	m^*_x	m^*_y	m^*_z
e l e c t r o n s			
3.9	3.28	2.85	0.61
h o l e s			
7.5	0.47	0.15	0.45

From the previously discussed carrier effective masses in $CrSi_2$, $MoSi_2$ and WSi_2 [4.30, 4.40], one knows that the effective masses for states dominated by d character are of the order of the free-electron mass. In the case of Ru_2Si_3 there is one exception. The highest occupied valence band exhibits a strong dispersion due to the overlap of delocalized silicon p states and the corresponding hole masses are about one order of magnitude smaller than the values typical for silicides [4.127]. They are close to the values known for silicon.

OsSi (5d). There are neither theoretical nor detailed experimental data describing band structure and related electronic properties of this silicide. It has been identified as a semiconductor with $E_g = 0.34$ eV only by the experimentally measured temperature dependence of resistivity [4.132].

Os_2Si_3 (5d). This compound has an orthorhombic *Pbcn* unit cell isostructural with Ru_2Si_3. According to the experimentally observed temperature dependence of resistivity it was concluded to be a semiconductor with a fundamental energy gap of 2.3 eV [4.132]. This gap value is the highest among the semiconducting silicides, but is almost twice the theoretically predicted gap.

The first theoretical study of electronic properties of Os_2Si_3 has reported it to be a direct-gap semiconductor [4.129]. Its band structure is indeed similar to that of orthorhombic Ru_2Si_3 (see Fig. 4.10a). It is not surprising in view of the identity of the crystal structures and valence-electron compositions of both silicides. A direct transition occurs in the Γ point and has the value of 0.95 eV. According to the orbital composition data, eigenstates of the conduction band minimum are mainly determined by osmium d with admixture of silicon p electrons. The valence band maximum is characterized by largely different electron states formed by silicon p and osmium d electrons with a small contribution of silicon s and osmium p electrons. The contribution of p states is more essential than that in the case of Ru_2Si_3. It may cause a larger underestimation of the band gap energy. The striking feature is that in osmium silicide the 112th band moved down to the 111th band so that the difference between the first and the second direct transitions constitutes less than 0.1 eV.

Theoretical study of the isostructural osmium germanide Os_2Ge_3 has shown it to be a quasi-direct-gap semiconductor [4.129]. The indirect gap, as in case of Ru_2Ge_3, is located between the Γ and Y points of the Brillouin zone, while the direct one is in the Γ point. The band gap values of orthorhombic Me_2Si_3 silicides and related germanides summarized in Table 4.8 clearly demonstrate high prospects of ternaries and quaternaries within this family of semiconductors for effective energy gap engineering.

$OsSi_2$ (5d). This compound, having the orthorhombic *Cmca* unit cell, is isostructural to β-$FeSi_2$. Optical and electrical measurements clearly indicate a semiconducting behavior of this disilicide in a wide temperature range. The derived band gaps vary from 1.4 to 2.0 eV [4.132, 4.134]. The first theoretical calculations performed in [4.76] have found it to be a semiconductor with a very small indirect gap value of 0.06 eV, which strongly differs from available experimental data. One of the possible reasons is that the LMTO method used for the calculations is very sensitive to a choice of the atomic sphere radii ratio. Moreover, it is necessary to note that the energy gap value of 0.44 eV for β-$FeSi_2$ obtained within the same approach [4.76] is about half the typical experimental values obtained in electrical and optical measurements. This indicates the inadequate choice of the parameters introduced into the calculations.

A more correct electronic property simulation of $OsSi_2$ with the use of the LMTO method was performed in [4.135]. A fragment of the calculated band structure along the high-symmetry directions of the appropriate Brillouin zone is shown in Fig. 4.12. The total number of valence electrons is equal to 128 per unit cell. A gap appeared between the 64th and 65th bands. There are well resolved global extrema for both valence and conduction bands at the Λ and Y points, respectively, thus indicating $OsSi_2$ to be an indirect-gap semiconductor. Calculated and experimentally observed gap energies are summarized in Table 4.10.

In $OsSi_2$ the regions of bonding silicon p and osmium d states and nonbonding osmium d states are shifted to higher energies with respect to those in β-$FeSi_2$. This is expected to be related to the stronger overlap of osmium d and silicon p wave functions implying a greater bonding/antibonding splitting.

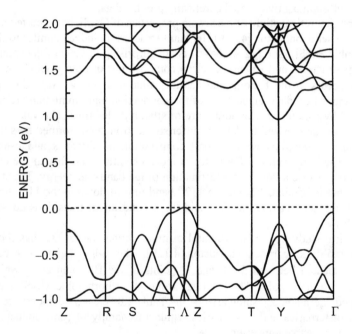

Fig. 4.12. Band structure of $OsSi_2$ calculated by the LMTO method [4.135]

Table 4.10. Band gaps in $OsSi_2$ (d – direct, in – indirect)

LMTO calculations		Optical measurements at room temperature		Resistivity measurements (temperature dependence)	
E_g (eV)	Refs	E_g (eV)	Refs	E_g (eV)	Refs
0.06 (in)	4.76	2.0	4.134	1.8	4.134
0.95 (in)	4.135			1.4	4.132
1.14 (d)	4.135				

Eigenvalues near the gap energy region are mainly determined by d electrons of osmium atoms. However, orbital composition analysis shows strong admixture of other electron states, particularly silicon s and p states in the lowest conduction band. Unlike the situation in β-FeSi₂, where two direct transitions with low oscillator strength at the Λ and with an appreciable one at the Y points exist, there is only one direct transition at the Y point in OsSi₂. The corresponding local minimum of the conduction band is moved from the Λ to the Γ point. Moreover, it is located higher in energy than the minimum at the Y point.

Unfortunately, there is a lack of optical data for OsSi₂. Only reflectance measurements performed in [4.134] and roughly estimating the gap to be 2 eV can be mentioned. However, the indirect gap theoretically predicted in [4.135] is somewhat smaller than the experimental one. It is well known that a band gap calculated within the LDA scheme is usually underestimated with respect to the ones derived from experiments. The value of such an underestimation depends on the shift of eigenvalues in the energy range analyzed due to so-called correlation effects. Nevertheless, in the case of isostructural β-FeSi₂ rather good agreement between theoretical and experimental gap energies was obtained [4.73]. This effect can be ascribed to the fact that corresponding eigenfunctions at extremum points of the band structure are mainly composed by d-electron states of iron atoms. Hence, they undergo almost the same shift. For OsSi₂ larger admixtures of other electron states to the eigenfunctions in the energy region near the gap occur.

Ir₃Si₅ (5d). The crystalline structure of this compound (monoclinic $P2_1/c$) is rather complex for use in theoretical calculations of its fundamental electronic properties. This silicide has been found to be a semiconductor experimentally [4.136–4.138]. Electrical property measurements yielded the band gap of 1.2 eV [4.139, 4.140]. Optical measurements indicated direct-gap transitions corresponding to E_g = 1.56 eV [4.141].

The UPS spectrum of Ir₃Si₅ is characterized by the absence of electronic states near the Fermi energy [4.142] pointing to an energy gap in this material. The characteristic d-band feature is shifted to higher binding energy compared to the $3d$ silicides and has a marked peak at −4.2 eV with shoulders at −6.2, −2.5, and −1.1 eV. The shift to higher energies is in agreement with the general expectation that the substitution of $3d$ metals by $4d$ and $5d$ metals results in a larger splitting of the bonding and antibonding metal d and silicon p states. Due to the larger spatial extent of iridium d wave functions and the larger overlap of the silicon p wave function the hybridization between these states increases.

No detailed theoretical simulation of electronic band structure of Ir₃Si₅ has been performed.

4.2.5 General Trends

Most semiconducting silicides are formed by transition metals. These are chromium, molybdenum, and tungsten belonging to group VI of the Periodic Table of elements, manganese and rhenium from group VII, and iron, ruthenium, osmium, and iridium from group VIII. While they have a different number of valence electrons, all of them occupy d states. This results in some common features in the fundamental electronic properties of the semiconducting silicides produced and also of the transition metal silicides with metallic properties. They can be summarized as follows.

The electronic structure of the transition metal silicides is dominated by the metal d states. The common characteristics of the valence band density of states are: nonbonding d states near the Fermi level E_F with the energy close to that for the d band in the corresponding metal; the bonding pd hybridized states, which dominate the stability of the silicide and extend to about 6 eV below E_F; and silicon s states about 14–16 eV below E_F. Compared to the elemental transition metals the nonbonding d band is narrowed. This is explained by the fact that the atomic d bands are driven further away and the hopping integrals decrease due to the presence of silicon atoms in the lattice. With the increase of the atomic number in the series of the $3d$-$5d$ transition metal silicides the d-derived states enhance the binding energy. They appear first as a broad band of mostly empty states at the beginning of the transition metal rows and transform into a narrow band of mostly occupied states at the end of the rows. Near the middle of the d-transition metal rows the motion is relatively slow as the density of d states is high at E_F. Due to the hybridization of metal d states with the silicon $3p$ states, the center of gravity of the silicon $3p$ states moves to a higher binding energy as the atomic number of the d metal is increased, following the shift of the metal d states. Whereas the occupied d-derived states move to greater binding energies within the transition metal rows, the antibonding pd combinations appear mostly above the Fermi level and their energetic position remains nearly invariant in the series of silicides. Silicon s-derived states overlap relatively little with the silicon p and metal d states and are weakly involved in chemical bonding. The transition metal d states form a dominant nonbonding d state between bonding and antibonding states for the $3d$ transition metal silicides, whereas the majority of nd states are bonding for the $4d$ and $5d$ transition metal silicides [4.143].

The main chemical bonds in the transition metal silicides are formed by basically covalent interactions between silicon $3p$ electrons and metal d electrons. Detailed investigations for the $3d$ transition metal disilicides have shown that charge flows occur mainly between silicon sp and transition metal $4sp$ bands [4.144]. Therefore the ionic contribution to bonding is small. Silicon–silicon and transition metal–silicon bondings dominantly contribute to the covalent bonds. The strength of the silicon–silicon bonds decreases and that of transition metal–silicon bonds increases with increasing atomic number of the transition metal. The main participant of the transition metal-silicon bonding shifts from the silicon $3d$ bonding to the silicon $4sp$ bonding when passing from the middle to the end of the $3d$ transition metal row.

The vast majority of band structure calculations performed for semiconducting transition metal silicides applied the LDA theory including exchange and correlation terms. Such calculations for semiconductors are usually known to yield gap energies which are 50–100% smaller than experimental ones. Nevertheless, for the semiconducting silicides the calculated gaps are in surprisingly good agreement with the experimental data. The value of the underestimation of energy gaps depends on the shift of energy values due to the so-called correlation effects. One reason for the agreement between experimental and theoretical gaps in semiconducting transition metal silicides might be the fact that the corresponding eigenfunctions at the extremum points of the band structure are mainly composed of metal d electron states, which undergo almost the same shift due to correlation effects. Another remarkable fact is a very good correlation between experimental and LDA calculated interband optical spectra of semiconducting silicides, as has been demonstrated for $CrSi_2$, $ReSi_{1.75}$, β-$FeSi_2$, and Ru_2Si_3. Thus, LDA seems to be a useful approximation when it is correctly applied to semiconducting silicides.

Small changes in atomic positions of constituents in a lattice induce relatively large shifts in the band edge states of semiconducting silicides. This stresses that atomic positions must be well defined if a reliable determination of the band-edge states is required. It also illustrates that band-gap tuning of semiconducting silicides can be achieved by a controlled distortion of their lattices conveniently provided by impurity doping or by strains.

An experimental gap energy determination from optical measurements has to take into account that the oscillator strengths for optical across-gap transitions. They are generally small due to the dominant d character of the states at the extremum points of the valence and conduction bands. As far as semiconducting silicides can have defects of different origin which produce extended gap tails, weak optical interband features might easily escape detection. Thus, an adequate identification of the nature and energy of lowest interband gaps being derived from optical measurements requires samples of high purity and good structural quality.

Calculated effective electron and hole masses in semiconducting transition metal silicides are of the order of the free electron mass. Experimentally determined effective masses are much larger. In particular, electron effective masses are an order of magnitude higher. The reasons for this discrepancy are not very clear. Nevertheless, one can note that the experiments performed used thermopower measurements subsequently fitted by standard simplified analytical relations. Eventually, more refined models for evaluating these measurements as well as other experimental approaches are necessary to implement for semiconducting silicides.

4.3 Interband Optical Spectra

The photon energy dependence of real ε_1 and imaginary ε_2 parts of the dielectric function most profoundly characterizes the interband optical properties of semiconductors. For semiconducting silicides it is discussed first in this section. Regarding the importance of optical absorption data at the interband edge for understanding the character of the band gap, they are analyzed in the second part of the section. The presentation is arranged according to the similarities traced for specific families of semiconducting silicides, in particular with respect to their

d-electron states. In this way semiconducting Mg_2Si and $BaSi_2$ without d electrons are different from those produced by metals with d valence electrons belonging to the family of transition metal silicides.

Reflectance measurements for **Mg₂Si** have been performed in [4.145] at 77 K over the energy range from 0.6 to 11 eV. The results are analyzed by the Kramers–Kronig relations to yield the dielectric function presented in Fig. 4.13. The experimental ε_2 curve starts with an indirect tail at 0.6 eV. At about 2 eV ε_2 sharply rises to a dominant peak at 2.70 eV. Above the main peak the next structure is a peak of lower intensity at 3.7 eV. Beyond this energy ε_2 continuously decreases exhibiting a small bump around 5 eV. From the band structure result [4.12] the optical constants have been calculated, which could successfully reproduce the experimental data.

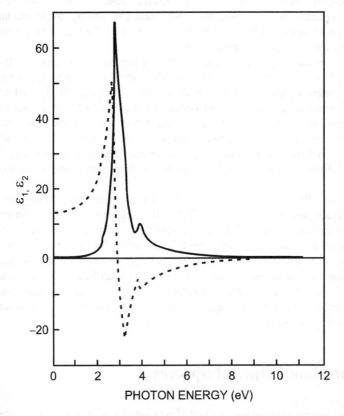

Fig. 4.13. Photon energy dependence of real (solid lines) and imaginary (dashed lines) of the dielectric function for Mg_2Si calculated from experimental reflectance data [4.145]

The main structures in the experimental spectra could be explained in terms of critical points of the joint density of interband states. The sharp rise at 2 eV has been attributed to optical transitions in the center of the Brillouin zone with an associated critical point of M_0 type. The main peak at 2.7 eV (theoretically at 2.6 eV) is ascribed to three transitions at the boundary of the Brillouin zone, all being M_0 type. The weaker peak at 3.7 eV has its origin in transitions along symmetry lines in the Brillouin zone and at its boundary. The contributing critical points have M_2 and M_3 symmetry. Above this peak a large number of different regions contribute to ε_2.

Semiconducting **BaSi$_2$** has been not optically characterized yet.

Most intensive studies of the interband optical properties of semiconducting silicides have been performed for the $3d$ transition metal silicides **CrSi$_2$, MnSi$_{2-x}$,** and **β-FeSi$_2$** over a broad energy range extending up to 24 eV [4.146–4.148]. Epitaxial layers and polycrystalline films yielded spectra of similar shape above 3 eV. Towards lower energy the spectra of these films agree in their gross features but differ in the fine structure. The imaginary parts of the dielectric function are compared in Fig. 4.14 for the $3d$ transition metal silicides CrSi$_2$, MnSi$_{2-x}$, and β-FeSi$_2$ as well as for the semiconducting $5d$ transition metal silicide Ir$_3$Si$_5$. There is a close resemblance in the spectra for the silicides of the $3d$ series in the high-energy region.

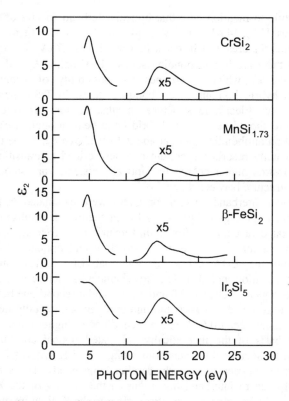

Fig. 4.14. Imaginary part of the dielectric function at high photon energies for CrSi$_2$, MnSi$_{2-x}$, β-FeSi$_2$, and Ir$_3$Si$_5$ [4.148]

The spectral details observed in the optical constants for the $3d$ transition metal silicides can be directly related to their band structures on the basis of experimental [4.49] and theoretical [4.47, 4.49] information for $CrSi_2$ and β-$FeSi_2$. It can be assumed that these considerations are also principally applicable to the neighboring $3d$ silicide $MnSi_{2-x}$. In these materials the conduction band density of states in the vicinity of the Fermi energy shows a relatively smooth variation with energy. Therefore, ε_2 being proportional to the joint density of interband states, reflects major features of the valence band density of states distribution (VBDOS). The VBDOS for $CrSi_2$ and β-$FeSi_2$ can be divided into three main parts. The first one near the Fermi energy is predominantly the region of nonbonding metal d states with an admixture of metal and silicon p states. Transitions from these states into empty conduction band states correspond to the first (low-energy) peak in ε_2. The second due to strong bonding between metal d and silicon p states extends over a region of several eV. Transitions from these states into empty conduction band states are responsible for the broad maximum in ε_2 between 4 and 5 eV. The third VBDOS region extending down to 14 eV below E_F corresponds to silicon s-like states. Optical transitions from these states are reflected in the small peak in ε_2 at 14.5 eV.

It is worthwhile to mention here that optical interband spectra of the metallic early $3d$ silicides $TiSi_2$, [4.416] and VSi_2 [4.150, 4.151] also fit to this scheme. The first interband features in VSi_2 occur between 1 and 1.7 eV. Theory states that in this material there are hybrid bonding states -1 eV below E_F and antibonding states just above E_F which determines the possibility of strong interband transitions from hybrid bonding to antibonding states. In $TiSi_2$ main structures occur around 2 eV. Electronic structure calculations for this silicide [4.152] indicate that there should be a d_{Ti}-p_{Si} hybrid bond peak at about 1 eV below the Fermi level and an antibonding peak at about 1 eV above E_F. Thus, the structure of the experimentally recorded spectrum can be ascribed to transitions between these states. Towards higher energy the optical constants vary smoothly with a second weaker structure between 4 and 5 eV.

Detailed optical interband spectra for **CrSi$_2$** were calculated in [4.31, 4.37, 4.43] for photon energies up to 6 eV. In [4.43] the theoretical results are compared with experimental data obtained from ellipsometric measurements on poly- and single-crystalline samples. The electronic structure and optical properties were calculated within the LDA using the semirelativistic LMTO method with exchange and correlation included. The first dominant structure in the theoretical optical constants appears between 0.8 and 1.5 eV and a second one between 4 and 5 eV. In this energy range ε_1 passes through zero. Experimentally determined ε_2 exhibits two maxima: one at 0.6 eV and a second one of higher intensity at 1.1 eV [4.43]. On the basis of the theoretically calculated optical constants the first maximum could be attributed to transitions along the M-K symmetry direction of the Brillouin zone between two nearly parallel lying bands. The second one at 1.1 eV is mainly due to interband transitions in the vicinity of the K point and along the K-Γ and M-K symmetry directions close to the K point involving several conduction and valence bands. Similar theoretical results were obtained in [4.37] where the first dominant peak in the theoretical optical conductivity (defined as

$\sigma = \omega\varepsilon_2$) occurring at 0.85 eV was found to relate to transitions from pd hybridized states along the M-K and M-L lines. Further peaks in the optical conductivity are at 1.25, 1.52, 1.85, and 2.1 eV. All theoretical calculations for CrSi$_2$ reveal a broad maximum in ε_2 between 4 and 5 eV to which transitions between numerous bands contribute.

A comparison of the experimental ε_1 and ε_2 spectra for thin film samples with the theoretical data is shown in Fig. 4.15.

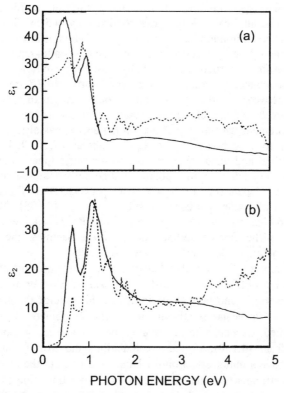

Fig. 4.15. Experimental (*solid lines*) and theoretical (*dashed lines*) photon energy dependence of real (a) and imaginary (b) parts of the dielectric function for CrSi$_2$ [4.43]. The calculated curves are shifted to lower energies by 0.4 eV

After a shift of the theoretical spectra by 0.4 eV to lower energies there is very good agreement with the experiment in the low-energy region. A possible reason for the discrepancy with respect to energetic position might be the density functional approximation used which is well fulfilled for states around the gap having predominantly chromium d character, but becomes less valid for states originating from chromium states hybridized with silicon p states. In the experimental spectra the theoretically predicted strong interband transitions around 5 eV are absent. According to the theoretical calculation [4.43], the anisotropy in the optical spectra is relatively weak. The experimental spectral dependencies of optical constants agree well with those recently obtained on polycrystalline $CrSi_2$ films [4.153]. The ε_2 spectrum for these layers well reproduces the dominant double structure around 1 eV (see Fig. 4.15) and, in addition, shows a pronounced peak around 4 eV. The latter peak has been also observed in [4.148] on polycrystalline $CrSi_2$ layers (see Fig. 4.14).

The real and imaginary parts of the dielectric function of **$MnSi_{2-x}$** in the above-gap region up to 2 eV exhibit most prominent peaks at 1.25 eV with a shoulder at 1.5 eV [4.147]. The spectral distribution of the absorption coefficient shows a clear minimum at about 2 eV and then rises continuously to 5 eV. The reflectance exhibits a pronounced minimum between 2 and 3 eV, which is at variance with the results for $CrSi_2$ and β-$FeSi_2$. Due to the complicated crystallographic structure of this material there are no particular calculations available up to now.

The most pronounced interband feature for nearly all $3d$ transition metal silicides occurs between 1 and 2 eV. It exhibits some fine structure, which usually can be only observed on material of high crystalline perfection.

Precise measurements of the reflectance on **β-$FeSi_2$** epitaxial layers above the interband gap up to 2 eV revealed a series of peaks at 0.9, 1.19, 1.5, 1.62, and 1.8 eV [4.154]. Real and imaginary parts of the dielectric function are shown in Fig. 4.16 for polycrystalline films, polycrystalline bulk material and epitaxial films. The spectral dependence of ε_1 and ε_2 for polycrystalline films is in good correspondence with other published results, see e.g. [4.155]. Whereas the variation of the optical constants for polycrystalline films is relatively smooth and featureless due to the fine-crystalline nature of the samples, the spectra for epitaxial films contain detailed band structure information.

The first attempt of theoretical calculations of interband optical spectra of β-$FeSi_2$ was carried out in [4.74, 4.156] for photon energies up to 5 eV. In the simulation of electronic and optical properties of this silicide the self-consistent LMTO method in its scalar relativistic form with combined correction terms [4.157] was applied. Exchange and correlation potentials were included using the approximation from [4.158]. In these calculations the iron f functions were included which have a minor effect on the band structure, but the $d \rightarrow f$ oscillator strength is much stronger than that of the $p \rightarrow d$ or $s \rightarrow p$ transitions. Knowing the one-electron energy spectrum and the momentum matrix elements, the interband contribution to the imaginary part of the dielectric function ε_2 has been calculated for three different light polarization along the principal axes of the crystal, and the real part ε_1 has been obtained from ε_2 via the Kramers–Kronig relation. The spectroscopic ellipsometric measurements were carried out on high-quality MBE-grown β-$FeSi_2$ layers on Si(100) substrates. Fig. 4.17 shows a comparison of the experimental and theoretical ε_1 and ε_2 spectra for light polarization along the a axis and proves the agreement in detail between experiment and theory.

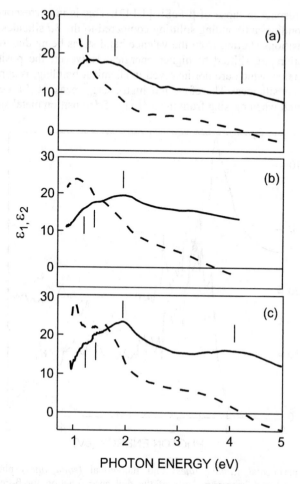

Fig. 4.16. Photon energy dependence of real (*solid lines*) and imaginary (*dashed lines*) parts of the dielectric function for β-FeSi$_2$ polycrystalline films (a), polycrystalline bulk material (b), and epitaxial films (c) [4.154]

Strong absorption starts at about 0.8 eV. The main feature of ε_2 is a pronounced maximum at 1.65 eV in the theoretical spectra, which is slightly shifted towards higher energy in the experiment. It is mainly associated with transitions from hybridized pd states of iron and silicon into conduction band states. Both theory and experiment show a noticeable drop of ε_2 in the higher energy region with a minimum followed by a second maximum at 4.1 eV. Anisotropy of the optical functions in the energy range of 1–3.5 eV is relatively weak. At the energies around the second maximum some anisotropy occurs. This peak is absent for light polarizations along b and c axes.

For the $5d$ semiconducting transition metal silicide **Ir$_3$Si$_5$** there are also three interband structures observed in ε_2, but the low-energy ones are shifted to higher energies with respect to those of β-FeSi$_2$ [4.147]. This is in correspondence with the stronger bonding-antibonding splitting compared to the 3d silicides. Therefore all optical transitions starting from the valence band states being due to the metal-silicon interaction are shifted to higher energies. However, the position of the silicon s-like states, which are not involved in chemical bonding, is approximately the same in the silicides. Therefore the high-energy peak at 14.5 eV remains almost unaffected when passing from the 3d to the 5d transition metal silicides.

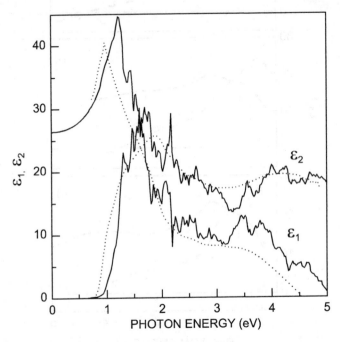

Fig. 4.17. Experimental (*dotted lines*) and theoretical (*solid lines*) photon energy dependence of real and imaginary parts of the dielectric function for β-FeSi$_2$ for light polarization along the a axis [4.156]

The same trend in interband optical spectra is observed for the 5d semiconducting silicide ReSi$_{1.75}$ [4.63]. This is evident from the spectra presented in Fig. 4.18. The most pronounced low-energy interband feature in ε_2 has its maximum at 3.3 eV, which is due to transitions from the hybridized d_{Re}-p_{Si} states into the conduction band. Towards higher energies ε_2 smoothly decreases up to 12 eV with weak shoulders appearing between 5.5 and 7.2 eV. The real part of the dielectric function ε_1 passes through zero around 5 eV. Towards lower energies a shoulder becomes apparent at 0.6 eV which probably arises from transitions starting from the rhenium d manifold into the conduction band.

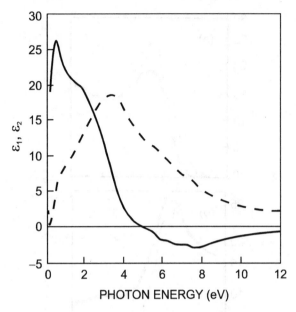

Fig. 4.18. Photon energy dependence of real (*solid lines*) and imaginary (*dashed lines*) parts of the dielectric function for ReSi$_{1.75}$ [4.63]

First results of optical interband investigations for semiconducting **Ru$_2$Si$_3$** for photon energies up to 5 eV are reported in [4.128]. The optical constants including real and imaginary parts of the dielectric function, refractive index and absorption coefficient were obtained by ellipsometry and reflectance measurements on polycrystalline films. These spectra are compared with the electronic structure and optical properties calculated with the LMTO method with correlation and exchange terms in LDA. For the same reason, as mentioned above for β-FeSi$_2$, it was necessary to include ruthenium f functions in the calculations. Fig. 4.19 shows a comparison of the experimental and theoretical optical data. The agreement is excellent up to 4 eV. At higher energies it is likely that roughness of the samples and surface layers influence the experimental spectra. The small value of the oscillator strength from the energy gap to approximately 1 eV is due to the fact that the states at the bottom of the valence band and at the top of the conduction band consist mainly of ruthenium d states with only a small admixture of ruthenium and silicon p states. This conflicts with the theoretical prediction made in [4.127, 4.129]. ·

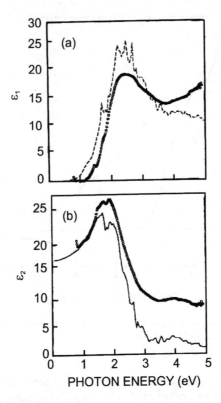

Fig. 4.19. Experimental (*dotted lines*) and theoretical (*solid lines*) photon energy dependence of real (a) and imaginary (b) parts of the dielectric function for Ru_2Si_3 [4.128]

The optical interband spectra of the isostructural compounds Ru_2Si_3 and Ru_2Ge_3 have been recently compared in [4.159]. Reflectance and ellipsometric measurements have been performed over the spectral range from 0.8 to 10 eV on polycrystalline films prepared by metal deposition on silicon and germanium substrates with subsequent annealing. The spectra showed great similarity with dominant interband features at about 2 and 5 eV. Above this energy the optical constants exhibited a smooth variation. In comparison with the silicide, the main structures are shifted slightly to lower energies for the germanide.

Interband optical spectra revealing the dielectric function of other transition metal semiconducting silicides are not available at the moment, but they are still desired to confirm general trends in this family and to search for individual peculiarities.

Extensive optical studies have been performed on the interband edges of semiconducting silicides in order to reveal their gap energies and the gap nature. In all cases this information has been derived from conventional square or square-root plots of the absorption coefficient versus photon energy according to (4.15)–(4.17) for allowed direct or phonon-assisted indirect optical transitions. However, the following critical comments should be made for this kind of data handling in the case of semiconducting silicides:

(i) The theoretical calculations have shown that the across-gap oscillator strength $f_{cv} = 4V_{cv}^2/(E_c - E_v)$ (where V_{cv} is the optical interband matrix element for transitions between valence and conduction band, $E_{c,v}$ are the conduction and valence band energies, respectively) for CrSi$_2$ and β-FeSi$_2$ are 0.02 [4.31] and 0.01 [4.78]. It is also very low for Ru$_2$Si$_3$ [4.128]. These are orders of magnitude smaller than those for the zincblende-type semiconductors. This is due to the predominant metal d character of the states on both sides of the interband gap for most semiconducting transition metal silicides.

(ii) A pronounced absorption tail exists for all semiconducting silicides studied so far, which amounts to 10^3–10^4 cm^{-1}, depending on the structural film quality, and extends far into the gap region, see e.g. [4.32, 4.41, 4.77, 4.79, 4.91, 4.95, 4.100, 4.107, 4.155]. This strong absorption tail evidently masks interband features having a low oscillator strength.

(iii) The interband matrix elements are expected to vary appreciably over a narrow energy range in the edge region due to the strong change in the character of band states when going deeper into the bands, and this will also be reflected in the energy dependence of the absorption coefficient.

The importance of these comments is well illustrated with results obtained for β-FeSi$_2$ and CrSi$_2$.

Optical absorption data for high crystalline perfection epitaxial films of **β-FeSi$_2$** have been interpreted in the light of recent band structure calculations [4.73]. The plot of the absorption coefficient as square and square root versus photon energy, presented in Fig. 4.20, yields a direct gap value of 0.87 eV. This value agrees well with the theoretical prediction of the interband transition of 0.825 eV at the Y point (see Fig. 4.8), which has an appreciable oscillator strength because of the considerable weight of silicon and iron p states at the valence band maximum. In the lower energy region the situation is more complicated. The linear segments of the $(\alpha^*E)^{1/2}$ versus E plots are indicative of indirect transitions with a phonon absorption and phonon emission. An indirect edge energy of 0.765 eV with energy of the participating phonon of 35 meV is derived. The latter value corresponds very well to the results of IR studies on phonon spectra of epitaxial layers. The theory predicts indirect transitions between 0.738 and 0.750 eV starting from the valence band maximum at the Λ point. Besides, the theory also yields direct transitions with an energy of 0.742 eV occurring at the Λ point having much lower oscillator strength than those at the Y point. Thus, some deviations from the linear approximation in the energy region below 0.8 eV might be due to a superposition of both indirect and direct low-oscillator strength transitions.

Experimental evidence for the existence of an indirect absorption edge is also found in [4.83, 4.84, 4.88, 4.117] with indirect transition energies $E_g^{ind} + hv_p = 0.78$ eV [4.83, 4.84] and 0.79 eV [4.88] at room temperature. This agrees fairly well with the above result considering a phonon energy hv_p of about 35 meV. At 77 K the indirect transition was found to be 0.86 eV in [4.83], and 0.83 eV in [4.117]. In all optical investigations of β-FeSi$_2$ the direct interband edge is clearly identified irrespective of the differences in the structural quality of the films, which seems to affect the direct gap value only a little.

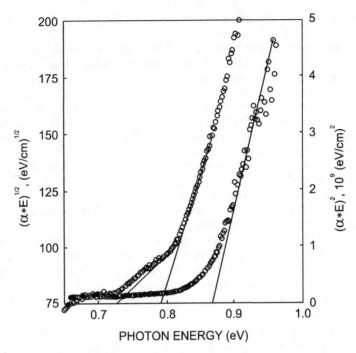

Fig. 4.20. Photon energy dependence of the absorption coefficient in β-FeSi$_2$ in coordinates appropriate for estimation direct and indirect optical transitions [4.73]

The temperature dependence of the direct edge measured between 4 and 300 K [4.100, 4.117] shows two contributions to the variation of the interband gap with temperature. These are thermal dilation of the lattice and electron–phonon interaction. The first can usually be neglected, therefore the second one is dominant. According to the thermodynamical model [4.88, 4.100] it is described by

$$E_g(T) = E_g(0) + S(h\nu)(\coth(<h\nu>/2k_BT) - 1),$$ (4.18)

where $E_g(0)$ is the band gap at the absolute zero temperature, $<h\nu>$ is the average phonon energy, and S is a dimensionless coupling parameter. The experimental data could be fitted by $E_g(0) = 0.901$ eV, $<h\nu> = 34.5$ meV, and $S = 2.15$ [4.100]. The zero-gap value agrees very well with that obtained in [4.117], however, the $<h\nu>$ and S values differ strongly (71 meV and 6.22, respectively). At 10 K an experimental direct energy gap of 0.93 eV has been determined from transmission measurements on epitaxial β-FeSi$_2$ films [4.161]. The temperature coefficient of E_g at a constant pressure is estimated to be 3.23×10^{-4} eV/K around room temperature. This value is in good correspondence with that obtained from temperature-dependent photoconductivity studies on β-FeSi$_2$ single crystals [4.161], while previous investigations [4.117] yielded a considerably higher value of 1.2 meV/K.

The first evidence of excitonic transitions in optical absorption spectra of β-FeSi$_2$ has been revealed recently from low-temperature transmission measurements on epitaxial films of high structural perfection grown on Si(100) substrates [4.161]. The related photon energy dependence of the absorption coefficient is shown in Fig. 4.21.

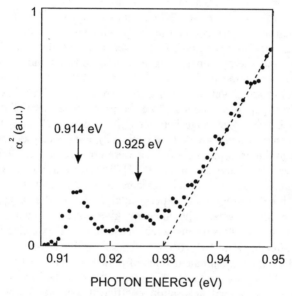

Fig. 4.21. Square of the absorption coefficient versus photon energy calculated from transmission measurements on β-FeSi$_2$/Si(100) epitaxial structures at 10 K [4.161]

The two peaks resolved below 0.93 have been ascribed to the ground and first excited state of a Wannier–Mott exciton from which a binding energy of the exciton of 16 meV is determined. Estimated values of the excitonic Rydberg range are between 4 and 21 meV, taking into account the presently known parameter range of reduced effective masses and static dielectric constants in this material.

It is frequently observed that there is considerable absorption in the below-gap region of the films [4.77, 4.100, 4.107, 4.155], even in the case of epitaxial films of high structural quality [4.73]. In [4.100] it was pointed out that band tail states and impurity states give rise to absorption below the fundamental edge in ion-beam synthesized films. The band tails could be fitted by an exponential Urbach edge. Both static and dynamic disorders contribute to the Urbach tail. At room temperature structural disorder was found to make dominant contributions. A critical study of the influence of defect-induced absorption on band gap determinations of β-FeSi$_2$ films was performed in [4.107]. High defect densities of the order of 10^{19} cm^{-3} and above, which are often met in polycrystalline thin films, can produce band tails with an apparent energy width close to the energy difference between indirect and direct energy gaps. It is suggested that the existence of band tails in the density of states may prevent an assignment of the indirect band gap nature in polycrystalline β-FeSi$_2$ and may account for the range of energy gap values reported in the literature.

An influence of the third component on the position of the direct band gap in β-FeSi$_2$ was studied on buried layers of $(Fe_{1-x}Co_x)Si_2$ synthesized by subsequent implantation of iron and cobalt into Si(100) [4.162]. The orthorhombic structure of β-FeSi$_2$ remains stable up to the cobalt content corresponding to $x = 0.2$. With an increase of the cobalt fraction from zero the energy of the direct gap continuously decreases down to 0.7 eV for $x = 0.15$. Simultaneously, the absorption background increases considerably with cobalt addition because of increasing fine-grained film structure. This makes it impossible to obtain a reliable determination of the energy gap for higher cobalt contents and does not allow indirect transitions to be traced.

All theoretical investigations of the electronic structure of **CrSi$_2$** have predicted it to be an indirect-gap semiconductor (see the previous section). The calculated energies of the indirect gap range between 0.21 and 0.38 eV and those of the direct gap between 0.37 and 0.48 eV (see Table 4.2). For polycrystalline CrSi$_2$ thin films two steps in the absorption coefficient were observed experimentally in the range of the lowest interband transitions [4.32]. The derived indirect gap value at room temperature of 0.35 eV is in fair agreement with the band structure calculations. In the high-energy region the experimental data plotted as a square root of the absorption coefficient versus photon energy give a direct gap value of 0.67 eV. For these polycrystalline films a considerable absorption tail extended into the gap region, which amounted to 10^3–10^4 cm^{-1}. These results were later questioned [4.147] because all theoretical calculations of optical spectra for CrSi$_2$ suggested that up to about 0.7 eV, i.e. in the energy range covering the lowest indirect and direct transitions, the optical absorption coefficient was very low due to the predominant d character of the states on both sides of the gap. Therefore, by comparing the theoretically calculated absorption coefficient for CrSi$_2$ with the experimental absorption spectrum in the photon energy range of the lowest interband transitions, it was concluded [4.147] that the experimentally observed spectral features had to be attributed not to the lowest indirect and direct energy gaps, but to higher-energy interband transitions.

A quite different conclusion on the nature of the lowest gap is drawn on the basis of optical investigations in the energy range from 0.09 to 1.1 eV of high-quality epitaxial CrSi$_2$ films of A-type grown on Si(111) substrates [4.163]. The background absorption in these films was considerably lower than that in the polycrystalline as a result of strongly reduced free carrier concentration in the films and negligible defect-related absorption. The absorption coefficient versus photon energy is plotted in Fig. 4.22. The data fitted by the straight line in the plot of the square of the absorption coefficient versus photon energy yield a direct character of the involved interband transitions. The lowest direct-gap energy obtained is 0.37 eV, which belongs to the range of theoretically determined direct energy gaps. Two additional direct interband gaps are derived with energies of 0.83 and 0.93 eV. The former is M$_1$-type and the latter is M$_0$-type critical points of the joint density of interband states.

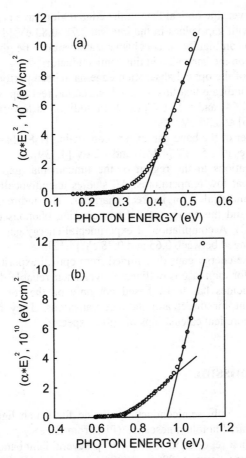

Fig. 4.22. Photon energy dependence of the absorption coefficient in A-type epitaxial CrSi$_2$ films for two energy ranges: 0.1–0.6 eV (a) and 0.6–1.1 eV (b) [4.163]

It was concluded that the previous investigations [4.32, 4.43] were hampered by a high background absorption connected with free carriers and lattice defects, which prevented the authors from a reliable identification of the interband gap. However, it should be added that: (i) in the high-quality epitaxial films a small absorption background also exists, and (ii) the oscillator strength of the indirect transition is very small due to the above-mentioned factors. Therefore, the indirect edge preceding the direct one might have escaped detection, being masked by the absorption tail. This means that the results for A-type epitaxial CrSi$_2$ films [4.163] may not necessarily contradict the conclusion derived from theoretical band structure calculations.

Gap nature optical investigations of the other semiconducting silicides are relatively scarce. The data obtained for **MnSi$_{2-x}$** [4.58, 4.148] seem to indicate this material to be an indirect-gap semiconductor with a fundamental gap energy of 0.45–0.47 eV and a relatively high energy of the participating phonons of

60–70 meV. However, the optical data in the edge region have been interpreted by direct transitions with gap values in the interval 0.78–0.83 eV [4.59] and 0.66 eV [4.164]. Due to this ambiguity no conclusive statement can be given at present on the gap nature and on the gap value in this semiconductor.

Several studies of the optical absorption edge have been performed for $ReSi_{1.75}$ [4.64, 4.66]. They indicate that this material has an indirect gap with reported gap energies of 0.12 [4.64] and 0.15 eV [4.67]. In addition, a lowest direct edge has been found in [4.66] at 0.36 eV.

A direct character of the band gap in semiconducting Ir_3Si_5 has been concluded with the gap energies of 1.57 eV [4.148] and 1.2 eV [4.139].

Optical investigations in the region of the fundamental gap of Mg_2Si were mainly carried out at low temperatures. All studies unequivocally stated that the band gap of this material is of indirect character. The indirect gap energy of 0.623 eV at 90 K and the energy of the participating phonons of 30 meV have been obtained [4.23]. A compilation of experimental energy gaps extrapolated to 0 K gives values ranged between 0.65 and 0.78 eV [4.12].

In conclusion, the energy gaps determined from optical experiments have to be critically assessed for the reasons outlined above. Their reliable determination for semiconducting silicides has to be based not only on the conventional energy power laws relevant for direct and indirect transitions. They have also to be substantiated by theoretical calculations of optical spectra.

4.4   Light Emission

Iron disilicide is the only semiconducting silicide for which light emission has been observed so far. Photoluminescence (PL) from β-FeSi$_2$ polycrystalline thin films at 1.8 K was first reported in 1990 [4.102]. Several faint bands were found in the wavelength range from 1400 to 1700 nm whose intensity and energetic positions varied with the growth temperature of the films. Subsequently many efforts have been undertaken to produce this material for application in optoelectronic devices. However, light emission from β-FeSi$_2$ was especially found on the samples prepared by iron implantation into silicon [4.84, 4.91, 4.98, 4.165, 4.166]. All efforts to find a significant light emission from epitaxial layers of high structural perfection or from single crystals of β-FeSi$_2$ grown by chemical vapor transport under optical or electron-beam excitation failed [4.167]. Thus, it seems that the light emission observed for β-FeSi$_2$ is of extrinsic rather than of intrinsic origin. Because of the high defect density existing in ion beam synthesized β-FeSi$_2$ it cannot be excluded that some not yet identified defects in β-FeSi$_2$ or defect states at the β-FeSi$_2$/Si interface might be responsible for the light emission.

A PL signal was observed at 2 K with a maximum at 0.8 eV (1.55 μm) on buried Si/β-FeSi$_2$/Si heterostructures which were produced by high-dose implantation of iron into Si(111) substrates and subsequent two-step annealing [4.84]. However, the structures formed in optimized conditions had a decreased light intensity. Therefore it has been suggested that the observed PL might originate from some structural defects in the silicon substrate, in particular from dislocations giving the D1 line at 0.81 eV. A similar PL signal was observed at 5 K in samples fabricated by high-dose iron implantation followed by halogen lamp annealing [4.98]. Its intensity and linewidth increased with annealing time and annealing temperature (up to 950°C).

In [4.91] ion beam synthesis of β-FeSi$_2$ layers has been performed by a triple-energy ion implantation into Si(100) and Si(111) substrates followed by two-step annealing. The PL spectra recorded at 2 K show several emission peaks, with two dominant ones around 0.836–0.837 eV and 0.80 eV. Different interpretations are given for the origin of these peaks. The low-energy one is assigned to dislocation loops, respectively bound-exciton emission associated with residual phosphorus impurities in the silicon substrate or associated with trap(acceptor)-related emission from β-FeSi$_2$. The faint PL emission observed at 0.837 eV for some samples has been attributed by the authors to intrinsic free-to-free or free-to-bound transitions in β-FeSi$_2$.

The PL spectra recorded at 2 K for polycrystalline β-FeSi$_2$ layers formed by ion beam synthesis followed by two- and three-step annealing contain several emission peaks [4.96]. Their energies differ slightly from each other just as forbidden gap values depending on the annealing temperature of the samples. The main ones are located at 0.79, 0.80, 0.81 and 0.87 eV. The low-energy peaks are assumed to belong to intrinsic optical radiative transitions in β-FeSi$_2$.

Light emission at 1.5 μm from β-FeSi$_2$ layers fabricated by ion-beam synthesis followed by high-temperature annealing was studied in detail [4.166]. The authors tried to find arguments against the suggestion of [4.84] that this emission might be a D1 dislocation-related emission from silicon. For this purpose they introduced silicon filters into the excitation path and performed experiments using excitation energies below and above the silicon band gap. Additionally they carried out experiments on samples in which dislocations had been deliberately introduced by silicon implantation. Finally, the emission at 1.5 μm was concluded to relate to the silicide band edge. Later, the authors succeeded in realization of a light-emitting electroluminescent device operating at 1.5 μm in which β-FeSi$_2$ is incorporated into a conventional silicon p-n junction [4.165]. The p-n junction was formed by MBE and consisted of a boron doped p-region grown on an antimony doped n-region. These silicon epitaxial layers were grown on an n-type Si(100) substrate. The dose of iron implantation was chosen to produce single-crystalline β-FeSi$_2$ precipitates upon annealing (at 900°C for 18 h). The implant energy 950 keV was used to place the peak of the precipitate distribution above the depletion region of the silicon p-n junction. The diodes exhibited good current–voltage characteristics, which demonstrated that the diode integrity had been retained during the processing steps. Fig. 4.23a shows the electroluminescence spectrum measured at low temperature. It has a peak at 1.54 μm characterized by the full width at half height of 50 meV. The intensity of the electroluminescence is temperature-dependent, as is illustrated in Fig. 4.23b.

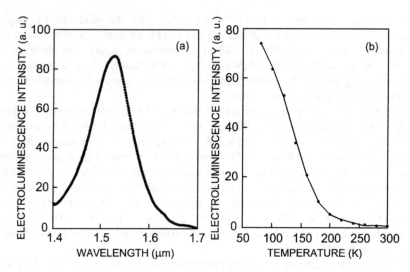

Fig. 4.23. Spectrum of electroluminescence measured at forward current of 15 mA at 80 K (a) and integrated electroluminescence intensity as a function of the measurement temperature (b) for silicon p-n junction incorporating ion-beam synthesized β-FeSi$_2$ [4.165] © *Nature*

The intensity reduces with rising temperature, but the light signal could still be seen clearly at room temperature. The shift in the peak energy with temperature between 80 and 300 K was 50 meV and followed the band gap shift determined from optical absorption measurements.

From the theoretical point of view there is a certain hope to use semiconducting silicides as light-emitting materials in optoelectronics. The calculations performed for strained β-FeSi$_2$ lattices show the possibility of an indirect-to-direct band gap transformation providing direct gaps at the Γ [4.80], Y [4.94] or Λ [4.82] points. This is very important for an effective light emission, but one should also remember that sizable oscillator strength is possible only in the case of different orbital compositions of the valence band maximum and the conduction band minimum.

Concerning other semiconducting silicides, many of them seem to have an indirect energy gap. Among silicides of group VIII metals only Ru$_2$Si$_3$ and Os$_2$Si$_3$ are well pronounced direct-gap semiconductors. However, no experimental investigations of luminescent properties of these compounds have been performed yet. According to the theoretical data of orbital composition of the band structure extremum points, the oscillator strength of direct transitions is expected not to be very high. This might relate to the fact that both valence band maximum and conduction band minimum are mainly formed by d states of the metal with only small admixture of metal and silicon p states. Nevertheless, one can expect quite favorable (for light emission) changes in the band gap structure of semiconducting silicides, when they are fabricated as nanodimensional or strained structures imbedded into silicon epitaxial structures.

4.5 Infrared Optical Response and Raman Scattering

Infrared optical spectroscopy and Raman scattering are powerful experimental techniques for investigation of lattice vibrations and related processes in semiconductors [4.168]. In a two-atomic crystal lattice composed of atoms of masses M_1 and M_2 the atoms can oscillate around their equilibrium positions with an amplitude which is small compared to the internuclear distance. Under this assumption the potential energy of this system depends quadratically on the displacements of atoms and the system behaves as if the atoms are coupled by harmonic springs. The equation of motion for this system yields normal modes with the dispersion relation $\omega(k)$, which are appropriate combinations of the individual displacements of atoms. In terms of these normal modes the total Hamiltonian is the sum of individual uncoupled single Hamiltonians.

For a crystal with z atoms per unit cell for each k-value there are $3z$ normal modes. These modes are either optical, where atoms of opposite charge move out of phase and produce an oscillating dipole moment, or acoustic with constant speed of motion for $k \rightarrow 0$. There are three acoustic and $3z - 3$ optical modes. Each mode can be either longitudinal (motion along an atomic chain) or transverse (motion perpendicular to an atomic chain). The vibrational frequency of a transverse optical (TO) mode in the above simple model is

$$\omega_{TO} = (2C/m)^{1/2}, \tag{4.19}$$

whose frequency lies in the infrared region, C is the force constant of the harmonic spring, $m = M_1 M_2 / (M_1 + M_2)$ is the reduced mass. Absorption of infrared radiation occurs if the frequency of the incoming light wave $\omega = \omega_{TO}$. Although the normal modes can be treated within the classical model, they can be also regarded as quantum-mechanical quantities, i.e. phonons which are quanta of mechanical lattice vibration waves. As the wave vector of the electromagnetic radiation is small, infrared techniques measure the $k = 0$ phonons. Since the harmonic oscillations in the crystal lattice are damped due to anharmonic coupling to other excitations in the crystal, the solution of the equation of motion of damped (damping constant γ^*) simple N harmonic oscillators per volume V of dynamic charge Q leads to the dielectric response

$$\varepsilon = \varepsilon_\infty + (4\pi Q^2 N/mV)[1/\{(\omega_{TO}^2 - \omega^2) - i\gamma^*\omega\}], \tag{4.20}$$

where ε_∞ summarizes the dielectric response of all excitations which occur at much higher frequencies. This expression can be easily extended to the case of superposition of several TO modes. The Lyddane–Sachs–Teller equation provides a useful relationship between the transverse ω_{TO} and longitudinal ω_{LO} frequencies

$$\varepsilon/\varepsilon_\infty = [(\omega_{LO}^2 - \omega^2)/(\omega_{TO}^2 - \omega^2)]. \tag{4.21}$$

This expression shows that the longitudinal frequency is given by the zero of ε and the transverse frequency by the pole of ε. As for (4.19), this expression can also be extended to the case of several oscillations in crystals with symmetry lower than cubic.

The standard technique in infrared optics is Fourier transform infrared (FTIR) spectroscopy. It is superior to infrared grating spectroscopy, which does not make the most effective use of the limited power of the light sources. The basic process in FTIR spectroscopy is light interference. FTIR spectroscopy replaces the dispersive grating by a Michelson interferometer.

If light is incident on a semiconductor, a small part of it interacts inelastically with the phonon modes. The scattered light contains frequencies, which are shifted from that of the incident light. This happens due to a phonon absorption (anti-Stokes process) or phonon emission (Stokes process). The above described process is called Raman scattering. As in infrared spectroscopy, Raman scattering probes the $k = 0$ phonons only. The Raman scattering is theoretically understood by a phonon-induced change in lattice polarizability, which is a change in dielectric susceptibility.

The intensity of the scattered light is very low as it is proportional to the square of the small polarization. In crystals having a center of symmetry infrared-active modes cannot be simultaneously Raman-active. In non-centrosymmetric crystals the lattice modes can be simultaneously Raman- and infrared-active. Because of the weakness of the scattered light intensity in Raman measurements powerful light sources (Ar^+, Kr^+, or tunable dye lasers) operating in the visible range are required. Since the Raman shift is small, the scattered light remains in the visible range and standard detectors and grating monochromators can be used. Operation of Raman systems is also possible in the infrared by using Nd:YAG lasers and incorporating highly sensitive Fourier methods.

The practical effectiveness of the infrared optical spectroscopy and Raman scattering in the study of semiconducting silicides has been demonstrated with $CrSi_2$, β-$FeSi_2$, $MnSi_{1.73}$, $ReSi_{1.75}$, $OsSi_2$, and Ir_3Si_5 Among these, most efforts on understanding of the phonon spectra have been undertaken for β-$FeSi_2$.

Systematic investigations of β-$FeSi_2$ epitaxial films grown by MBE on Si(111) and Si(100) substrates have been performed by using FTIR and Raman measurements [4.169]. In particular, the dependence of the phonon spectra on substrate orientation, film thickness, and growth and annealing temperature has been investigated. Preliminary studies of the phonon spectra on small β-$FeSi_2$ single crystals grown by a chemical vapor transport method displayed a sizeable anisotropy of the infrared spectra with minor effects in the Raman spectra [4.160]. The first calculations of the vibrational spectra of β-$FeSi_2$ by molecular dynamics simulations with a tight-binding potential provided a quantitative confirmation of the anisotropic behavior of the infrared measurements [4.160]. Compared to β-$FeSi_2$ the knowledge gained on the infrared properties for the other semiconducting silicides is quite scarce. Investigations have been made only on polycrystalline samples. $CrSi_2$ has been investigated in some detail, both experimentally and theoretically [4.170]. For semiconducting manganese, rhenium, and iridium silicides the measured infrared spectra [4.171] have a rather complex structure, which awaits a theoretical interpretation.

The first infrared optical measurements for β-FeSi$_2$ have been carried out on sintered ceramics [4.172], while later IR and Raman investigations were performed on polycrystalline β-FeSi$_2$ layers [4.173]. The polycrystalline layers and ceramics exhibited a characteristic five-peak infrared pattern in reflectance and absorbance spectra, as is illustrated in Fig. 4.24.

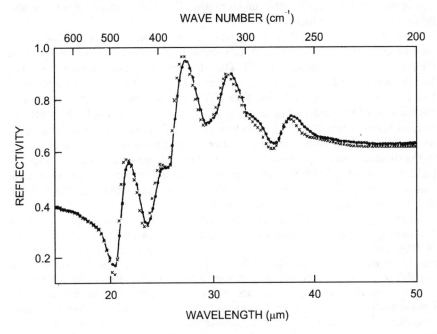

Fig. 4.24. Lattice mode reflectance spectra of β-FeSi$_2$ ceramics at 300 ($\bullet$) and 85 K ($\times$) [4.172]

A tentative interpretation of the spectra has been performed in terms of the factor-group analysis of the four atomic sites Fe1, Fe2, Si1, Si2 in the orthorhombic lattice [4.173]. This gives totally 36 expected Raman-active and 28 infrared-active modes. In order to simplify the analysis, these modes have been differentiated into internal modes where the molecules (or crystallographic groups) move as a unit. Meanwhile, in the internal modes the atoms move differently inside the molecules. Internal modes would lead to higher-energy vibrations, while the external modes occur at lower energies as they correspond to the motion of the heavier molecule as a whole.

In order to distinguish between internal and external modes the atoms have been grouped into molecules as Fe1+2Si2 or Fe2+Si1+Si2 and the symmetry of these units has been examined. This leads to 24 Raman-active and 18 infrared-active external lattice modes and 10 infrared-active and 12 Raman-active internal modes. From the similarity of the two iron sites it is further inferred that the internal modes of the iron center will display an overlap. Thus, the five experimentally observed IR peaks have been ascribed to the five IR-active close-lying modes.

As the crystallographic cell of the orthorhombic lattice has a symmetry center group theory forbids the IR modes to be simultaneously Raman-active and conversely. In the Raman spectra 14 lines have been observed, which are considered to be mainly attributable to the 12 Raman-active internal modes predicted by the group theory.

Measurements of the IR transmittance and reflectance on β-FeSi$_2$ films with increasing crystallite size revealed an increasing complexity of the spectra with the growing layer perfection [4.171]. The films grown epitaxially on Si(111) and Si(100) substrates have even richer phonon spectra [4.169, 4.171] than the five-line pattern of the polycrystalline films. This can be obviously understood as a convolution of the contribution of many individual grains having fairly small size and, probably, a low degree of crystalline order and perfection. Epitaxial samples have different absorbance spectra for different orientation of the film with respect to the substrate, which probably have their origin in stress effects at the interface [4.169]. The use of β-FeSi$_2$ single crystals allowed the observation of a clear anisotropy in the IR measurements [4.160], which is characterized by the different positions and relative intensities of the peaks.

Lattice dynamics in β-FeSi$_2$ was studied by estimating suitable autocorrelation functions during the molecular dynamics evolution of a well equilibrated sample [4.160]. The basic idea underlying this approach is that each atomic displacement can be represented by a superposition of the normal modes. Therefore it is possible to obtain the vibrational density of states as well as the absorption and the Raman scattering spectra by Fourier transforming from time to frequency the autocorrelation functions of the atomic velocity, the derivative of the cell dipole moment, and polarizability. The simulations have been performed for a nonprimitive orthorhombic unit cell containing 48 atoms with periodic boundary conditions. Strong IR modes are expected around 300 to 350 cm^{-1}. They are due to a counterphase motion of iron and silicon atoms. Similarly, the main Raman mode is expected around 250–270 cm^{-1}. Here only iron is moving. The features above 400 cm^{-1} involve only silicon displacements. The sharp structures between 260 and 280 cm^{-1} are produced by iron displacements.

A comparison between the calculated and measured absorbance spectra is shown in Fig. 4.25. For the light polarization along the a axis the agreement is quite satisfactory for the peaks located at 310, 340–350, and 390 cm^{-1}, while the sharp peak at 266 cm^{-1} is theoretically underestimated by 22 cm^{-1}. For the b polarization the anisotropic trend is quite well reproduced. By moving from a to b polarization the peaks above and below 400 cm^{-1} are enhanced, the most intensive feature is shifted from 310 to 303 cm^{-1}. The central peak in the triplet structure around 340 cm^{-1} as well as the sharp shoulder at 290 cm^{-1} are missed. This might be due to the high sensitivity of the spectra with respect to the actual atomic configuration and that the equilibrium situation in the experiment was possibly slightly different from that used in the theoretical simulation.

The experimental Raman spectra taken on β-FeSi$_2$ single crystals for two light polarizations are presented in Fig. 4.26. They appear to be quite similar to each other. The most important differences concern the relative intensities of the peaks.

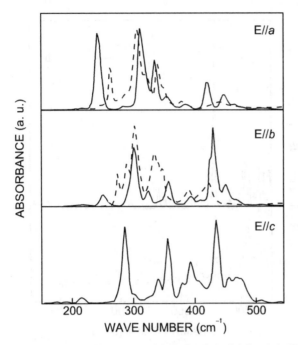

Fig. 4.25. Experimental (*dashed lines*) and calculated (*solid lines*) infrared absorbance spectra of β-FeSi$_2$ single crystals for a, b, and c light polarization [4.160]

A series of structures is seen between 150 and 200 cm^{-1} in both spectra with slightly different intensities for both polarizations. The highest-intensity peak occurs at about 250 cm^{-1} for the polarization parallel to the b axis exhibiting a shoulder at slightly higher frequencies, but it is no longer the largest one for the polarization perpendicular to this axis. Above 340 cm^{-1} only some small peaks can be discerned. The theoretical simulation of the Raman spectra also showed main features at 200 and 250 cm^{-1} which are rather permanent for the different polarizations. In agreement with the experiment, no structures appear for wave numbers larger than 340 cm^{-1}.

The IR reflectance of polycrystals of the isostructural silicide OsSi$_2$ bear great similarities to those of polycrystalline β-FeSi$_2$ [4.134]. The spectra exhibit a six-line pattern. The main IR-active lattice modes are situated in the same wave number range for both compounds.

Another picture is typical for CrSi$_2$ [4.170]. Far-infrared reflectance and absorbance spectra of polycrystalline films of this semiconductor deposited on oxidized silicon substrates are shown in Fig. 4.27. The higher-frequency modes at 600 and 450 cm^{-1} are due to silicon and SiO$_2$. There are four main peaks at 351, 295, 250, and 228 cm^{-1} and a shoulder at 375 cm^{-1} in the spectra. The performed theoretical analysis of the lattice modes shows that in the point group D$_2$, which is isomorphous with the space group of CrSi$_2$, there are four one-dimensional representations A$_1$, A$_2$, B$_1$, B$_2$ and two two-dimensional representations E$_1$ and E$_2$.

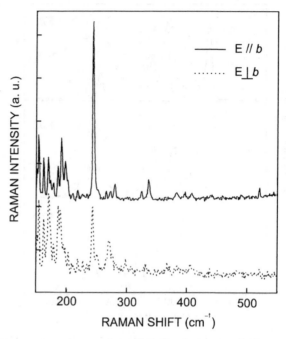

Fig. 4.26. Raman spectra measured on β-FeSi₂ single crystals for two different light polarizations [4.160]

Those modes, which transform according to the representations E_2 and A_2 are infrared-active, whereas those modes transforming according to A_1, E_1 and E_2 are Raman-active.

Among the normal modes of $CrSi_2$ there are two vibrations, which transform according to A_2 and four, which transform according to E_1. These are also simultaneously Raman-active. Altogether, there are six infrared-active modes. A lattice-dynamical calculation of the relative intensities of the infrared-active modes gave the result that one of the six vibrations, namely the lowest-frequency A_1 mode, has only a very small dipole moment. The IR reflection spectra measured in [4.41] were fitted with a dielectric function consisting of the high-frequency contribution of all valence electrons, six phonon oscillator terms, and a free carrier Drude term. From this fitting six infrared-active phonon modes were determined at 231, 253, 295, 342, 362, and 381 cm⁻¹. The underlined modes had the highest intensity. The intensity of the mode at 342 cm⁻¹ had more than an order of magnitude weaker intensity.

Raman measurements performed in [4.41] and [4.160] demonstrate that strong structures at about 295–310 cm⁻¹ dominate the spectra. This might correspond to the high-intensity infrared-active lattice mode at 295 cm⁻¹. Further bands appear at 350 cm⁻¹ and at about 400–410 cm⁻¹.

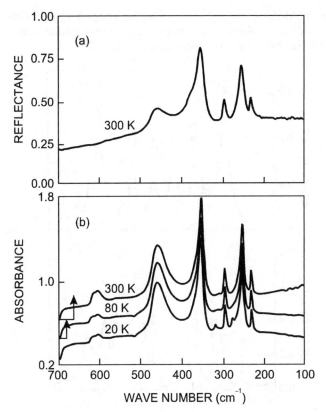

Fig. 4.27. Reflectance (a) and absorbance (b) spectra of 200 nm thick CrSi$_2$ film on oxidized silicon [4.170]

Later reexamined infrared and Raman spectra of CrSi$_2$ single crystals and polycrystalline films [4.174] demonstrated six IR-active modes. They have been ascribed to $2A_1 + 4E_2$ modes. The observed wave numbers (symmetries) are: 229 (E_1), <u>252</u> (A_2), 290 (E_1), 297 (E_1), <u>355</u> (E_1), 382 cm^{-1} (A_2). From the nine Raman-active modes predicted by the group theory, eight have been observed in Raman measurements. The A_1 mode involves only pure silicon motions. The A_2, E_1, and E_2 modes are due to displacements of silicon and chromium atoms. However, only one of the four E_1 and E_2 modes has a predominant contribution of chromium atom motion, whereas the three others involve mainly silicon atom motions.

From IR reflectance measurements on semiconducting rhenium silicide six phonon bands could be identified with energies of 35, 41, 44, 49, 58, and 68 meV (282, 331, 355, 395, 468, 548 cm^{-1}, respectively) [4.63], the one of highest intensity being located at 41 meV. The theoretical analysis assuming this silicide to have the orthorhombic lattice gives also six infrared-active bands. IR measurements on the semiconducting rhenium silicide performed in [4.65] also show particular features in the reflectance and transmittance spectra in between 400 and 200 cm^{-1} with the one of main intensity occurring at about 300 cm^{-1}.

IR spectra of $MnSi_{1.73}$ and Ir_3Si_5 are of higher complexity than those of $CrSi_2$ and β-$FeSi_2$. Typical absorbance spectra of polycrystalline semiconducting $MnSi_{1.73}$ and Ir_3Si_5 thin films are demonstrated in Fig. 4.28.

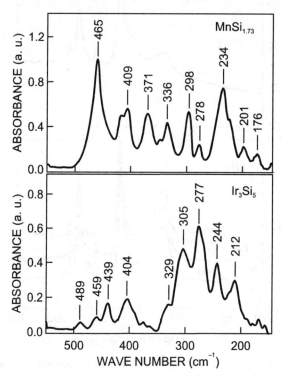

Fig. 4.28. IR absorbance spectra of $MnSi_{1.73}$ and Ir_3Si_5 thin films on Si(111) substrates [4.171]

The measurements on higher manganese silicide crystals grown from the melt gave similar spectra with dominant bands at 250 and 350 cm^{-1} [4.164]. The spectra are consistent with the seven-oscillator model. This is not surprising because of the large number of atoms per unit cell in these compounds. Due to the complicated structure of these silicides a factor-group analysis predicting the number of IR-active modes is difficult. More promising seems to be a lattice-dynamical calculation along the lines outlined in [4.160] for β-$FeSi_2$. From IR spectra fittings by superposition of phonon oscillators rather high static dielectric constants between 23.5 and 12.0 for β-$FeSi_2$ [4.88, 4.148, 4.172], 43 and 30 for $CrSi_2$ [4.148], 70 for $ReSi_{1.75}$ [4.63], and 34 for $MnSi_{2-x}$ [4.164] have been obtained.

They are comparable to those of narrow-gap semiconductors. Direct measurements give values of 31 for β-$FeSi_2$ [4.114] and 28 for $CrSi_2$ [4.32]. A much lower dielectric constant of 19 has been obtained by direct measurements on epitaxial $CrSi_2$ layers [4.163]. For Mg_2Si a static dielectric constant of 12.9 has been reported in [4.145].

The fact that the lattice vibrations of all semiconducting silicides studied are observed in the same frequency range is an indication that the phonon spectra are mostly influenced by bonding properties (force constants), which are very similar for these compounds. In all silicides studied the shortest metal–silicon distances drop below those for metals and silicon. Greater mass differences, for instance between iridium and manganese, obviously do not play an essential role.

Band structure calculations for semiconducting transition metal silicides imply that there is only a small charge transfer between silicon and the metal, i.e. ionicity contributes little to bond formation. Nevertheless, the IR spectra of all semiconducting silicides show high absorbance in the range of lattice vibrations, which is typical of ionic compounds. One explanation which has been put forward in [4.170] is the occurrence of dynamical charges arising when the atoms vibrate and relating to their large polarizability.

4.6   Concluding Remarks

The actual basic knowledge gained on the electronic structure and related optical properties of semiconducting silicides is rather valuable. Narrow-gap and wide-gap semiconductors with an indirect or direct type of gap can be found among the silicides investigated. This forms an appropriate background for attempts of their practical application in silicon integrated microelectronics, optoelectronics and thermoelectrics. Nevertheless, their study is far from being completed. The problems requiring a solution are discussed in this section.

There is still uncertainty regarding an exact description of the band gaps in the semiconducting silicides. Even for the best investigated Mg_2Si, $CrSi_2$, β-$FeSi_2$ one can find wide range scattering in the energy gaps both theoretically predicted and experimentally observed. Also, the gap nature and its origin need more detailed understanding. As for the theoretical simulation, the problem is mainly inherent in the local-density approximation typically used in the calculations. In fact, *ab initio* calculations particularly performed by methods which are free from any shape approximation for the charge or potential, such as FLAPW, full potential LMTO or PP, predict the electronic band structure of silicides quite reliably. The energy gaps are described much better than those in group IV or $A^{III}B^{V}$ semiconductors. The energy gap underestimation appears to be larger for semiconductors with a different symmetry of the valence-band and the conduction-band states, which are *s* and *p*, respectively in the group IV or $A^{III}B^{V}$ semiconductors. In transition-metal silicides the bands on both sides of the gap are predominantly formed by the states of the same symmetry (typically *d* symmetry). This reduces the underestimation of the calculated energy gaps and provides their acceptable correlation with the experimental data.

However, some semiconducting compounds, such as rhenium, ruthenium and osmium silicides have different symmetry of the valence-band and the conduction-band states (mainly *p* and *d* symmetry). One of the solutions to this problem within the density functional theory can be the corrections by the use of Hedin's

GW approximation [4.175] in which an energy-dependent, nonlocal self-energy replaces the LDA for the exchange-correlation energy, and true quasiparticle eigenvalues are then found by correcting the LDA eigenvalues through perturbation theory. As shown in [4.176], band structure calculations of group IV or $A^{III}B^V$ semiconductors, for which LDA underestimation is 2 times and more, such an approach has an accuracy of 0.1 eV. Nevertheless, the proposed theoretical approaches should go beyond the local-density approximation and take into account nonlocality effects. Generalized gradient approximations [4.177, 4.178] could become an appropriate alternative for this. As for experiments, one should take care of the material purity and lattice perfection. Scattering in the experimental results obtained can be explained by the influence of these factors. A systematic study of the impurity effect on electronic and optical properties of semiconducting silicides could resolve the problem.

The luminescence of semiconducting silicides promises impressive progress in silicon-based optoelectronics, but it needs deeper understanding. At the moment experimental observations of this phenomenon are limited with quasidirect β-FeSi$_2$. The luminescence from this disilicide is preferentially registered in layers which have been synthesized by iron implantation into silicon substrates. It might be suggested that it is mainly defect-related. However, there is evidence that an intrinsic signal also exists and it does not completely overlap the defect-induced one. From the theory it is known that the direct gap at the Y point can be made the lowest by deformation in the lattice of β-FeSi$_2$. Thus, small imbedded silicon precipitates of this silicide may exhibit electronic properties differing from those of the bulk material. This explanation, as well as the role of quantum confinement and interface effects, has to be tested.

Luminescence properties of other theoretically predicted direct-gap semiconducting silicides, Ru$_2$Si$_3$ and Os$_2$Si$_3$, should also be experimentally investigated. Remembering their identical crystalline lattices and relevant isostructural quasidirect germanides Ru$_2$Ge$_3$ and Os$_2$Ge$_3$ one can expect unique opportunities for energy gap engineering in ternaries and quaternaries synthesized within this family. The oscillator strength might also be enhanced by an appropriate choice of the atoms substituting for the main components in these compounds.

The theoretical understanding of the infrared optical spectra of semiconducting transition metal silicides is only at the beginning. A large number of atoms per unit cell gives problems in the simulation of the vibrational modes, which cannot be accomplished by the usual methods of lattice dynamics. Therefore, alternative strategies have to be developed for the calculation of the infrared and Raman spectra for large structures.

Practical applications of semiconducting silicides suppose an existence of the possibility to modify their fundamental electronic properties by technological methods. In semiconductors this is known to be achieved usually by impurity doping, multicomponent compound (ternaries and quaternaries) formation, or scaling down to nanodimensions. The problems related to the first two approaches have already been mentioned. As for the use of changes in electronic properties of solids under quantum confinement in nanosize structures, it is quite new in physics and the technology of semiconductors. Theoretical and experimental results already obtained for silicon, germanium, $A^{III}B^V$ and $A^{II}B^{VI}$ compounds show dramatic changes of their bulk properties when nanostructures from these materials are fabricated. A transformation from an indirect to direct character of the gap can take place. For semiconducting silicides such regularities remain unknown. Thus, theoretical and experimental investigation of electronic and optical properties of semiconducting silicide-based quantum dots, quantum wires and quantum films are one of the hot problems. Moreover, controlled lattice distortions in strained nanostructures buried into the crystalline silicon matrix can provide an additional, very effective instrument in property regulation.

Progress in the resolution of the above mentioned problems will promote semiconducting silicides to effective applications in novel generations of electronic, optoelectronic, and thermoelectric devices.

5 Transport Properties

Ludmila Ivanenko

Belarusian State University of Informatics and Radioelectronics
Minsk, Belarus

Horst Lange

Hahn-Meitner-Institute
Berlin, Germany

Armin Heinrich

Institute of Solid State and Materials Research Dresden
Dresden, Germany

CONTENTS

In this chapter we present the effect of external electric and magnetic fields on free carriers in semiconducting silicides. The case of weak fields producing changes in the distribution of electron velocities, which are a small perturbation of the equilibrium distribution, is considered. The behavior of carriers in this case can be described by Ohm's law. The electrical conductivity σ in an isotropic semiconductor in which both free electrons and free holes contribute to the flow of current, can be expressed as

$$\sigma = e(n_e \mu_e + n_h \mu_h). \tag{5.1}$$

Here e is the electron charge, μ_e and μ_h are the mobilities of electrons and holes, respectively, and n_e and n_h are the corresponding charge carrier concentrations. The inverse conductivity $1/\sigma = \rho$ is called resistivity. Practically important, σ is closely connected with the fundamental electronic properties of the material, which are n and μ. Thus, the transport properties of an electron gas in a semiconductor can be described most conveniently in terms of these quantities.

The chapter is organized in the following manner. In the beginning the basic concepts and theoretical analysis of scattering mechanisms in semiconductors are presented. An examination of these mechanisms for semiconducting silicides is discussed. The available resistivity data for each semiconducting silicide are next described in detail. The experimental carrier mobilities are then presented and analyzed in the light of the particular role of different carrier scattering mechanisms. In the final section the thermoelectric properties will be discussed with special emphasis on the thermoelectric figure of merit.

5.1 Free Charge Carriers and Their Mobility in Semiconductors

Electrons in the conduction band and holes in the valence band of a semiconductor are referred to as free carriers since they are certainly free to move. The states in the conduction band can be populated at thermal equilibrium by thermally excited electrons from lower (impurity and/or valence band) states. In an intrinsic semiconductor free carriers can be created only by electronic excitations from the valence band to the conduction band.

The observable transport phenomena for free charge carriers in a semiconductor depend strongly on the carrier concentration. So, we start with some important relations for the free charge carrier concentration in a semiconductor in intrinsic and extrinsic conditions.

5.1.1 Free Charge Carrier Generation

The number of free charge carriers in a semiconductor depends on temperature and on the presence of defects or impurities. At thermodynamic equilibrium the energy distribution of both free electrons and free holes is characterized by the Fermi level E_F. The dependence of the concentration of conduction band electrons n_e on the temperature T for the case of nondegenerate electron gas is given by [5.1]:

$$n_e = N_c \exp\frac{E_F - E_c}{k_B T}, \tag{5.2}$$

where E_c is the energy of the bottom of the conduction band, k_B is the Boltzmann constant. The quantity $N_c = 2 \left(\dfrac{2\pi m^*_e k_B T}{(2\pi \hbar)^2} \right)^{3/2}$, where m^*_e is the effective electron mass and $\hbar$ is Planck's constant, is often called the effective density of conduction band states. In a nondegenerate electron gas $n_e \ll N_c$.

Identically, the number of holes in the valence band with the effective mass m^*_h, is

$$n_h = N_v \exp \frac{E_v - E_F}{k_B T}, \tag{5.3}$$

where N_v is the parameter analogous, for the valence band, to N_c in the conduction band.

In an intrinsic semiconductor n_e and n_h are equal, as far as conduction band electrons can be derived only by excitation from valence band states. The intrinsic concentration of electron–hole pairs $n_i = \sqrt{n_e n_h}$ is:

$$n_i = \sqrt{N_c N_v} \exp \frac{-E_g}{2k_B T}, \tag{5.4}$$

where E_g is the energy gap corresponding to $E_c - E_v$.

For quite pure materials and at relatively high temperatures, the intrinsic excitation determined by valence-to-conduction band transitions, dominates in a semiconductor. Particularly at low temperatures, the intrinsic concentration of electron–hole pairs n_i is negligibly small compared to the density of one kind of free carrier thermally excited from the impurity or defect-related levels in the forbidden band of a semiconductor. Thus, a real semiconductor inevitably becomes either n-type or p-type extrinsic, when it is cooled.

In an n-type extrinsic semiconductor the concentration of free electrons excited from a set of N_d donor levels with the ionization energy E_d, is [5.2]:

$$n_e = \sqrt{\frac{N_c N_d}{2}} \exp \frac{-E_d}{2k_B T}, \text{ if } n_e \ll N_d. \tag{5.5}$$

Fig. 5.1 gives a schematic diagram of the carrier concentration versus the inverse temperature. Three distinct regions are evident. At low temperatures, the thermal energy $k_B T$ is not sufficient to permit thermal excitation of electrons from the valence band to the conduction band, but it is sufficient for extrinsic excitation. Nevertheless, a large number of donors (in the case of an n-type semiconductor) still retain their valence electrons, i.e. are not ionized, and there is carrier freeze-out at the donors.

At higher temperatures all donors are thermally ionized and the concentration of conduction electrons is equal to that of donors.

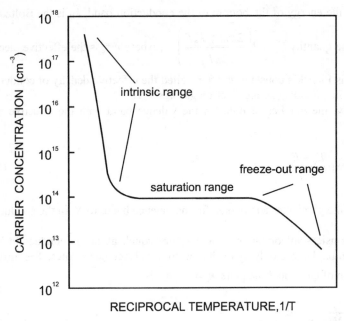

Fig. 5.1. Typical variation of charge carrier concentration in semiconductors with temperature

This temperature region is commonly called the saturation range of the system, since no further change in the electron density can occur and $n_e \approx N_d$. With the subsequent increase of the temperature the thermal energy becomes sufficient to start active intrinsic excitation. In this case the free carrier concentration is defined by (5.4).

The exponential factor in (5.4) controls the entire temperature dependence for the intrinsic carriers. A semilogarithmic plot of n_i versus $(1/T)$ has an average slope, which is a tolerable first approximation to the value of $(-E_g/2k_B)$. The temperature dependence of intrinsic carrier concentration, as well as the temperature dependence of conductivity in the intrinsic regime, is often used for the energy gap determination.

5.1.2 Charge Carrier Scattering Mechanisms and Related Mobilities

The carrier mobility is one of the most important parameters characterizing transport properties of semiconductors. It can be calculated if the band structure of the material and the main scattering mechanisms of carriers are known. Carriers in a semiconductor are scattered by their interaction with both acoustic and optical phonons, ionized impurities, neutral defects and other lattice imperfections (dislocations, grain boundaries, etc.).

If one assumes contributions of different scattering mechanisms to be independent, the total mobility according to Mathiessen's approximation is $\mu^{-1} = \sum_i \mu_i^{-1}$, where i stands for different scattering mechanisms under consideration. In our case one has

$$\mu^{-1} = \mu_{AC}^{-1} + \mu_{NPO}^{-1} + \mu_{PO}^{-1} + \mu_{IM}^{-1} + \mu_O^{-1}, \tag{5.6}$$

where μ_{AC}, μ_{NPO}, μ_{PO}, μ_{IM}, μ_O, are the carrier mobilities controlled by scattering within acoustic lattice mode, nonpolar and polar optical modes, charged and neutral impurities, respectively.

In general, the carrier mobility can be expressed in the form

$$\mu = e\langle\tau\rangle/m^*, \tag{5.7}$$

where m^* is the carrier effective mass. The average relaxation time $\langle\tau\rangle$ for the case of a nondegenerate electron gas is given by [5.3]:

$$\langle\tau\rangle = \frac{\int_{E_c}^{\infty} f_0'\tau v^2 N(E)\,dE}{\int_{E_c}^{\infty} f_0'v^2 N(E)\,dE}, \tag{5.8}$$

where E is the carrier energy, E_c is the bottom of the conduction band, f_0' is the derivative of the distribution function with respect to energy, $N(E)$ is the density of states, and v is the carrier velocity.

In the case of parabolic band and nondegenerate charge carriers for each separate mechanism one has [5.4]:

$$\mu_i = \frac{4e}{3m^*(k_B T)^{5/2}\sqrt{\pi}} \int_0^{\infty} E^{3/2}\tau_i \exp\left(-\frac{E}{k_B T}\right) dE, \tag{5.9}$$

where τ_i is the momentum relaxation time for the i-th mechanism.

The momentum relaxation time related to acoustic phonon scattering is [5.3]

$$\tau_{AC} = \frac{\pi\hbar^4 \rho^* v_s^2}{\sqrt{2m^{*3}}\, E\, \Xi^2 k_B T}, \tag{5.10}$$

where ρ^* is the material density, v_s is the mean longitudinal sound velocity, and Ξ is the constant defined by components of the deformational potential tensor. Unfortunately, the latter is usually unknown for semiconducting silicides. For tetrahedrally coordinated semiconductors its value is estimated to vary from 5 to 15 eV.

Other important centers of scattering in semiconductors are optical phonons of higher energy. At low temperatures the contribution from this mechanism is negligible, since there are no electrons with sufficient energy to emit optical phonons. The momentum relaxation time due to the polar optical phonon scattering is given by [5.3]

$$\tau_{PO} = \sqrt{\frac{2^5 E}{m^*}} \frac{\pi \hbar \varepsilon_p}{e^2 \omega_0} \left[n(\omega_0) \sqrt{1 + \frac{\hbar \omega_0}{E}} + (n(\omega_0 + 1)) \sqrt{1 - \frac{\hbar \omega_0}{E}} + \right.$$
$$\left. + \frac{\hbar \omega_0}{E} \left((n(\omega_0 + 1)) \operatorname{arcsin} h\left(\sqrt{\frac{E}{\hbar \omega_0} - 1} \right) - n(\omega_0) \operatorname{arcsin} h\left(\sqrt{\frac{E}{\hbar \omega_0}} \right) \right) \right]^{-1}, \tag{5.11}$$

where ω_0 is the optical phonon frequency defined by the Debye temperature Θ ($\omega_0 = k_B \Theta / \hbar$), $\varepsilon_p = (1/\varepsilon_{in} - 1/\varepsilon_0)^{-1}$, ε_{in} and ε_0 are the high frequency and the static dielectric constants, and $n(\omega_0) = (\exp(\hbar \omega_0 / k_B T) - 1)^{-1}$.

The appropriate formula for the momentum relaxation time in the case of nonpolar optical phonon scattering has been obtained in the form [5.3]:

$$\tau_{NPO} = \frac{\sqrt{2} \pi \hbar^3 \omega_0 \rho^*}{D_0^2 m^{*3/2}} (n(\omega_0)(E + \hbar \omega_0)^{1/2} + (n(\omega_0) + 1)(E - \hbar \omega_0)^{1/2})^{-1}, \tag{5.12}$$

where D_0 is the optical deformation tensor.

Charge carriers can also be scattered by ionized impurities. This mechanism becomes dominant at low temperatures in the freeze-out regime. For the ionized impurity scattering there are many models suitable for different conditions [5.3, 5.5]. For the case of mobility estimation it seems to be reliable to deal with the Brooks–Herring approach. On the one hand, it is more consistent than the others and includes the Coulomb screening. On the other hand, it is quite simple compared, for example, to the method of phase shifts. In this case one has [5.3]

$$\tau_{IM} = \frac{\sqrt{2 m^* E} \varepsilon_0}{\pi e^4 N_i \varphi(x)}, \tag{5.13}$$

where N_i is the ionized impurity or defect concentration and $\varphi(x) = \ln(1+x) + x/(1+x)$. Here $x = 8 m^* E \Gamma_0 / \hbar^2$, $\Gamma_0 = (\pi/3n)^{1/3} \varepsilon_0 \hbar^2 / 4 m^* e^2$ is the reciprocal square of the Debye length, $n = (2 m^* k_B T)^{3/2} F_{1/2}(E_F)/2 \pi^2 \hbar^3$ is the carrier concentration. $F_{1/2}(E_F)$ and E_F are the Fermi integral and the Fermi energy, respectively.

For the neutral impurity scattering mechanism Erginsoy's formula can be used [5.3]

$$\tau_0 = e^2 m^{*2} (80 \pi \varepsilon_0 \hbar^3 N_0)^{-1}, \tag{5.14}$$

where N_0 is the concentration of the neutral impurity.

Since the different scattering mechanisms have different dependencies on electron energy and temperature, they yield specific temperature dependencies of the carrier mobility. These are summarized in Table 5.1.

Table 5.1. Temperature dependence of the charge carrier mobility in semiconductors for different scattering mechanisms

Dominating scattering mechanism	Temperature dependence
acoustic phonons (deformation potential)	$\mu \sim T^{-3/2}$
ionized impurities	$\mu \sim T^{3/2}$
neutral impurities	$\mu \sim T^{0}$
nonpolar optical phonons, $k_B T \gg \hbar\omega_0$	$\mu \sim T^{-3/2}$
polar optical phonons, $k_B T \gg \hbar\omega_0$	$\mu \sim T^{-1/2}$

Comparing an experimentally observed temperature dependence of the carrier mobility with the theory makes it possible to evaluate contributions from different scattering mechanisms.

5.2 Experimental Resistivities

The most common quantity measured to characterize integrally transport properties of semiconducting silicides is electrical resistivity. Its systematic study started at the end of the 1950s. The majority of silicide phases was found to exhibit rather small electrical resistivity, namely in the range from 10 to 100 μohm cm. This is close to the values typical for transition metals. However, some silicides were found to have a resistivity of the order of 10^3–10^4 μohm cm. They are represented by CrSi$_2$, β-FeSi$_2$, ReSi$_{1.75}$ (ReSi$_2$), BaSi$_2$ and Mg$_2$Si. Their resistivities are too high for metals, being more characteristic of semiconductors. For the manganese silicon-rich silicide MnSi$_{2-x}$ the value of resistivity is somewhat lower but still high (about 460 μohm cm at room temperature). The semiconducting properties of these silicides have been established by investigation of the temperature dependence of their electrical resistivity, and also by studies of their thermoelectric properties [5.6–5.9].

With the exception of Mg$_2$Si single crystals, the measurements were first made on the samples obtained by sintering of the constituent elements at high temperatures. Such powder metallurgical techniques generally lead to a large concentration of impurities, which in some cases stabilize silicide phases [5.10]. Thus, most of these results are more of historical importance and the resistivity values determined from these measurements are questionable.

New semiconducting silicides of ruthenium, iridium, osmium and tungsten were reported in the 1980s [5.11–5.15]. Their semiconducting behavior was also confirmed by measurements of the temperature dependence of their resistivity in the intrinsic regime.

At present there are a number of the resistivity data recorded for each semiconducting silicide. The most reliable of them are summarized in Table 5.2.

The resistivity values were measured on both single crystals and thin films. The polycrystalline thin film silicides were formed either by diffusion synthesis in thin metal films on monocrystalline or polycrystalline silicon or cosputtered (metal+silicon) onto monocrystalline silicon substrates. The electrical resistivity was measured with a conventional four-probe technique using direct current.

It is apparent that there is scatter in the resistivity values reported by different researchers. This can be attributed to the quality of the samples investigated and to the presence of impurities in the silicides, which are known to form energy levels in the forbidden band of semiconductors as well as being the scattering centers for free charge carriers. Hence, the intrinsic properties of the silicides can be obscured. In the following sections we present and discuss available resistivity data for particular semiconducting silicides with emphasis on the impurity effects.

5.2.1   Silicides of Group II Metals

Among the silicides of group II metals of the Periodic Table two compounds are clearly known to be semiconductors: Mg_2Si and $BaSi_2$ (the orthorhombic phase). The cubic phase of barium disilicide also shows a negative temperature dependence of resistivity, like a semiconductor, but its value is about one order lower. There is another silicide in this group, namely Ca_2Si, which has been calculated to be a semimetal or an extremely narrow-gap semiconductor [5.16], but no experimental data on its electrical resistivity confirming this result are available in the literature.

Mg_2Si. Quite reliable data electrically characterizing this silicide were obtained as early as the 1950s. The electrical resistivity of single crystals of Mg_2Si has been studied by Morris et al. [5.9], by Whitsett [5.17], and by LaBotz et al. [5.18]. Polycrystalline samples of Mg_2Si were measured by Winkler [5.19] and Bose et al. [5.20].

The most comprehensive analysis of the resistivity data on both n-type and p-type Mg_2Si single crystals was carried out by Morris et al. [5.9]. Electrical resistivity was measured from 77 to 1000 K. The p-type material was prepared by crystal doping with silver or copper. An attempt to dope the material with aluminum or iron did not result in p-type samples. All undoped crystals of Mg_2Si were n-type. The resistivity versus temperature is shown in Fig. 5.2. At room temperature the resistivity of Mg_2Si is quite high and ranges from 0.08 to 0.2 ohm cm for n-type single crystals and from 0.2 to 0.7 ohm cm for p-type samples. The temperature dependence of the resistivity follows a classical semiconductor behavior, i.e. it decreases as the temperature increases. On the basis of studies on the electrical conductivity and Hall effect, the most reliable value of 0.78 eV for the energy gap of Mg_2Si was obtained.

Table 5.2. Resistivity of semiconducting silicides at room temperature

Silicide	Sample characterization	Conductivity type	Resistivity (ohm cm)	Refs
Mg$_2$Si	single crystal	n	$0.08 - 0.2^a$	5.9
	single crystal	p	$0.2 - 0.7^a$	5.9
	single crystal	n	0.09	5.18
	polycrystalline bulk	n	0.06	5.18
	polycrystalline bulk		35	5.20
BaSi$_2$	polycrystalline bulk	p	0.4	5.21
orthorhombic	polycrystalline bulk		7	5.22
	polycrystalline bulk	n	100	5.23
CrSi$_2$	single crystal	p	$1 \times 10^{-3} (//c)$	5.24
	single crystal	p	$0.5 \times 10^{-3} (\perp c)$	5.24
	single crystal		0.74×10^{-3}	5.25
	single crystal	p	0.9×10^{-3}	5.26
	polycrystalline film		0.6×10^{-3}	5.10
	polycrystalline film		3.6×10^{-3}	5.27
	polycrystalline film	p	7×10^{-3}	5.28
	polycrystalline film	p	8×10^{-3}	5.29
	polycrystalline film		10×10^{-3}	5.30
	epitaxial film	p	20×10^{-3}	5.31
WSi$_2$ hexagonal	polycrystalline film		$0.4 \times 10^{-3\ b}$	5.15
MnSi$_{2-x}$	single crystal	p	$1.1 \times 10^{-3} (//a)$	5.32
	single crystal	p	$5.5 \times 10^{-3} (//c)$	5.32
	polycrystalline bulk	p	2×10^{-3}	5.33
	polycrystalline film		4.1×10^{-3}	5.34
	polycrystalline film	p	7.7×10^{-3}	5.35
	polycrystalline film	p	13×10^{-3}	5.36
ReSi$_{1.75}$	single crystal	p	0.005	5.37
	polycrystalline film	p	0.018	5.38
	polycrystalline film		0.018	5.39
	epitaxial film	p	0.1	5.40
β-FeSi$_2$	single crystal	p	0.1	5.41
	single crystal	p	2	5.42
	single crystal	n	20	5.43
	polycrystalline film	p	0.3	5.44
	polycrystalline film	p	1	5.45
	polycrystalline film	p	0.5	5.46
	epitaxial film	p (on p-type Si)	0.3	5.47
	epitaxial film	p (on p-type Si)	0.3	5.47
	epitaxial film	p	0.005	5.48
Ru$_2$Si$_3$	single crystal	p	0.01	5.49
	single crystal	n	2	5.50
	single crystal	p	5	5.51
	polycrystalline film		7	5.52
	polycrystalline bulk	p	7.5×10^{-3}	5.53
OsSi$_2$	polycrystalline bulk	p	8.3	5.13
Ir$_3$Si$_5$	single crystal	p	0.01	5.54
	polycrystalline bulk	p	0.06	5.55
	polycrystalline bulk	p	0.2	5.56
	polycrystalline film	p	6.7	5.12

[a] The ranges in the resistivity have been assigned because of the various saturation carrier concentration, [b] Calculated from the data presented in [5.15].

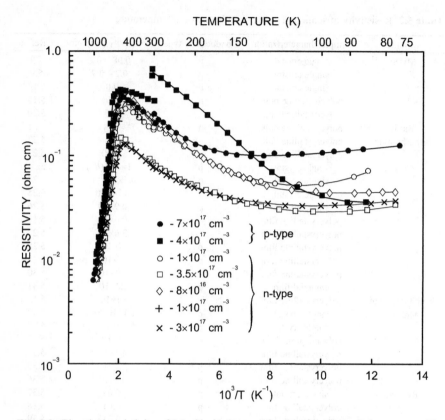

Fig. 5.2. Electrical resistivity of Mg$_2$Si single crystals as a function of temperature. The different curves correspond to the samples with various saturation carrier concentrations [5.9]

Later, polycrystalline Mg$_2$Si was characterized in detail by studying its structural and electrical properties by Bose *et al.* [5.20]. The results indicate that the material is a nondegenerate semiconductor. The variation of electrical conductivity with temperature was measured for the sintered pellets of Mg$_2$Si of various grain-size materials. From the study of a particular specimen (0.086 μm grain size) it was observed that at 300 K the resistivity of the material was 35 ohm cm. The electrical conductivity was found to be smaller in the samples consisting of smaller grains. The variation of the electrical conductivity with particle size of the material is well explained by the model proposed by Petritz for polycrystalline semiconductors [5.20]. According to this model the electrical conductivity varies inversely with the number of crystallites per unit length. Therefore, the results on the higher conductivity for larger grain size of the material can be understood.

BaSi$_2$. Due to the extraordinary difficulties in preparation of this disilicide, its electrical properties are rather poorly studied [5.21–5.23, 5.57]. As mentioned in

Chapt. 1, $BaSi_2$ has three crystallographic modifications under normal conditions. All of them were fabricated and electrically characterized by Imai and Hirano [5.23, 5.58]. The orthorhombic phase sample was fabricated by melting of starting material (orthorhombic $BaSi_2$ with purity 98%) in an argon arc furnace. The metastable cubic and trigonal phases were obtained by subjecting the orthorhombic $BaSi_2$ to high-pressure and high-temperature processing. The resistivity of $BaSi_2$ as well as its temperature variation strongly depend on the crystal structure. The changes of the structure from orthorhombic to trigonal through the cubic phase cause a considerable decrease in the resistivity. The orthorhombic phase has the highest resistivity among the three phases and a large negative temperature coefficient like a semiconductor. The dominant carriers are electrons, which is consistent with the earlier results [5.22]. The cubic phase is also characterized by a negative temperature coefficient of the resistivity with increasing temperature but its value is much lower than that of the orthorhombic phase. The trigonal $BaSi_2$ demonstrates the lowest resistivity and a positive temperature coefficient that is typical for a metal. A similar jump in electrical resistivity of about three orders of magnitude during structural transformations in $BaSi_2$ was also observed by Evers [5.57].

The conductivity σ of orthorhombic $BaSi_2$ is shown in Fig. 5.3 as a function of temperature [5.23]. Two linear relations fitting the conductivity data signify the electrical conduction to be controlled by a thermally activated mechanism in both high and low temperature regions. An activation energy of 0.049 eV was observed for temperatures above 230 K. The energy gap obtained in the previous study in the temperature range from 293 to 1073 K on polycrystalline samples of high purity is 1.3 eV [5.22]. In this case the activation energy is expected to be 0.65 eV. Thus, the small activation energy observed by Imai and Hirano [5.23] is conditioned by the temperature range used, which is too low to measure the exact activation energy in the intrinsic range. Since the sample contained impurities and defects, the small activation energy observed in [5.23] seems to represent the energy gap between the conduction band edge and the donor level related to the impurities, defects or their combination.

Taking into account the negative temperature coefficient of the resistivity of the cubic phase and its relatively large value, 5.6 ohm cm at room temperature, this phase may also be a semiconductor. As in the orthorhombic $BaSi_2$, the dominant carriers are electrons.

In contrast to the former two phases, the trigonal phase is a hole metal that shows superconductivity with an onset temperature of 6.8 K [5.59].

Thus, the electrical properties of $BaSi_2$ depend on the crystal structure. Imai and Hirano consider that the structural change in $BaSi_2$ is related to the variation of interatomic distance [5.23]. The point is that Ba–Ba and Ba–Si distances decrease with the structural change, while the Si–Si distance of the three phases, which is slightly longer than that of diamond-type structure silicon, is almost unchanged. Because the electronic structure of $BaSi_2$ near the Fermi level is due to the interaction between barium and silicon atoms these changes of the Ba–Ba and Ba–Si distances are reflected in the electrical properties.

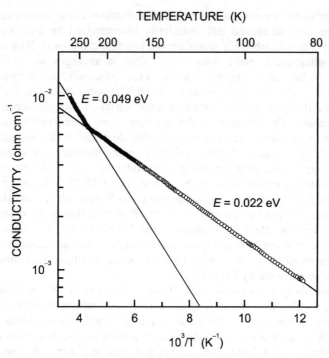

Fig. 5.3. Electrical conductivity of orthorhombic $BaSi_2$ as a function of temperature. The *solid lines* are fits with $\sigma = \sigma_0 \exp(-E/k_B T)$ [5.23]

So far, the electronic structure of $BaSi_2$ has not been calculated, and hence it is impossible to discuss in detail. However, the results obtained by Fahy and Hamann [5.60] on the electronic structure of $CaSi_2$, whose structure is the same as that of trigonal $BaSi_2$, provide some evidence for the correctness of the above hypothesis.

5.2.2 Silicides of Group VI Metals

Three silicides in this group are reported to be semiconductors. These are hexagonal phases of $CrSi_2$, WSi_2 and $MoSi_2$. While the conductivity mechanisms of $CrSi_2$ have been extensively studied, there are only a few reports on the electrical properties of hexagonal WSi_2 and no experimental resistivity data for the hexagonal phase of $MoSi_2$.

$CrSi_2$. The electrical parameters of chromium disilicide, including resistivity, strongly depend on the crystalline structure of the sample (see Table 5.2). The resistivity of $CrSi_2$ single crystals ranges from 0.5 to 1 mohm cm [5.24–5.26]. It is one order of magnitude higher and ranges from 3.6 to 20 mohm cm in thin films [5.27–5.30]. There is only one exception for the films obtained by Murarka [5.10], which had a resistivity of about 0.6 mohm cm.

Generally, the high resistivity of thin films can be attributed to grain boundaries, stacking faults and impurities in the polycrystalline samples. The discrepancies in the resistivity values among single crystals and thin films can be related to the formation technique and, in the case of thin films, to the substrates used. The resistivity variation due to phase transformation also has to be considered, since a sputtered silicide is usually amorphous at room temperature and crystallizes at a certain elevated temperature.

In pure $CrSi_2$ holes are the major charge carriers [5.26, 5.28, 5.61, 5.62]. Manganese atoms introduced into $CrSi_2$ act as donors. Their activation energy was estimated to be 0.01 eV [5.26, 5.62]. Excess chromium atoms act as acceptors [5.24].

Electrical investigations of $CrSi_2$ single crystals as a function of temperature were carried out by Shinoda et al. [5.26] yielding the energy gap of 0.30 eV. The temperature dependence of resistivity for $CrSi_2$ thin films was determined by Nava et al. [5.28] and Bost and Mahan [5.30]. The energy gaps obtained from their measurements is about 0.27 and 0.35 eV, respectively. The energy gap values derived from electrical measurements for single crystals and thin films are close to each other. However, the result obtained by Bost and Mahan seems to be more accurate, since it was confirmed by optical measurements [5.30, 5.63].

Experimentally recorded conductivity of $CrSi_2$ thin films as a function of temperature is presented in Fig. 5.4.

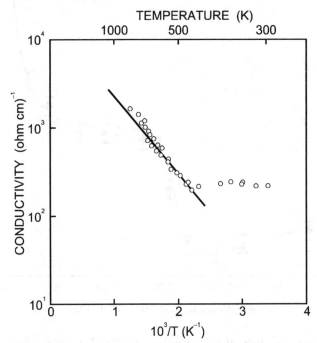

Fig. 5.4. Electrical conductivity of $CrSi_2$ polycrystalline thin films formed by the diffusion synthesis as a function of temperature. A least-squares-fit line corresponds to the activation energy of 0.175 eV [5.30]

Two distinct temperature regions with different conduction mechanisms are evident. At low temperatures, the conductivity is essentially independent of temperature due to the saturation of shallow impurities. At high temperatures ($T > 500$ K), the conductivity is thermally activated. The data in this region have been fitted with a straight line corresponding to an activation energy of 0.175 eV. Under the assumption that conductivity is intrinsic in this temperature region the energy gap is estimated to be 0.35 eV.

A detailed study of the electrical and structural properties of Cr/Si thin films as a function of annealing temperature was performed by Nava *et al.* [5.28] and Gong *et al.* [5.27]. The room temperature resistivity of the annealed films was found to be strongly dependent on heat treatment. The highest resistivity, about 10 mohm cm, was observed after annealing for 1 h at 900°C or above [5.28]. Fig. 5.5 shows the resistivity versus annealing temperature for the 62 nm thin film of sputtered CrSi$_2$ annealed in an argon atmosphere between room temperature and 600°C for 60 min.

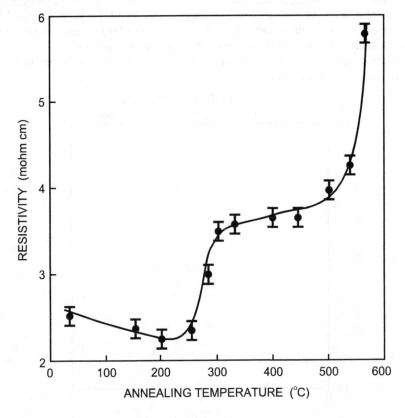

Fig. 5.5. Electrical resistivity of the sputtering deposited CrSi$_2$ film versus annealing temperature [5.27]

Below 200°C the resistivity decreases slightly with annealing temperature owing to a structure relaxation in the film. It was amorphous as-deposited and became short-range ordered upon annealing. However, above 200°C a sharp increase in the resistivity in the narrow temperature interval is observed. Since a crystalline silicide generally has lower resistivity than amorphous material, the increase in the resistivity has been associated with the microstructure of the film, in which boundaries surrounding columnar grains were observed. The boundaries, which scatter charge carriers thus reducing the current through the film, may be formed due to volume shrinkage during crystallization of the amorphous film. Upon annealing at temperatures higher than 550°C the boundaries become rather wide to disconnect adjacent columnar grains. This results in a sharp increase in the electrical resistivity (Fig. 5.5). Since the resistivity of a $CrSi_2$ thin film after crystallization is mainly determined by the grain boundaries and the columnar grains do not tend to grow at a temperature below 600°C, only a short annealing time is needed to reach a certain resistivity value.

Although a thin film with a relatively stable resistivity of 3.6 mohm cm was obtained by heat treatment, the observed microstructure of the films indicates that this value does not represent intrinsic properties of the material. Note that similar anomalous electrical behavior of coevaporated Cr-Si thin films upon crystallization was also observed in [5.28].

In spite of $CrSi_2$ being a semiconductor, Hirano and Kaise [5.25] and Lasjaunias et al. [5.64] observed a positive temperature dependence of resistivity for $CrSi_2$ single crystals in the range from room to liquid helium temperature. The reason is that, according to Shinoda [5.26] and Nishida [5.62], $CrSi_2$ is a degenerate semiconductor and the temperature interval from room temperature to 4.2 K corresponds to its extrinsic region.

The resistivity of $CrSi_2$ is likely to show a crystallographic orientation dependence. The resistivity measured along the $[2\bar{1}\bar{1}0]$ direction is higher than that for the [0001] one [5.25]. A large anisotropy with $\rho_{//c}/\rho_{\perp c} \cong 2$ at room temperature was also observed on single crystals [5.24]. The resistivity in the intrinsic region obeys an exponential temperature dependence giving the energy gap of 0.32 eV and pre-exponential factors of 3×10^{-4} and 1.6×10^{-4} for measurements along and perpendicular to the c axis, respectively.

WSi_2. Resistivity measurements of coevaporated amorphous WSi_{2+x} thin films were performed by Nava et al. [5.15]. A metastable, semiconducting hexagonal phase WSi_2 appears during crystallization of the amorphous film at around 400°C. Its formation is associated with a sharp increase in resistivity. The hexagonal phase is stable up to 600°C, then it transforms into the tetragonal phase with metallic properties.

In order to electrically characterize the hexagonal WSi_2 phase, the temperature dependence of the sheet resistivity of $WSi_{1.79}$, WSi_2, $WSi_{2.18}$ films was measured from 555 to 900 K. The results are presented in Fig. 5.6. Two stages may be seen in the electrical behavior of these films. The first high-temperature stage (>800 K) is characterized by linear relationships, especially pronounced for the stoichiometric WSi_2 specimens. The second low-temperature stage (<800 K) demonstrates a nonlinear correlation. The linear behavior is typical for semiconductors in the intrinsic region. From the slope of the linear dependence the energy gap of 0.40 eV, was determined by assuming that the sheet resistivity is proportional to $\exp(-E_g/2k_B T)$.

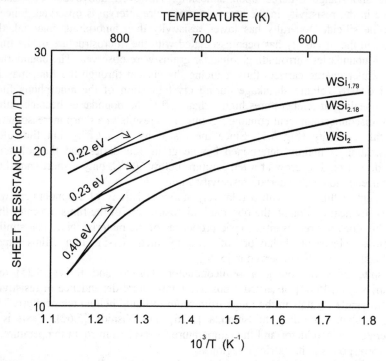

Fig. 5.6. Sheet resistance of coevaporated $WSi_{1.79}$, WSi_2, and $WSi_{2.18}$ thin films as a function of temperature [5.15]

5.2.3 Silicides of Group VII Metals

Silicon-rich silicides of manganese and rhenium are known to have semiconducting properties among silicides of group VII metals. The conductivity mechanism of these silicides has not been clarified for a long time because of the complexities of their composition and the crystal structure. The stoichiometric disilicides of manganese and rhenium were first reported to be semiconductors [5.8, 5.38, 5.39]. In later studies Mn-Si and Re-Si alloys have been investigated by many researchers to confirm this composition [5.32, 5.33, 5.37, 5.65–5.67]. They have concluded that the silicon-rich silicides of manganese and rhenium are not the stoichiometric disilicides but $MnSi_{-1.73}$ and $ReSi_{1.75}$, whose semiconducting behavior has been examined. Nevertheless, the definitions from the original papers are used for convenience.

$MnSi_{2-x}$. There are many different reports on the transport properties of the higher manganese silicide because of the reasons mentioned above. The resistivity values of this silicide obtained by different investigators range from 1.1 to

5.5 mohm cm for single crystals and from 4.1 to 13 mohm cm for thin films at room temperature (see Table 5.2). The value of 4.1 mohm cm [5.34] is considerably lower than that mostly reported for $MnSi_{\sim 1.73}$ thin films. The reason seems to be the phase composition of the silicide films investigated. The films formed at 500°C were reported to be $MnSi_{1.7}$ without any accurate determination of their structures. Probably, the films of low resistivity consisted of MnSi, since it was found that $MnSi_{2-x}$ samples with $x > 0.28$ were two-phase alloys composed of $MnSi_{1.72}$ and MnSi [5.33]. The monosilicide of manganese behaves electrically as a metal. Its presence in a film can reduce the resistivity.

The resistivity of $MnSi_{2-x}$ with x ranging from 0.25 to 0.28 was examined by Nishida [5.33] and Kawasumi et al. [5.32] for bulk and single crystals, respectively. The absolute resistivity of $MnSi_{2-x}$ decreases with increasing x, while the general trend in its temperature-dependent resistivity has been found to be independent of the composition, as illustrated by the experimental data presented in Fig. 5.7. The compound $MnSi_{2-x}$ was confirmed to be a p-type degenerate semiconductor with the energy gap of about 0.4 eV [5.33] and 0.7 eV [5.32], estimated from the temperature dependence of the resistivity. The difference in the values of the energy gap may indicate an effect of impurities. The intrinsic resistivity of $MnSi_{2-x}$ varies with temperature as $3.6 \times 10^{-4} \exp(2320/T)$ [5.33].

An anisotropy between the a and c axis directions for the $MnSi_{\sim 1.73}$ single crystals has been found. The resistivity measured along the c axis is about 5 times higher than that measured along the a axis [5.32]. Even larger anisotropy was found in [5.68].

Semiconducting properties of $MnSi_{1.73}$ thin films in the temperature range of 100–850 K have been studied by Krontiras et al. [5.36]. A degenerate semiconducting behavior was observed below 450 K. This is confirmed by the data shown in Fig. 5.7. As in the case of bulk $MnSi_{2-x}$, the electrical resistivity increases with the temperature rise up to 500 K. Above 500 K the resistivity decreases with temperature according to $\rho \sim \exp(-E_g/2k_BT)$ with $E_g \approx 0.45$ eV. It well agrees with $E_g = 0.4$ eV found for bulk $MnSi_{1.73}$ by Nishida [5.33].

$ReSi_{1.75}$ ($ReSi_2$). Structural and transport properties of silicon-rich rhenium silicide single crystals were studied by Gottlieb et al. [5.37, 5.66]. The studies show that only $ReSi_{1.75}$ exhibits semiconducting behavior. The resistivity of $ReSi_{1.75}$ single crystals measured on two different samples for two crystallographic axes in the wide temperature range 4–800 K is displayed in Fig. 5.8. The electrical measurements indicated thermally activated behavior soon above room temperature, as expected for a semiconductor with a small band gap. The room temperature resistivity is about 5 mohm cm. The resistivity for the current I parallel to the a direction is higher than for $I//c$, but the general behavior of the two samples is the same. Below 10 K the resistivity measurements indicate dominant carrier transport via localized states [5.37]. In the extrinsic region at low temperatures ($10\,K \leq T \leq 200\,K$) the resistivity increases slightly with the temperature up to the exhaustion range. Unfortunately, it is difficult to distinguish different effects in the conductivity mechanism in this region. At high temperatures in the intrinsic region $ReSi_{1.75}$ shows thermally activated behavior

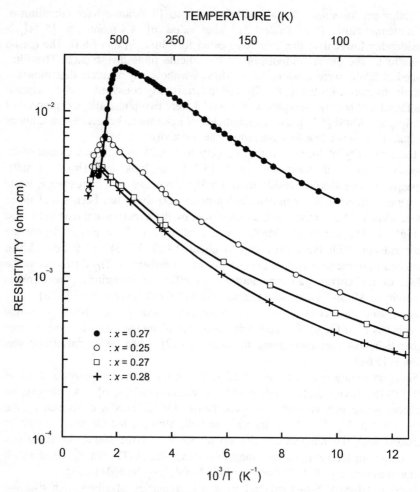

Fig. 5.7. Electrical resistivity of $MnSi_{2-x}$ as a function of temperature for bulk material ($\circ$, $\square$, $+$) [5.33] and thin films ($\bullet$) [5.36]

with one (or two) energy gap $E_g = 0.16$ (0.30 eV). The values are quite close to those previously reported by Long *et al.* for thin films for the phase denominated $ReSi_2$ [5.39].

Electrical property measurements on $ReSi_2$ thin films in a wide temperature range were performed by Krontiras *et al.* [5.38] and Long *et al.* [5.39]. The films show semiconducting behavior, consistent with the data for single crystals. The room temperature resistivity of thin films is about 20 mohm cm. The temperature dependence of the conductivity in the range of about 300–900 K is plotted in Fig. 5.9. The apparent thermal activation energies for the two linear regions are 0.045 eV for low temperatures and 0.14 eV for the high temperatures. Since the

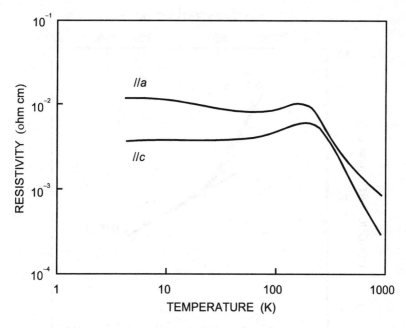

Fig. 5.8. Electrical resistivity of ReSi$_{1.75}$ single crystals as a function of temperature for two different crystallographic axes measured on two different samples [5.37]

material is assumed to be intrinsic over the entire temperature range, it is difficult to apply the traditional interpretation that each of the two activation energies should be equal to half of the corresponding energy gap at 0 K [5.39]. The observed temperature dependence of the conductivity was explained within a band model considering the results of both optical and electrical measurements. Accounting for the optical data there are two transitions, a low energy indirect transition through the gap of 0.12 eV and a higher energy direct transition at 0.36 eV. Band structure calculations [5.69] have shown that the material is indeed a narrow-gap semiconductor with an indirect gap of 0.16 eV. The first direct transition with an appreciable oscillator strength at 0.30 eV is predicted.

A resistivity behavior of ReSi$_2$ thin films grown epitaxially on Si(111) was studied by Ali *et al.* [5.40] below 300 K. The carrier transport seems to take place in the valence and impurity bands.

5.2.4 Silicides of Group VIII Metals

Among the twelve semiconducting silicides reported up to this time, six of them are formed by group VIII metals. These are silicides of iron (β-FeSi$_2$), ruthenium (Ru$_2$Si$_3$), osmium (OsSi, Os$_2$Si$_3$, OsSi$_2$) and iridium (Ir$_3$Si$_5$). β-FeSi$_2$ and OsSi$_2$, as well Ru$_2$Si$_3$ and Os$_2$Si$_3$ are isostructural orthorhombic compounds. This provides the possibility of solid solution formation between them in the couples over the

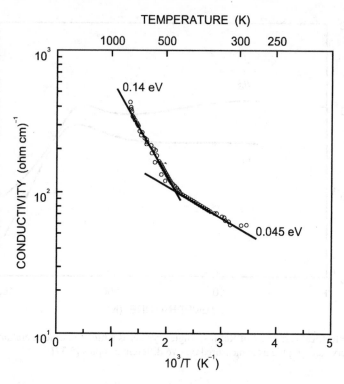

Fig. 5.9. Electrical conductivity of ReSi₂ thin films as a function of temperature. Solid lines correspond to fitting with the activation energies indicated [5.39]

whole composition range. Most attention has been devoted to the study of the electrical properties of semiconducting iron disilicide because of its potential use in optoelectronics and thermoelectrics. Progress in the material synthesis, especially in the growth of high-purity single crystals of β-FeSi₂, Ru₂Si₃, Ir₃Si₅, has allowed detailed information on the electrical behavior of these semiconductors only in the 1990s.

β-FeSi₂ is the most studied material among the semiconducting silicides. The above mentioned availability of high quality single crystals has provided detailed studies of its electrical parameters in a wide temperature range. Undoped semiconducting β-FeSi₂ is usually p-type. A conversion to n-type has been observed by doping with cobalt, nickel, platinum, palladium [5.70–5.73] or boron [5.74], while a replacement of iron in β-FeSi₂ by manganese, chromium, vanadium, titanium [5.71, 5.72] and aluminum [5.41, 5.73] gives p-type conductivity. The conductivity can also be changed from p-type to n-type if the impurity concentration is reduced [5.75]. The activation energies of most n- and p-type dopants vary between 0.07 and 0.14 eV. They decrease with an increase of dopant concentration [5.73, 5.76]. Occasionally n-type conductivity is observed in β-FeSi₂ epitaxial thin films with an excess of silicon [5.77]. Polycrystalline β-FeSi₂ layers were also reported in [5.78] to be n-type in unintentionally doped samples. Semiconducting β-FeSi$_2$ crystals grown from high-purity starting materials were found to be n-type [5.43, 5.79]. The nature of point defects and impurities, which could be important in the electrical behavior of β-FeSi$_2$, has been investigated in single crystals by electron paramagnetic resonance spectroscopy [5.80]. Intentional doping of the crystals during the growth undoubtedly proved the presence of chromium, manganese, cobalt, and, with less certainty, nickel and copper. The impurities substitute iron in its two inequivalent lattice positions and produce energy levels within the forbidden band.

A typical temperature dependence of conductivity of β-FeSi$_2$ thin films is shown in Fig. 5.10. The material is extrinsic at room temperature, where the value of ρ is about 1 ohm cm [5.81]. The same value of resistivity for β-FeSi$_2$ thin films grown at different temperatures was also reported in [5.45]. Above 500 K the material becomes intrinsic and the conductivity is thermally activated with an activation energy of about 0.43 eV.

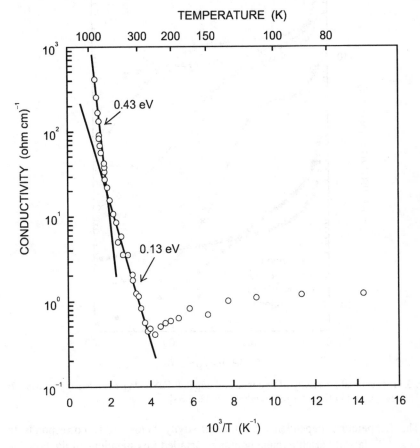

Fig. 5.10. Electrical resistivity of β-FeSi$_2$ thin films as a function of temperature. The solid lines correspond to fitting with the activation energies indicated [5.81]

The effect of doping on the electrical resistivity of β-FeSi$_2$ is demonstrated by the data shown in Fig. 5.11. For comparison, the results for bulk single crystals and thin films are also plotted. All the samples revealed semiconducting behavior. Absolute values of the resistivity at room temperature are in the range between 0.2 and 100 ohm cm. The films show higher conductivity (ρ < 1 ohm cm at 300 K) than single crystals. In single crystals the difference between doped and undoped material at room temperature is more pronounced. Above 80 K the resistivity behavior of these samples is strongly influenced by doping. In the chromium-doped samples a transition between the defect-band conduction and valence-band conduction was observed in the temperature range of 50–100 K [5.82].

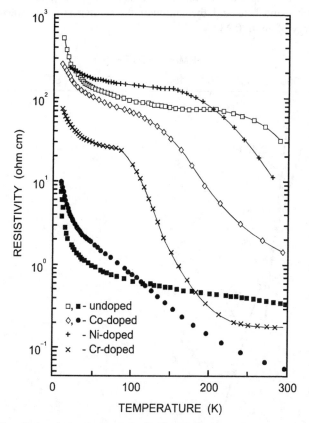

Fig. 5.11. Electrical resistivity of undoped and doped β-FeSi$_2$ single crystals (*open symbols and crosses*) and thin films (*closed symbols*) [5.42, 5.83]

The temperature dependence of the resistivity below 50 K corresponds to Mott's $T^{-1/4}$ law of variable range hopping. Detailed investigations of the hopping conductivity of p-type β-FeSi$_2$ single crystals were performed in [5.84]. The experimental data have been analyzed using the Mott [5.85] and Shklovskii–Efros

[5.86] models. Values of the characteristic temperatures of variable range hopping and widths of the Coulomb gap have been determined as well as densities of states and localization radii. At low temperatures ($T < 200$ K), a variable range hopping conduction was also observed in polycrystalline β-FeSi$_2$ thin films [5.87].

The investigation of resistivity and thermopower of chromium-doped β-FeSi$_2$ single crystals in the temperature range of 20–950 K [5.88] provides more details. Fig. 5.12 shows the results obtained.

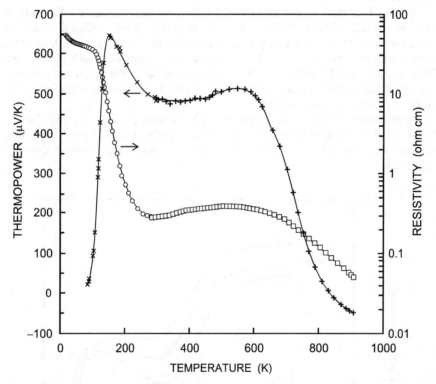

Fig. 5.12. Temperature dependence of the thermopower and resistivity of chromium-doped (~ 0.1 at %) β-FeSi$_2$ single crystals [5.88]

Four temperature regions are apparently distinguished. The strong increase of the thermopower at low temperatures occurs when the resistivity passes from variable range hopping to conduction in the valence band. Above 100 K carriers are thermally activated into the valence band and both parameters show the usual E_a/T behavior. The increase in the thermopower above 300 K correlates with the increase of the resistivity. The slight growth of the resistivity in this temperature range probably means that the increase of electron–phonon scattering in the valence band exceeds a further increase of the carrier concentration. The decrease of the resistivity and thermopower above 600 K reflects the onset of intrinsic conduction with $E_a = E_g/2 = 0.4$ eV. The change in sign of the thermopower from positive to negative above 850 K supports this interpretation. Both resistivity and thermopower of doped sintered β-FeSi$_2$ exhibit, above 600 K, a temperature dependence similar to those observed on single crystals [5.73]. Below 600 K the transport behavior strongly depends on the preparation conditions. Whereas selected samples also show the four characteristic temperature ranges, others behave in a completely different manner. These data were explained by polaron effects [5.73], which are not confirmed by the results on single crystals. Further investigations are necessary in order to obtain a quantitative understanding of the role of polarons in β-FeSi$_2$.

Ru$_2$Si$_3$. Electrical properties of this silicide were reported for the first time at the beginning of the 1980s for bulk samples. Above 1200 K electrical measurements indicated a phase change from the low-temperature orthorhombic phase with the energy gap of 0.7 eV to the high-temperature tetragonal phase, having a gap of 0.44 eV [5.11]. Later investigations of this material were focused on the orthorhombic semiconducting phase Ru$_2$Si$_3$ because of its potential use for thermoelectric applications [5.50].

High quality single crystals were grown by Vining *et al.* [5.51] and Gottlieb *et al.* [5.49, 5.89] in order to study the semiconducting behavior of orthorhombic Ru$_2$Si$_3$. The resistivities measured along three different orientations are shown in Fig. 5.13.

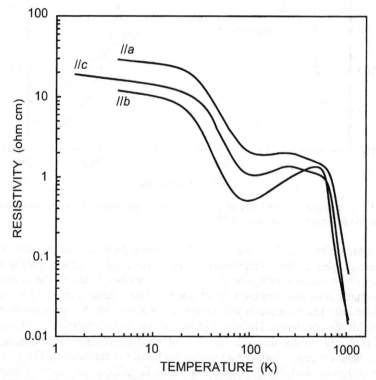

Fig. 5.13. Electrical resistivity of Ru$_2$Si$_3$ single crystals along main crystallographic axes as a function of temperature [5.89] (high temperature data from [5.51])

In the low temperature range the resistivity of Ru_2Si_3 single crystals is anisotropic with $\rho_a/\rho_b \approx 2$ and $\rho_c/\rho_b \approx 1.5$ roughly independent of temperature. A resistivity anisotropy between the b and c axis was also found above 300 K [5.51]. The variation of the resistivity of Ru_2Si_3 with temperature is typical for a doped semiconductor with an extrinsic regime below 300 K and the intrinsic regime at higher temperatures. At low temperatures ($T \le 20$ K) the carrier transport occurs via localized states (impurity band conduction). The activation energies of impurities measured in the extrinsic region vary between 19 and 24 meV [5.89]. In the intrinsic region the energy gap value calculated from the resistivity data was found to be 1.08 eV [5.51] which is considerably higher compared to 0.70 eV found in [5.11].

The room temperature resistivity of the Ru_2Si_3 single crystals ranges from 1 to 5 ohm cm. These are very close to the resistivity of polycrystalline thin films which is about 7 ohm cm [5.52].

OsSi, Os$_2$Si$_3$, OsSi$_2$. There are very few electrical data characterizing semiconducting osmium silicides. All three compounds have a negative temperature coefficient of the resistivity in the high temperature region, as is evident from the experimental results presented in Fig. 5.14.

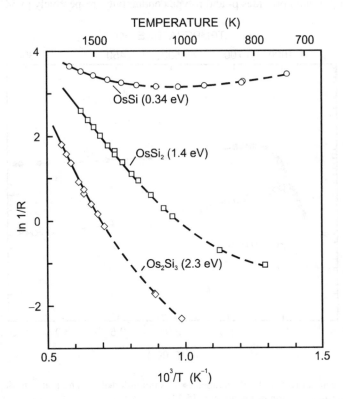

Fig. 5.14. High temperature resistance behavior of osmium silicides [5.14]

The energy gaps of about 0.34 eV for OsSi, 2.3 eV for Os_2Si_3, and 1.4 eV for $OsSi_2$ were estimated from these measurements [5.14]. The last value is noticeably smaller than that of 1.8 eV reported for $OsSi_2$ in [5.13]. This can be explained by the inclusion of 1.5 at % of aluminum in the samples investigated in [5.13].

More investigations are necessary on these semiconducting silicides in order to obtain adequate information about their transport properties.

Ir_3Si_5. Resistivity measurements of this semiconducting silicide were carried out on both bulk material and thin films [5.12, 5.54–5.56, 5.90, 5.91]. The samples studied demonstrate p-type conductivity [5.55, 5.90]. For polycrystalline Ir_3Si_5 the room temperature resistivity is 0.06 ohm cm. It increases to 0.1 ohm cm at 750 K and then decreases to 0.012 ohm cm at 1300 K, which is typical for a doped semiconductor [5.55]. The temperature dependence of the resistivity for Ir_3Si_5 crystals in the high temperature region is shown in Fig. 5.15. The p-type undoped Ir_3Si_5 becomes intrinsic around 800 K, while the sample doped with osmium reaches this state above 1050 K. No anisotropic transport properties in the crystals have been reported. The energy gap of about 1.2 eV was estimated from the high temperature resistivity behavior of the undoped sample [5.54], which is equal to the value previously reported by Petersson et al. [5.12]. Doping of Ir_3Si_5 with osmium or platinum provides p- and n-type conductivity, respectively [5.54].

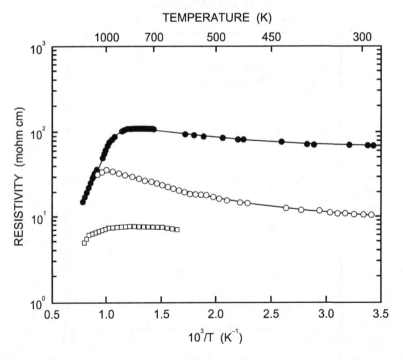

Fig. 5.15. Electrical resistivity of undoped (●), osmium-doped (○) and platinum-doped (□) Ir_3Si_5 crystals as a function of temperature [5.54]

The behavior of thin Ir_xSi_{1-x} films at low temperatures was investigated in [5.90]. Below 300 K the conductivity is extrinsic and the resistivity decreases rapidly with temperature. This results from the rapidly increasing number of charge carriers via impurity ionization.

5.3   Mobility of Charge Carriers

The charge carrier mobility can be calculated if the main scattering mechanisms of carriers are known. Experimentally, the carrier mobility can be obtained from the analysis of Hall effect and electrical conductivity measurements. The sample for these measurements is usually taken in the form of a rectangular bar. The electric field E_x is applied along the x axis while the magnetic field of the induction B is applied perpendicular to the xy plane of the sample. According to Lorentz's law electrons drift in the y direction in the presence of the magnetic field and accumulate on the sides of the sample to produce the so-called Hall voltage. An electric field E_y is induced by the presence of the magnetic field. The effect of the appearance of the electric field in the y direction is known as the Hall effect.

The Hall effect can be described with the help of the Hall coefficient, which is defined as [5.92]:

$$R_H = E_y \, (j_x B)^{-1}, \tag{5.15}$$

where j_x is the current density in the sample. It may be shown (see, for example [5.4]) that the Hall coefficient is equal to

$$R_H = -r/ne, \tag{5.16}$$

where n is the carrier concentration and r is called the Hall factor. The value of the Hall factor depends on a combination of the scattering processes. It is usually of the order of 1. Some criteria for its definition are discussed in [5.2, 5.4]. The sign of the Hall coefficient depends on the charge carrier type. Thus, the concentration and the type of charge carriers in a sample can be obtained by the measurement of the Hall coefficient. The carrier mobility is determined from the Hall coefficient and conductivity using the relation

$$\mu_H = \sigma |R_H| . \tag{5.17}$$

The mobility obtained with (5.18) is called the Hall mobility. It is not the same as the drift mobility defined by (5.7). They are connected by

$$\mu_H = r \mu. \tag{5.18}$$

For some silicides it was possible to measure the drift mobility by independent methods, however, one often has to rely only on Hall mobility data. The most representative data, as determined from the Hall effect and conductivity measurements, are listed in Table 5.3.

Table 5.3. Room temperature transport properties of semiconducting silicides

Silicide	Sample characterization	Conductivity type	Carrier concentration (cm^{-3})	Hall mobility $(cm^2/V\ s)$	Refs
Mg_2Si	single crystal	n	8×10^{16}	406	5.9
		p	4×10^{17}	56	
	single crystal	n	$10^{17}-10^{18}$	200	5.17
	single crystal	n	2.6×10^{17}	300	5.18
$BaSi_2$	polycrystalline bulk	n			5.23
$CrSi_2$	single crystal	p	4×10^{20}	14	5.26
	single crystal	p	$10^{21}-10^{23}$	40	5.94
	single crystal	p	$4\times10^{20}-2\times10^{21}$		5.61
	single crystal	p	8.5×10^{20}	$21(//c)$	5.64
				$10(\perp c)$	
	epitaxial film	p	10^{17}	2980	5.31
	polycrystalline film	p	$10^{21}-10^{23}$	0.1–10	5.94
	polycrystalline film	p	7×10^{18}	10–18	5.29
	polycrystalline film	p	4×10^{19}	10	5.28
$Cr_{1-x}Mn_xSi_2$	polycrystalline bulk	p	7.7×10^{20}	10	5.62
		(p $\rightarrow$ n with $\uparrow x$)			
$MnSi_{2-x}$	single crystal	p	2.1×10^{21}	$0.45(//c)$	5.32
				$2.9(\perp c)$	
	polycrystalline bulk	p	$1.8-2.3\times10^{21}$	1.3	5.33
	polycrystalline bulk		2×10^{19}	230	5.95
	polycrystalline film	p	7.1×10^{20}	0.6	5.36
	polycrystalline film	p	2×10^{20}	1.4	5.35
$ReSi_{1.75}$	single crystal	p	3.7×10^{18}	370	5.37
	polycrystalline film	p	3.9×10^{19}	9.0	5.38
$\beta\text{-}FeSi_2$	single crystal	p		10–20	5.41
	single crystal	p (Cr-doped)	$10^{17}-10^{18}$	<2	5.42
		n (Co-doped)			
	epitaxial film	p		1–4	5.96
	epitaxial film	p	9×10^{18}	104	5.48
	epitaxial film	p	8×10^{18}	128	5.97
	epitaxial film	p	5×10^{18}	up to 200	5.98
	polycrystalline film	p	$10^{18}-10^{19}$	0.1–0.2	5.99
	polycrystalline film	p	7.8×10^{18}	2.3	5.44
	polycrystalline film	p	10^{19}	1	5.45
	polycrystalline film	p	2×10^{18}	3	5.81
	polycrystalline film	p	1×10^{17}	97	5.46
	polycrystalline film	p	1×10^{17}	100	5.100
	polycrystalline film	n	6×10^{15}	900	5.78
Ru_2Si_3	single crystal	p	1×10^{18}	3.5	5.49
	single crystal	p	1×10^{18}	2.3	5.51
	polycrystalline bulk	p	2.8×10^{19}	29	5.53
$OsSi_2$	polycrystalline bulk	p (Al-doped)			5.13
Ir_3Si_5	polycrystalline film	p	4×10^{17}	2.4	5.12

The temperature dependence of the Hall mobility of charge carriers in semiconducting silicides and related scattering mechanisms are presented and discussed below in this section.

5.3.1 Silicides of Group II Metals

The first studies of charge carrier mobility and related scattering mechanisms in semiconducting Mg_2Si came to light in the 1950s. The mobility values were obtained from Hall effect measurements in a wide temperature range and quantitatively explained. Most of these investigations were performed on single crystals [5.9, 5.17, 5.93]. Among the silicides of this group another compound is known to be a semiconductor: $BaSi_2$. For the latter there are only resistivity data which were analyzed in Sect. 5.2.1.

Mg_2Si. The most detailed analysis of transport properties of both undoped n-type and p-type Mg_2Si single crystals was performed by Morris *et al.* [5.9]. The Hall mobilities obtained are quite high for this semiconducting silicide and amount to 406 cm^2/V s for n-type and 56 cm^2/V s for p-type crystals at 300 K. Saturation carrier concentrations are as low as $8 \times 10^{16} cm^{-3}$ and $4 \times 10^{17} cm^{-3}$, respectively. The electron-to-hole mobility ratio was 5. In the intrinsic range (above 500 K) the Hall mobility shows $T^{-5/2}$ behavior for all samples.

The electron mobility μ_e, which was obtained from resistivity and Hall data in accordance with Shockley's theory for electrons and holes in semiconductors [5.101] is shown in Fig. 5.16. An attempt to explain the temperature dependence of the mobility has been made. At $T > 160$ K μ_e varies with temperature as $T^{-2.3}$. Acoustic phonon scattering ($\mu \sim T^{-3/2}$) is not dominant in this region and scattering of charge carriers by polar optical-mode lattice vibrations and by ionized impurity was also considered. As a result, the best mobility fits to the experimental curve from 77 to 400 K gave a combination of polar optical-mode and ionized impurity scattering. At higher temperatures electron–hole scattering was also accounted for. The electron mobility can be explained by a combination of electron–hole scattering with either acoustic phonon or optical phonon scattering (Fig. 5.16).

It appears that at high temperatures electron–hole scattering becomes important. The particular role of electron–hole scattering ($\mu \sim T^{-5/2}$) in the intrinsic region was confirmed by later investigations of the Hall effect in $Mg_2Ge_xSi_{1-x}$ ternary crystals [5.18]. In the extrinsic region an average Hall mobility in these crystals follows a $T^{-3/2}$ behavior indicating scattering caused by acoustic phonons as the dominant mechanism. This disagrees with previous treatments of the Hall mobility of Mg_2Si single crystals by Morris *et al.* [5.9], which emphasized the combination of optical-mode and ionized impurity scattering in the temperature range of 77–400 K. A quite satisfactory explanation is that the $Mg_2Ge_xSi_{1-x}$ crystals were not free from defects and impurities, which essentially influence the electrical measurements in the extrinsic region. The electron Hall mobility in $Mg_2Ge_xSi_{1-x}$ was almost the same as that in the pure compound. The maximum values obtained are slightly above 300 cm^2/V s at 300 K.

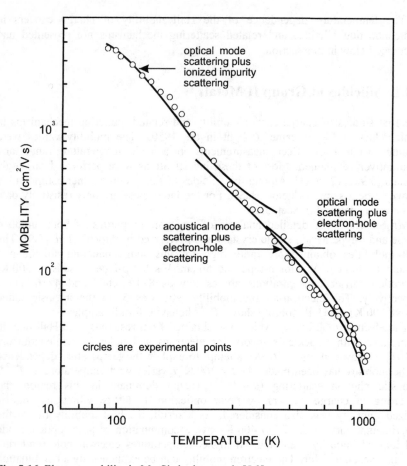

Fig. 5.16. Electron mobility in Mg_2Si single crystals [5.9]

5.3.2 Silicides of Group VI Metals

In this group of semiconducting silicides experimental mobility data have been obtained only for $CrSi_2$. Carrier transport in hexagonal WSi_2 and $MoSi_2$ was analyzed by theoretical simulation on the basis of their calculated electron band structure [5.102].

CrSi₂. Early carrier mobility investigations for this material were carried out by Shinoda *et al.* [5.26] and Nishida *et al.* [5.24, 5.62] on bulk samples. In pure $CrSi_2$ holes were found to be major charge carriers. Their concentration was rather high even in single crystals and amounted to 6.3×10^{20} cm^{-3} below 600 K [5.24]. The hole conductance can be compensated when $CrSi_2$ is doped with silicon or modified with manganese [5.26]. In all the above studies the Hall mobility is proportional to $T^{-3/2}$ above room temperature indicating acoustic phonon scattering to be dominant. Towards lower temperatures scattering processes become more complicated. At sufficiently low temperatures a $T^{3/2}$ variation of the Hall mobility of holes indicates ionized impurity scattering. The maximum mobility values obtained are about 40 cm^2/V s. From the analysis of Hall and thermoelectric power data for CrSi$_2$ single crystals in [5.24], the effective masses of holes parallel and perpendicular to the c axis were estimated to be five and three times as large as a free electron mass, respectively. The ratios of electron-to-hole mobility were $b_{//c} \ll b_{\perp c} = 0.01$. The anisotropy of hole mobility was found to be 0.5 irrespective of the temperature.

Later studies performed on polycrystalline CrSi$_2$ thin films in essence confirmed the results obtained on single crystals also showing acoustic phonon scattering as the dominant scattering mechanism above room temperature [5.28, 5.94]. Fig. 5.17 compares the temperature dependence of the mobility for thin films and bulk material.

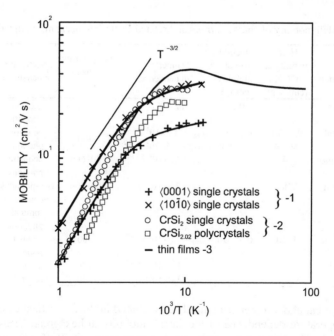

Fig. 5.17. Temperature dependence of Hall mobility of charge carriers in CrSi$_2$: 1-[5.24], 2-[5.26], 3-[5.28]

At variance with the above findings are the results obtained on A-type epitaxial layers of CrSi$_2$ grown on Si(111) by the template technique [5.31]. The transport measurements have shown that the layers have a hole concentration of 1×10^{17} cm^{-3} and Hall mobility of 2980 cm^2/V s at room temperature. The authors state that they took care to eliminate the contribution of the substrate. From the temperature dependence of the hole concentration an essentially higher energy gap than all previous studies of 0.42 eV has been derived. In the temperature range from 200 to 500 K the mobility obeys a T^{-3} law indicating a more complex scattering mechanism than by acoustic phonons. This has been tentatively attributed to two-phonon acoustic phonon scattering or scattering by optical phonons. The authors believe that the main reason behind these differences from the previous findings is the strongly reduced defect density in their epitaxial layers. Also, a very high electron mobility in $CrSi_2$, which is more than 1000 cm^2/V s at 77 K, has been estimated as a result of characterization of the electrical transport properties of $Si/CrSi_2/Si$ heterostructures fabricated by ion implantation [5.103]. The result seems to be reasonable in the light of theoretical calculations giving the value of the electron mobility of 10^7 cm^2/V s at 50 K [5.29].

The experimental data on the carrier mobility in $CrSi_2$ are collected in Table 5.4. Two mechanisms, namely scattering by acoustic phonons and by ionized impurities, are assumed to be dominant in this material. However, it can not be excluded that optical phonon scattering might contribute.

Table 5.4. Hall mobility of charge carriers and related scattering mechanisms in $CrSi_2$

Sample characterization	Hole mobility at 300 K (cm^2/V s)	Electron-to-hole mobility ratio, b	Temperature dependence	Temperature range (K)	Scattering mechanism	Refs
single crystal	14	0.01	$\mu_h \sim T^{-3/2}$	>300	acoustic phonons	5.26
single crystal	9.2(//c) 18($\perp c$)	$b_{//c} \ll b_{\perp c}$ $b_{\perp c} = 0.01$	$\mu_{h//c} = 4.8 \times 10^4\ T^{-3/2}$ $\mu_{h\perp c} = 9.4 \times 10^4\ T^{-3/2}$	>300	acoustic phonons	5.24
polycrystalline film	14		$\mu_h = 7.2 \times 10^4\ T^{-3/2}$	180–600	acoustic phonons	5.28
epitaxial film	2980		$\mu_h = 7.8 \times 10^{10}\ T^{-3}$	200–500	two-phonons, acoustic or optical phonons	5.31
polycrystalline bulk $Cr_{1-x}Mn_xSi_2$	1–10	0.01–0.1	$\mu_h = 7.0 \times 10^4\ T^{-3/2}$	>300	acoustic phonons	5.62
			$\mu_h = 1.1 \times 10^{-2}\ T^{3/2}$	<120	ionized impurities	
			$\mu_e = 3.0 \times 10^{-4}\ T^{3/2}$	<120		

Polar optical phonon scattering has been included in the theoretical calculations of the temperature dependence of the carrier mobility in hexagonal $CrSi_2$, $MoSi_2$ and WSi_2 [5.102]. For these calculations the expressions (5.7)–(5.13) were used. The mean longitudinal sound velocity and the deformational potential tensor were assumed to be 9×10^5 cm/s and 10 eV, respectively. As holes are the major carriers in the disilicides, their mobilities were calculated. The hole mobilities as a function of temperature with different ionized point defect concentrations are shown in Fig. 5.18.

According to the calculation acoustic phonon scattering practically dominates within the whole temperature range. The contribution of polar phonon scattering is within 10–15%. It should be noted that the influence of polar optical phonon scattering on the carrier mobility in the semiconducting disilicides is less

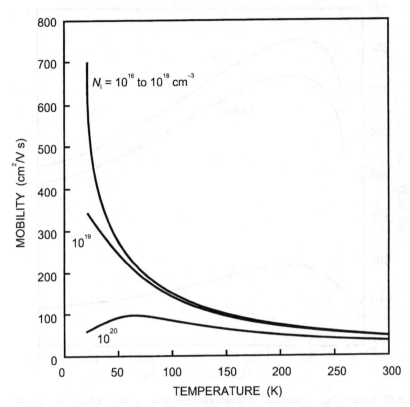

Fig. 5.18. Theoretically simulated temperature dependence of the hole mobility in CrSi₂ for different ionized impurity concentrations N_i [5.102]

important than in the case of III-V or II-VI compounds. The role of ionized point defect scattering in CrSi₂ becomes significant only at high concentrations $(N_i > 10^{20}$ cm$^{-3})$ and at low temperatures $(T < 50$ K$)$.

MoSi₂ and WSi₂. Hole mobilities versus temperature for the three hexagonal disilicides considered in the case of high defect concentrations are presented in Fig. 5.19. The carrier mobilities calculated for hexagonal MoSi₂ and WSi₂ appear to be four to five times higher than those in CrSi₂.

Comparison of the calculated mobilities with the experimental data for CrSi₂ thin polycrystalline films and crystals (6–40 cm²/V s in the temperature range 50–300 K) [5.24, 5.29, 5.62] shows good qualitative agreement between them. However, the calculated mobilities are somewhat greater than the experimental ones. The difference can be explained mainly by the difference in the calculated and measured values of the effective masses. More systematic studies are needed in order to clarify the scattering mechanism in CrSi₂ at low temperatures.

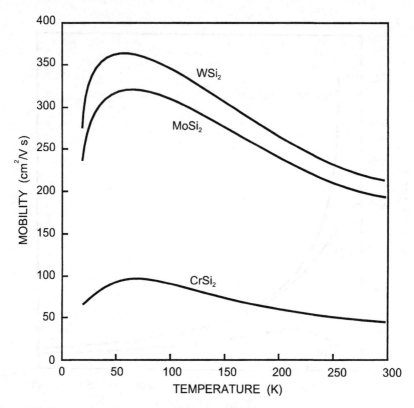

Fig. 5.19. Temperature dependence of hole mobilities in hexagonal $CrSi_2$, $MoSi_2$, and WSi_2 with $N_i = 10^{20}$ cm^{-3} [5.102]

5.3.3 Silicides of Group VII Metals

$MnSi_{2-x}$. Electrical properties of the most silicon-rich manganese silicide phase have been investigated on bulk material and polycrystalline thin films. This compound was found to be a highly degenerate p-type semiconductor. The hole concentration in the degenerate state increases with increasing x. Its room-temperature values for $MnSi_{2-x}$ bulk samples with x ranging from 0.25 to 0.28 are of the order of 10^{21} cm^{-3} [5.33]. Somewhat lower values of degenerate carrier concentration in the extrinsic range $(2-7.1)\times10^{20}$ cm^{-3} have been found in $MnSi_{1.73}$ thin films [5.35, 5.36].

The hole mobility varies with temperature above 250 K as $1.2\times10^4 T^{-3/2}$ indicating scattering caused by acoustic phonons to be a dominant mechanism in $MnSi_{2-x}$ [5.33]. Detailed investigations of the transport properties of $Mn_{15}Si_{26}$ ($MnSi_{1.73}$) single crystals grown by the Bridgman method were performed in [5.32]. While the Hall coefficient is isotropic over the whole temperature range from 77 to 1200 K, the Hall mobility, as well as the electrical resistivity (see

Sect. 5.2.3), measured in a and c directions show an anisotropy, see Fig. 5.20. As found in [5.32], the Hall mobilities decrease with increasing temperature and follow a $T^{-3/2}$ behavior above 600 K and 200 K for $\mu_{//c}$ and $\mu_{\perp c}$, respectively. In these temperature ranges acoustic phonon scattering is dominant and hole mobilities in c and a axis are expressed by $\mu_{h//c} = 0.62 \times 10^4 T^{-3/2}$ and $\mu_{h\perp c} = 1.2 \times 10^4 T^{-3/2}$, respectively. Scattering mechanisms at low temperature are not clear.

As in the case of crystals, the carrier mobility in thin $MnSi_{\sim 1.73}$ films follows a $T^{-3/2}$ dependence above 500 K, which suggests acoustic phonon scattering as the

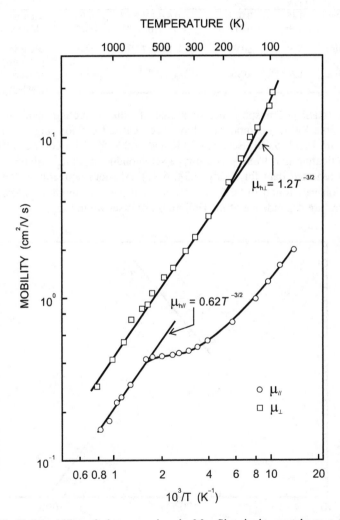

Fig. 5.20. Hall mobility of charge carriers in $Mn_{15}Si_{26}$ single crystals as a function of temperature. $\mu_{h//}$ and $\mu_{h\perp}$ indicate hole mobilities along [001] and [100] directions, respectively [5.32]

dominant mechanism [5.36]. The electron-to-hole mobility ratio was less than unity in all the above studies.

The mobility data are compared in Table 5.5. As one can see from the table, lattice phonon scattering is the dominant mechanism for all higher manganese silicide sample studied.

Table 5.5. Hall mobility of charge carriers and related scattering mechanisms in MnSi$_{2-x}$

Sample characterization	Hole mobility at 300 K (cm^2/V s)	Electron-to-hole mobility ratio, b	Temperature dependence	Temperature range (K)	Scattering mechanism	Refs
single crystal	0.45	0.023	$\mu_{h//c} = 0.62 \times 10^4\, T^{-3/2}$	>600	acoustic	5.32
	2.9	0.017	$\mu_{h\perp c} = 1.2 \times 10^4\, T^{-3/2}$	>200	phonons	
polycrystalline bulk	1.3	≤0.02	$\mu_h = 1.2 \times 10^4\, T^{-3/2}$	>250	acoustic phonons	5.33
polycrystalline film	0.6	0.36	$\mu_h \sim T^{-3/2}$	>500	acoustic phonons	5.36

ReSi$_{1.75}$ (ReSi$_2$). Transport measurements of semiconducting rhenium silicide with the stable ReSi$_{1.75}$ composition have been carried out for single crystals by Gottlieb *et al.* [5.37, 5.66]. Between 30 K and 660 K the Hall coefficient R_H was positive indicating that ReSi$_{1.75}$ is a p-type semiconductor, in accordance with the previously reported data for ReSi$_2$ [5.38, 5.95]. At room temperature the carrier concentration was found to be 3.7×10^{18} cm^{-3} and the Hall mobility 370 cm^2/V s. The temperature dependence of the Hall mobility is shown in Fig. 5.21.

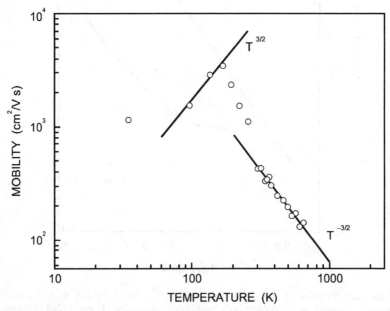

Fig. 5.21. Hall mobility of charge carriers in ReSi$_{1.75}$ as a function of temperature [5.37]

Above 300 K the hole mobility follows a $T^{-3/2}$ dependence indicating acoustic phonon scattering as the dominating mechanism. Between 100 and 200 K the mobility obeys a $T^{3/2}$ law pointing to ionized impurity scattering.

Electrical measurements on the thin layers grown epitaxially on Si(111) indicate that conduction takes place probably in two bands [5.40]. At low temperatures ($T < 180$ K) the conduction occurs in the impurity band with a low mobility. Above 180 K, thermal activation of carriers takes place from the impurity band to the valence band where the mobility is much higher. A considerable improvement of the room-temperature electrical parameters has been found after annealing of the samples in hydrogen at 850°C. After that the hole concentration decreased by one order of magnitude to 4.17×10^{17} cm^{-3} and the hole mobility in the valence band increased to 150 cm^2/V s.

5.3.4 Silicides of Group VIII Metals

The analysis of carrier mobility data and related scattering mechanisms has been performed only for semiconducting silicides of iron and ruthenium. No mobility data for the semiconducting silicides of osmium and iridium has been published.

β-FeSi₂. Studies of electrical parameters of β-FeSi$_2$ single crystals significantly extend our understanding of electronic properties of this material. The temperature dependence of the Hall coefficient in p-type crystals shows an exponential growth with decreasing temperature and a maximum at low temperatures [5.41]. A model proposed to describe the behavior of the Hall coefficient in p-type samples consists of an impurity band formed by shallow acceptor (activation energy 0.055 eV) and donor levels and an additional deep acceptor (activation energy 0.1 eV). The estimated hole effective mass of $1m_0$ is in satisfactory agreement with theoretical calculated $0.8m_0$ [5.104]. Below 100 K strong carrier freeze-out is observed in crystals and films as is illustrated by the experimental results presented in Fig. 5.22. Apparently, impurity conduction is the dominating conduction process at low temperatures. For chromium-doped crystals an activation energy of 75 meV was determined [5.83]. The temperature dependence of the Hall coefficient in p-type chromium-doped β-FeSi$_2$ single crystals is also explained in the limit of a two acceptor-one donor model [5.105]. The value of the activation energy of the deep acceptors, the concentration of the shallow and deep acceptors, as well as the concentration of the compensating donors were calculated.

The room temperature carrier concentration in β-FeSi$_2$ films and crystals could not be reduced below 10^{17} cm^{-3}. It usually ranges between 10^{18} and 10^{19} cm^{-3} for polycrystalline films and appears to be one to two orders of magnitude lower in single crystals.

A serious obstacle in the reliable determination of low-temperature transport parameters for β-FeSi$_2$ is the nonlinear variation of Hall voltage with magnetic field, which occurs in p- and n-type material. In n-type β-FeSi$_2$ crystals this nonlinear magnetic field dependence of the Hall voltage is considerably more pronounced. The effect is observed even at room temperature [5.107]. In addition,

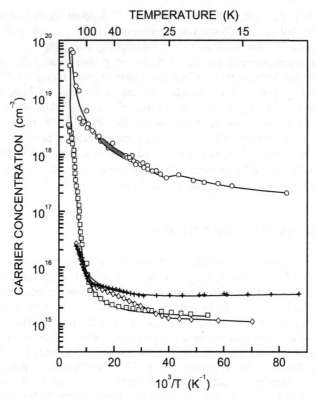

Fig. 5.22. Temperature dependence of carrier concentration in undoped (◊), chromium-doped (□) and cobalt-doped (+) β-FeSi$_2$ single crystals and polycrystalline films (○) [5.106]

a hysteresis in Hall voltage, as shown in Fig. 5.23, has been found for n- and p-type crystals at low temperatures, being also much stronger in n-type material [5.108]. The hysteresic character vanishes at 100–150 K.

Two models have been proposed for explaining the deviation of the Hall voltage from linearity. These are the occurrence of ferromagnetic order (or large magnetic moments) giving rise to an anomalous contribution to the Hall coefficient, or the presence of two types of charge carriers, light and heavy ones. The first contribution was rejected in [5.107] on the basis of magnetization measurements, which showed that the observed magnetization (less than 0.05 G) was too small to account for the observed nonlinearity. Moreover, neutron scattering experiments as well as the EPR measurements gave no indications of the existence of ferromagnetic order [5.109]. The strong decrease of the Hall coefficient with magnetic field was ascribed by Arushanov [5.107] to the presence of two types of charge carriers in the conduction band of n-type β-FeSi$_2$. These are heavy electrons with the mobility of 1.6 to 3.6 cm^2/V s and light electrons with the mobility of about 10^4 cm^2/V s at room temperature. The energetic separation of the two conduction bands was estimated to be 25 meV. These findings are in general correspondence to the band structure calculations of β-FeSi$_2$ [5.110]

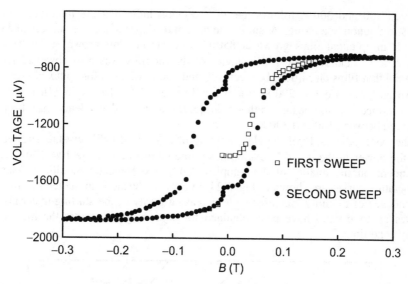

Fig. 5.23. Voltage between Hall contacts for a β-FeSi$_2$ crystal versus magnetic induction for first and second magnetic field sweep at 4.2 K [5.108]

predicting two groups of electrons, light ones at the Y point with effective mass of 0.49m_0 and heavy ones with considerably higher mass at the Λ point. The energetic separation between the two valleys is 8 meV. However, a nonlinear variation of the Hall voltage with magnetic field is also observed at low temperature in p-type material where this model is not applicable.

Later experiments with differently doped n- and p-type β-FeSi$_2$ single crystals and molecular-beam-epitaxy-grown films indicate that the anomalous contributions to the Hall resistivity have a magnetic origin. The measurements of magnetization and the result of neutron diffraction gave no indication of a magnetic phase transition of β-FeSi$_2$. It has been suggested that the cause of the anomalous contributions to the Hall resistivity found in most samples is of extrinsic nature. It has been proposed that the effects arise from cluster-like regions, which carry large magnetic moments and behave superparamagnetically [5.111]. Therefore, other models have to be considered which also give rise to an anomalous Hall effect, e.g. asymmetric scattering processes such as skew scattering or side jumps which always occur if the scattering center has nonspherical symmetry [5.112, 5.113].

Detailed studies of charge carrier scattering mechanisms in thin films and single crystals of β-FeSi$_2$ are scarce. The Hall mobility of holes continuously increases with decreasing temperature, reaches a maximum, and then decreases. The maximum mobility value of 1200 cm^2/V s was reported for p-type β-FeSi$_2$ single crystals at 74 K [5.41]. However, the results obtained by different researchers do not conform to each other. Due to the above-mentioned nonlinearity between the Hall voltage and magnetic field at low temperatures a reliable determination of transport parameters is generally limited to higher temperatures. According to [5.45] the mobility in p-type β-FeSi$_2$ thin films between 100 and 300 K follows the $T^{-3/2}$ relation indicating the importance of acoustic phonon scattering. A similar temperature dependence was observed in [5.114] in epitaxial films grown at 850°C, whereas in films grown at 750°C it follows a T^{-1} behavior. A T^{-1} behavior of the mobility was also observed for epitaxial thin films on Si(001) between 100 and 300 K whereas for epitaxial layers grown on Si(111) below 200 K mobility follows a $T^{-1/2}$ behavior [5.77]. In thin films formed by ion-beam synthesis in [5.96] a $T^{-1.9}$ dependence has been observed between 100 and 300 K.

The systematic study of the Hall mobility of holes in β-FeSi$_2$ crystals showed that the power in the $\mu \sim T^{-n}$ relationship in the temperature range 100–300 K is dependent on the quality of the samples [5.42]. Experimental mobility versus temperature curves are shown in Fig. 5.24. In crystals the slope in the temperature dependence of mobility increases with increasing purity of the starting materials. Slopes up to $n = 3.6$ have been obtained for crystals grown under the highest purity conditions.

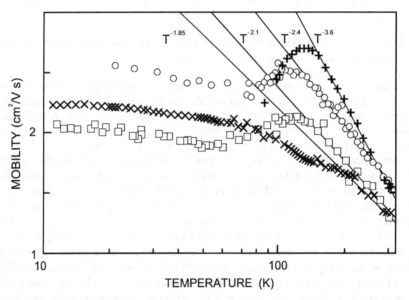

Fig. 5.24. Hall mobility of charge carriers in β-FeSi$_2$ single crystals grown from starting material of different purity as a function of temperature [5.42]. The slope of the curves near room temperature increases in samples made from purer material

The mobility data in the same temperature range for p-type material are approximated by superposition of acoustic and nonpolar phonon modes as well as scattering by charged and neutral impurities [5.115]. The changes in the mobility versus temperature when the material purity is reduced are traced. It is assumed that the purity of the crystal, among other reasons, is defined by the concentration of neutral impurities. The influence of this mechanism is negligible when the concentration of such impurities is less than $10^{17}\,\mathrm{cm}^{-3}$ and becomes more pronounced if the concentration is increased. This corresponds to the decreasing of the power in the $\mu \sim T^{-n}$ relationship, that is in good agreement with the experimental data. The data on charge carrier scattering mechanisms in β-FeSi$_2$ are collected in Table 5.6. The growth conditions of the silicide films are also presented.

Table 5.6. Hall mobility of charge carriers and related scattering mechanisms in β-FeSi$_2$

Sample characterization	Hole mobility at 300 K (cm^2/V s)	Electron-to-hole mobility ratio, b	Temperature dependence	Temperature range (K)	Scattering mechanism	Refs
single crystal	10–20	0.1				5.41
single crystal	<2		$\mu_h \sim T^{-3.6}$	100–300		5.42
epitaxial film	104		$\mu_h \sim T^{-3/2}$	100–300	acoustic phonons	5.48
epitaxial films: grown at 850°C			$\mu_h \sim T^{-3/2}$	100–200	acoustic phonons	5.114
grown at 750°C			$\mu_h \sim T^{-1}$	30–200		
epitaxial films: on Si(001)	20		$\mu_h \sim T^{-1}$	100–300		5.77
on Si(111)	>100		$\mu_h \sim T^{-1/2}$	<200	defect scattering	
epitaxial film	1–4		$\mu_h \sim T^{-1.9}$	100–300		5.96
polycrystalline film	1		$\mu_h \sim T^{-3/2}$	100–300	acoustic phonons	5.45

The room temperature mobility values for p-type polycrystalline and epitaxial β-FeSi$_2$ films usually range between 1 and 10 cm^2/V s. On the other hand, exceptionally high room temperature values between 80 and 100 cm^2/V s are also observed [5.46, 5.48, 5.77, 5.97, 5.100, 5.116, 5.117] for polycrystalline and epitaxial films fabricated by different techniques.

The Hall mobility of electrons of 900 cm^2/V s at room temperature in polycrystalline β-FeSi$_2$ thin films was reported [5.78]. Considering these differing data one has to remember that the silicon substrate markedly influences the experimental results. The influence of the substrate on the Hall-effect data has been investigated by Brehme *et al.* [5.82]. β-FeSi$_2$ thin films were grown by means of a template technique on high resistivity (001) and (111) silicon substrates. The near-room temperature part of the Hall coefficient was found to be significantly influenced by the substrate. At lower temperatures the influence is strongly reduced. On the other hand, the straightforward evaluation of the Hall-effect data for β-FeSi$_2$ is limited by the occurrence of the anomalous Hall coefficient and the transition to impurity-band conduction at low temperatures. These phenomena define a rather narrow observation window for Hall-effect studies of free carries in thin β-FeSi$_2$ films on insulating substrates. Resistivity measurements were found to be less influenced up to 300 K.

Ru$_2$Si$_3$. As in the case of β-FeSi$_2$, much interest is attracted by the electrical behavior of Ru$_2$Si$_3$, which is considered as a promising material for thermoelectric applications. Hall effect and electrical resistivity measurements on single crystals of this semiconductor were performed in the wide temperature range from 1.6 to 1300 K [5.50, 5.51, 5.89].

The Hall coefficient measured on the undoped material at $T \leq 300$ K is positive down to 10 K and becomes negative below [5.89]. Undoped single crystals obtained in [5.50] were n-type between 300 and 550 K; above 550 K the conduction was dominated by thermally activated holes. The temperature dependence of Hall mobility of charge carriers in Ru_2Si_3 single crystals is presented in Fig. 5.25. The high-temperature data were taken from [5.51].

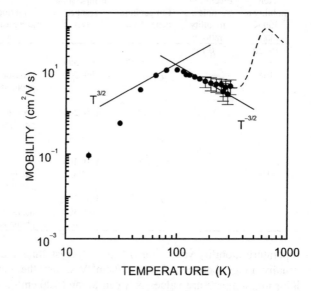

Fig. 5.25. Temperature dependence of Hall mobility of charge carriers in Ru_2Si_3 single crystals [5.89]

Below 100 K conduction in a semiconductor is normally dominated by scattering due to ionized impurities and the charge carrier mobility is proportional to $T^{3/2}$. In the case of Ru_2Si_3 this dependence has not been observed, at least over a large temperature region. Between 100 and 300 K the mobility follows a $T^{-3/2}$ law which suggests lattice phonon scattering as the dominant mechanism.

Measurements above room temperature up to 1300 K showed a strong increase of mobility with maximum values of 90 cm^2/V s around 700 K. Above 750 K the mobility consists of two contributions, a temperature-dependent hole mobility varying as $1.09 \times 10^5/T$ and a temperature-independent electron mobility of 77.2 cm^2/V s [5.51].

In order to realize the predicted potential of Ru_2Si_3 for thermoelectric applications suitable dopants have to be identified which are electrically active but without seriously reducing the mobility. In [5.118] manganese and iridium were found to be useful as p- and n-dopants, respectively, however, without achieving fundamental progress in raising the figure of merit. Further attempts have been made in [5.119, 5.120] to find proper dopants for Ru_2Si_3, which will be dealt with in Sect. 5.4.4.

5.4   Thermoelectric Properties

Semiconducting silicides were first considered as thermoelectric materials more than thirty years ago when the search became stronger for semiconductors with a high efficiency of thermoelectric energy conversion. Since that time the higher silicides $CrSi_2$, $MnSi_{1.73}$ and β-$FeSi_2$, the monosilicide $CoSi$ and Mg_2Si based solid solutions have been extensively investigated, see the early reviews [5.121, 5.18]. A wide spectrum of n-type and p-type single crystals and polycrystalline bulk samples and thin films have been grown and first prototypes of thermoelectric generator modules have been prepared and tested. In the beginning of the 1990s also Ru_2Si_3, Ir_3Si_5 and $ReSi_{1.75}$ received interest as possible high efficient thermoelectric materials. At present β-$FeSi_2$, Ru_2Si_3 and Mg_2Si based materials are regarded as the most promising silicides for thermoelectric applications. According to their energy gap as well as their thermal stability and corrosion resistance they are of interest mainly for generators working in the temperature range 500–900 K. Recent reviews are given in [5.68, 5.122, 5.123].

The general objective of thermoelectric studies is to develop materials with a high thermoelectric figure of merit $Z = S^2\sigma/(\kappa_{latt} + \kappa_{el})$ with S being the Seebeck coefficient, σ the electrical conductivity, κ_{latt} the lattice and κ_{el} the electronic components of the thermal conductivity. Fig. 5.26 shows schematically the dependence of the kinetic coefficients making up Z on the charge carrier concentration.

It is obvious that high Z values can only be achieved within a limited range of high carrier concentrations where the power factor is close to its maximum. Further, the lattice thermal conductivity κ_{latt} has to be minimized by incorporation of phonon-scattering centers whereas its electronic part κ_{el} is determined by the Wiedemann–Franz law $\kappa_{el} = L\sigma T$, where L is the Lorentz number and T the temperature. The opposite dependencies of the kinetic coefficients on the carrier concentration limit the possible Z values. As a result, all established thermoelectric materials such as $(Bi, Sb)_2(Te, Se)_3$, $Pb(Te, Se)$, SiGe and others fulfil the empirical bound of the dimensionless figure of merit $ZT \leq 1$ [5.124]. In a simple one-band model the figure of merit is approximately proportional to the parameter $\beta = T^{5/2}N_v m^{*\,3/2} \mu/\kappa_{latt}$ with μ being the charge carrier mobility, m^* the effective mass of the carriers and N_v the number of equivalent band valleys. Materials with high β-values are expected to be good thermoelectrics.

For most of the semiconducting silicides promising β-values have been estimated [5.122]. This is a result of low lattice thermal conductivities due to the complex lattice structure and of high effective masses that originate from the bonding and antibonding d states at the band edges of transition metal silicides.

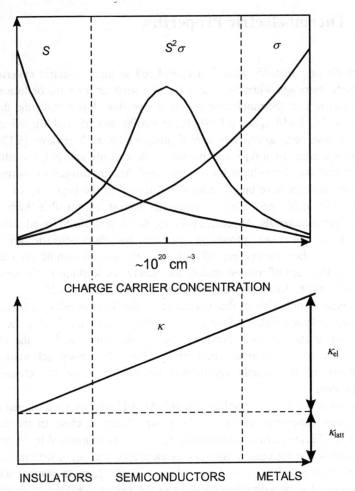

Fig. 5.26. Schematic dependence of the electrical conductivity σ, Seebeck coefficient S and thermal conductivity $\kappa = \kappa_{latt} + \kappa_{el}$ on the charge carrier concentration

bonding and antibonding d states at the band edges of transition metal silicides. Against this the charge carrier mobilities are rather small which limits the power factor. The semiconducting silicides are "classical" thermoelectrics in that their efficiencies are below the empirical limit $ZT = 1$. There are estimations of $ZT > 1$ for Mg_2Si based solid solutions [5.125] and for Ru_2Si_3 [5.126] but until now no experimental verification has been presented.

In order to achieve $ZT > 1$ new concepts have been recently developed. These are the phonon glass electron crystal (PGEC) concept for bulk materials [5.127] and the multilayer concept for thin films [5.128, 5.129]. In the multilayer concept an increase of Z can be achieved on the basis of the established thermoelectrics mainly by two effects, by quantum confinement in a quantum well structure [5.128] and by an enhanced scattering of phonons at the interfaces [5.129]. Also, semiconducting silicides should be of interest for the multilayer concept especially with silicon-based barrier layers.

In the following sections the thermoelectric properties of the semiconducting silicides are discussed in detail.

5.4.1 Silicides of Group II Metals

Among the semiconducting silicides of the Group II metals the compound Mg_2Si and its solid solutions with Mg_2Ge and Mg_2Sn are well known promising thermoelectric materials. According to the width of their energy gap they are candidates for thermoelectric energy conversion in the temperature range from 500 to 800 K [5.130]. $BaSi_2$ has also been considered with respect to its thermoelectric potential [5.122], but no attempts are known to verify experimentally the theoretically estimated figure of merit $ZT \approx 0.2$ at room temperature.

Mg_2Si. More investigations have been performed on $Mg_2Si_{1-x}Ge_x$ and $Mg_2Si_{1-x}Sn_x$ than on Mg_2Si because solid solutions have lower thermal conductivities due to an enhanced phonon scattering than their pure components. In the early work [5.18] it was shown that in the $Mg_2Si_{1-x}Ge_x$ system the lattice thermal conductivity has its minimum at $x = 0.4$. Mainly this composition was further investigated to optimize the figure of merit by appropriate doping. Antimony and silver were found to be an effective donor and acceptor of $Mg_2Si_{0.6}Ge_{0.4}$, respectively [5.125]. With these dopants room temperature carrier concentrations up to 5.4×10^{19} cm^{-3} were achieved which is in the desired range for highly efficient materials. The absolute thermopower reaches a maximum at about 600 K of 400 µV/K in n-type and 300 µV/K in p-type samples. In [5.125] the thermal conductivity was measured only at room temperature, but its temperature dependence was estimated with the help of the Wiedemann–Franz law for κ_{el} and under the assumption of a $T^{-1/2}$ dependence of κ_{latt}. The resulting figure of merit shows a maximum at about 650 K with ZT larger than unity: $(ZT)_{max} = 1.07$ for n-type and $(ZT)_{max} = 1.68$ for p-type material. The temperature dependence of κ_{latt} assumed in [5.125] was confirmed for antimony-doped $Mg_2Si_{0.6}Ge_{0.4}$ up to 400 K [5.131] but the verification of the estimated values of ZT is still open.

For solid solutions $Mg_2Si_{1-x}Sn_x$ $ZT = 0.8$ [5.132] and for the binary n-type compound Mg_2Si ZT-values up to 0.67 [5.133] have been reported.

For a practical use of Mg_2Si and $Mg_2(Si, Ge, Sn)$ solid solutions one has to solve preparation and handling problems caused by magnesium evaporation and oxidation as a result of its high vapor pressure and chemical reactivity. Spark plasma sintering with subsequent hot pressing was demonstrated in [5.133] to be successfully applied to produce Mg_2Si thermoelectric elements. Fig. 5.27 shows the resulting power factor and figure of merit of sintered antimony-doped n-type samples in their dependence on the sintering temperature. For silver-doped p-type material such high values are not yet achieved. In order to avoid the mentioned handling problems, alternative preparation methods such as mechanical alloying were also successfully applied [5.134].

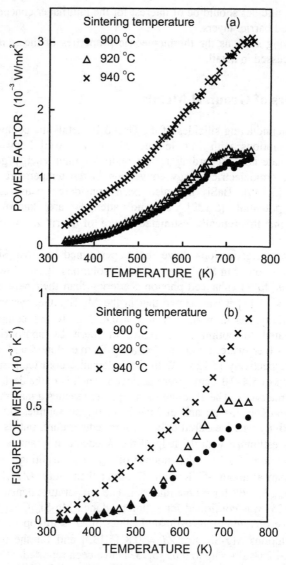

Fig. 5.27. Temperature dependence of power factor (a) and figure of merit (b) of 0.6 at %
antimony-doped Mg_2Si thermoelectric element for various sintering temperature [5.133]

5.4.2 Silicides of Group VI Metals

CrSi₂. $CrSi_2$ is the only semiconducting silicide of this group that has been
considered as a thermoelectric material [5.121]. It was found to be a degenerate
semiconductor with hole concentrations of about $5 \times 10^{20} \, cm^{-3}$. According to the
lattice structure, electrical resistivity and thermopower of single crystals are

anisotropic with respect to the crystallographic c axis. This anisotropy was observed in the temperature range from 85 to 1100 K with ratios of $S_{//}/S_\perp = 1.7$ and $\rho_{//}/\rho_\perp = 1.9$ at room temperature. This was explained by anisotropic carrier scattering [5.24, 5.135].

In the case of the thermopower parallel to the c axis mixed acoustic phonon and ionized impurity scattering have to be taken into account, but perpendicular to the c axis only acoustic phonon scattering has to be accounted for.

Two ways were attempted to modify the thermoelectric properties of chromium disilicide. These are a change of the Si/Cr ratio [5.121] and doping with $3d$ metals. In the latter approach titanium and vanadium have been used as acceptors while manganese and iron as donors. In the $Cr_{1-x}Ti_xSi_2$ and $Cr_{1-x}V_xSi_2$ mixed crystals a continuous change from semiconducting to metallic behavior was found with increasing x without any improvement of the power factor [5.136, 5.137]. In contrast, $Cr_{1-x}Mn_xSi_2$ alloys show for $x < 0.12$ an increase of the thermopower and the power factor [5.62].

Only for large manganese contents was a change from p-type to n-type conduction observed. The kinetic properties of the $Cr_{1-x}Mn_xSi_2$ alloys have been explained by different two-band models. One model includes hole and electron conduction in one valence and one conduction band [5.62].

In this model a large effective electron mass ($m^* \approx 20m_0$ at $x = 0$) and a low electron mobility are responsible for the dominance of p-type conductivity up to the high manganese contents of $x = 0.12$. In another model [5.138], two valence bands with different carrier mobilities and effective masses are expected.

Sintered $CrSi_2$ was used as the p-leg in modules for thermoelectric energy generation [5.121]. In these modules CoSi was applied as the n-leg. The best figure of merit of $CrSi_2$ was achieved in samples with a small excess of chromium, as illustrated in Fig. 5.28, but the derived value of $ZT = 0.24$ at room temperature is too small for large-scale applications. In [5.139] the slip casting technique was successfully applied to make $CrSi_2$ legs of a complex shape. Thereby, the addition of iron somewhat improves the thermoelectric properties.

The thermoelectric properties of $CrSi_2$ have also been investigated in thin films [5.140]. It was shown that in magnetron sputtered $CrSi_{2+x}$ films with an excess of silicon the power factor can be much larger in comparison with single phase films.

5.4.3 Silicides of Group VII Metals

The semiconducting silicides of this group $MnSi_{2-x}$ ($MnSi_{\sim1.73}$) and $ReSi_{1.75}$ have found very different levels of interest as possible thermoelectric materials owing to their different energy gaps of about 0.70 eV and 0.15 eV, respectively. While p-type $MnSi_{\sim1.73}$ is considered as an efficient material for thermoelectric power generation between 500 and 900 K the thermoelectric properties of the narrow-gap semiconductor $ReSi_{1.75}$ have found interest only recently. Its application would be restricted to the room temperature range.

$MnSi_{2-x}$. Reviews on the higher manganese silicides and their thermoelectric properties are given in [5.68, 5.121]. According to the lattice structure the kinetic properties of single crystals show a strong anisotropy over a wide temperature

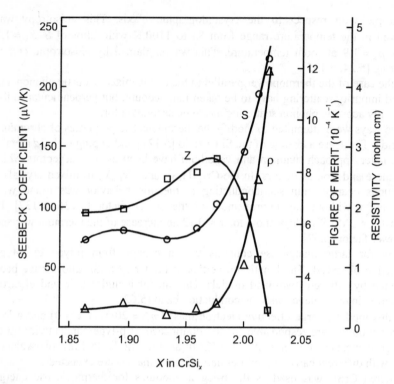

Fig. 5.28. Thermoelectric properties of CrSi$_x$ as a function of the silicon content [5.121]

range [5.141]. The thermopower anisotropy reaches values above 100 μV/K between room temperature and 600 K. This strong anisotropy makes this compound a promising candidate for transverse thermoelectric effects that can be used in one leg thermoelements [5.142].

Single crystals of MnSi$_{2-x}$ grown from the melt are always inhomogeneous. They exhibit stripes of metallic MnSi which lead to a decrease of the Seebeck coefficient. Therefore, polycrystalline samples can have a larger efficiency than single crystals [5.143]. Fig. 5.29 shows experimental temperature dependencies of thermal conductivity and figures of merit for polycrystalline undoped and doped p-type MnSi$_{\sim1.73}$ samples. The maximum figure of merit for the complex doped MnSi$_{\sim1.73}$ is the highest value for semiconducting silicides that has been published.

In [5.144] a thermoelectric module is described made of p-type MnSi$_{\sim1.73}$ and Co$_{1-x}$Me$_x$Si solid solutions (with Me = Ni, Fe) as the n-leg. For this module a rather high efficiency of 8% was estimated for a temperature difference of about 700 K. In [5.145] a module is described that consists of p-type MnSi$_{\sim1.73}$ and n-type β-FeSi$_2$. Both legs have been prepared by a common powder metallurgical process. The efficiency of this generator reaches 3% at a temperature difference of about 600 K.

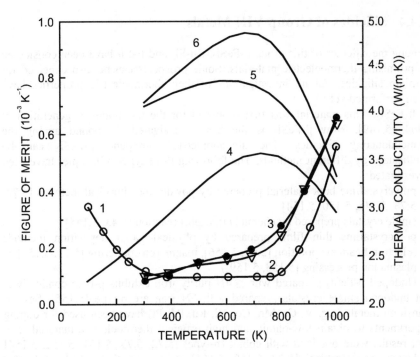

Fig. 5.29. Temperature dependence of thermal conductivity (1–3) and figure of merit (4–6) for undoped (1,4), complex doped (2,6) and germanium-doped (3,5) MnSi$_{1.73}$ [5.68]

ReSi$_{1.75}$. This compound became of interest recently within an approach to develop new heterogeneous thermoelectric materials. Two kinds of heterogeneity are under investigation. These are quasi two-dimensional multilayer systems made of conventional thermoelectric materials and barrier layers [5.128, 5.129] and composite systems with random distribution of the phases involved. In such composites the figure of merit can usually not exceed the best value of its components [5.146], but in nanodisperse composites the situation can be different. This was shown with nanocrystalline thin films made of ReSi$_{1.75}$ grains and a second amorphous Re-Si phase [5.147]. Amorphous Re-Si films were found [5.147] to form a nanocrystalline structure whose electrical conductivity and thermopower do not fulfil the effective medium equations.

As a consequence, the nanocrystalline Re-Si films are not subjected to the restrictions of classical composite systems. It is still an open question whether this effect can be used to obtain materials with higher figure of merits.

There is also an interest to use thermoelectric ReSi$_{1.75}$ thin films in silicide/silicon multilayers due to the small lattice mismatch between ReSi$_{1.75}$ and Si(100) [5.148]. Such multilayers with a quasi two-dimensional behavior are expected to exhibit a much larger figure of merit than the corresponding bulk materials [5.128, 5.129].

5.4.4 Silicides of Group VIII Metals

Among the silicides of this group β-FeSi$_2$, Ru$_2$Si$_3$ and Ir$_3$Si$_5$ have been considered as promising thermoelectric materials. Some prospects can be also expected for osmium silicides, but no investigations of their thermoelectric properties have been performed yet.

$\boldsymbol{\beta}$-**FeSi$_2$.** This material was first proposed for the thermopower generation in 1964 [5.149]. Now β-FeSi$_2$ is the best investigated compound among the semiconducting silicides. The stoichiometric compound β-FeSi$_2$ can be synthesized by different methods. The following three types of samples have been investigated:

1. polycrystalline bulk material prepared by powder metallurgical methods [5.72, 5.73, 5.145, 5.150–5.154]
2. single crystals prepared by chemical transport reaction [5.43, 5.155]
3. polycrystalline thin films prepared by physical vapor deposition methods (electron beam evaporation [5.99, 5.156], magnetron sputtering [5.71, 5.157] or plasma ion processing [5.158, 5.159]).

Undoped β-FeSi$_2$ prepared with $\leq$ 4N purity iron exhibits p-type conductivity but undoped single crystals prepared with 5N iron are n-type [5.155]. Several transition metals (e.g. V, Cr, Mn, Co, Ni, Ru, Pd, Pt) have been used for doping experiments to obtain low-ohmic and high efficient thermoelectric material. The best results were obtained with cobalt (n-type) [5.72, 5.73, 5.151, 5.152, 5.154] and manganese (p-type) [5.72, 5.150, 5.151] at concentrations of about 1 at %. Efficient p-type β-FeSi$_2$ was also obtained by doping with aluminum, which replaces silicon [5.73, 5.152] and in thin films by doping with chromium [5.71, 5.157]. An improvement of the efficiency was achieved for cobalt-doped n-type β-FeSi$_2$ by partial compensation with aluminum [5.145]. Table 5.7 gives a summary of the thermoelectric room temperature data for the doped polycrystalline bulk and thin film materials.

Table 5.7. Room temperature electrical and thermoelectric properties of doped polycrystalline β-FeSi$_2$

Dopant	σ (ohm cm)$^{-1}$	S (μV/K)	μ (cm^2/V s)	$S^2\sigma$ (μW/cm K^2)	κ (W/mK)	Refs
Co	40 – 150	−150 − −450	0.3	2.2 – 3.5	4 – 10	5.72, 5.73, 5.151, 5.152, 5.154
Co	500	−100 − −270		11	5	5.71, 5.157
Co+Al	230	−190		8.3	4.5	5.145
Ni	13	−110		0.17		5.71
Pt	110	−200		4.9		5.71
Mn	5 – 10	280 – 450	8	0.4 – 1.5	6	5.71, 5.72, 5.150, 5.151, 5.153, 5.157
Cr	30 – 35	220 – 300		1.4 – 3.0	12	5.71, 5.153, 5.157
Al	190	170	1.6 – 4	6	6	5.73, 5.152

The temperature dependence of the electrical conductivity and thermopower are characterized by four ranges with different carrier transport mechanisms: impurity band conduction, extrinsic range, saturation of band conduction and intrinsic range. Fig. 5.12 shows as an example the resistivity and thermopower, characterized by the Seebeck coefficient, of chromium-doped β-FeSi$_2$ single crystals in the temperature range from 4 K (70 K) to 950 K. The different behavior of single crystals and polycrystalline bulk material (ceramics) is shown in Fig. 5.30 for cobalt-doped samples. The different magnitudes and temperature dependencies shown are not only the result of the different structure but also of a different content of unknown impurities. Their influence on the thermoelectric properties makes it difficult to compare results from various authors. As is evident from Table 5.7 the published data scatter within a broad range.

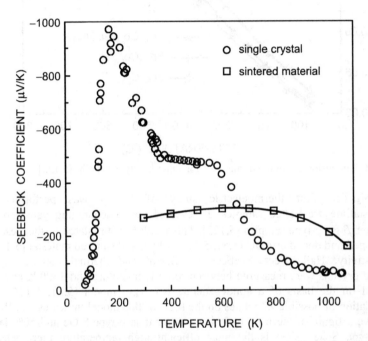

Fig. 5.30. Comparison of the Seebeck coefficient of about 1 at % cobalt-doped β-FeSi$_2$ single crystals [5.88] and sintered bulk material [5.151]

Therefore, the numerous attempts to optimize the thermoelectric properties of β-FeSi$_2$ ceramics [5.72, 5.73, 5.145, 5.150–5.154] can only be evaluated in connection with the preparation techniques.

In [5.123, 5.152] cobalt-doped n-type and aluminum-doped p-type β-FeSi$_2$ ceramics were hot pressed from source materials of 4N purity. Fig. 5.31 presents the figure of merit, which was achieved for n-type material with different cobalt concentrations and compensation degrees. This ceramic was applied for a generator module working up to 900°C. The efficiency of the p-type β-FeSi$_2$ ceramic is smaller in comparison with MnSi$_{-1.73}$. But the high chemical and thermal stability, the low price of the source materials and a simple preparation method are advantages in favor of the iron disilicide.

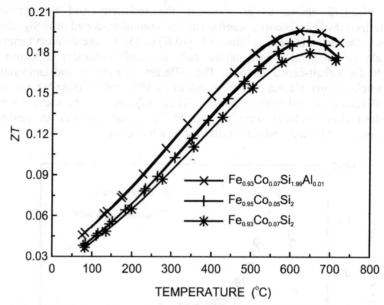

Fig. 5.31. Dimensionless figure of merit ZT for sintered n-type β-FeSi$_2$ [5.123]

Ru$_2$Si$_3$. The first thermoelectric study of Ru$_2$Si$_3$ was performed on polycrystalline arc-melted samples. It revealed very encouraging values of the parameter β for p-type samples [5.122]. This result has stimulated further research on undoped and doped crystals [5.51, 5.118, 5.119] and sintered material [5.120].

Resistivity, Hall effect, Seebeck coefficient and thermal conductivity of undoped crystals were measured between room temperature and 1300 K and then analyzed on the basis of a nondegenerate two-band model [5.51, 5.126]. The extrapolation of possible ZT values on the basis of this model indicates that Ru$_2$Si$_3$ may have a figure of merit up to 3 times larger than p-type SiGe and 50% larger than n-type SiGe (SiGe is the most efficient high temperature thermoelectric material). The model also suggests that carrier concentrations are required in excess of $10^{20}\,\text{cm}^{-3}$ to realize such high values. Further, the mobility must not degrade too much with the increasing doping level [5.126].

The efforts to find effective dopants for low-ohmic p-type and n-type Ru$_2$Si$_3$ were up to now only partially successful. Manganese was identified as an effective acceptor [5.118] and iridium, gallium and rhodium were found to be useful dopants for n-type Ru$_2$Si$_3$ [5.119, 5.120].

It was also suggested that dopants replacing silicon might be more useful than dopants replacing ruthenium. Meanwhile, the required carrier concentration has not been achieved yet. But the estimation of the figure of merit for rhodium-doped samples [5.119] shows that Ru$_2$Si$_3$ should be further considered as a promising thermoelectric material.

Ir$_3$Si$_5$. Thermoelectric studies of this semiconducting silicide started in 1992 with the investigation of the silicon-rich part of the Ir-Si phase diagram [5.55]. Among the five phases found in this part Ir$_3$Si$_5$ was identified as a semiconductor.

The preparation of single phase, polycrystalline Ir_3Si_5 samples was successfully performed with arc melting. Single crystals were also grown in the Bridgman-like furnace [5.54]. The kinetic properties of osmium-doped (p-type) and platinum-doped (n-type) samples were measured between room temperature and 1300 K and then compared to those of SiGe alloys. The thermal conductivity was found to be smaller with respect to SiGe but the power factor remained clearly below that of SiGe. This is illustrated in Fig. 5.32.

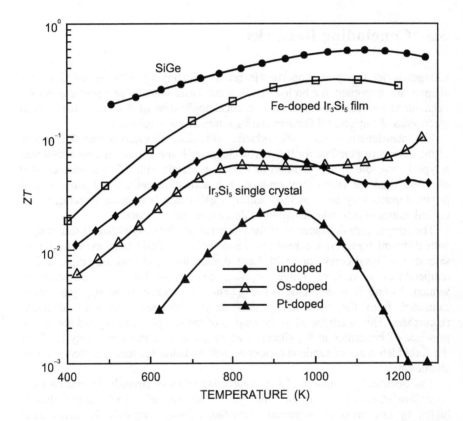

Fig. 5.32. Dimensionless figure of merit ZT for Ir_3Si_5 single crystals [5.54] and thin films [5.160]

The resulting maximum ZT is close to 0.1, which is about a factor of five below the best SiGe values. Optimum doping and alloying would give better results, but an appearance of ZT values exceeding those typical for SiGe looks unlikely [5.54].

Thermoelectric properties of polycrystalline Ir_xSi_{1-x} films have been investigated in the composition range from $0.3 \leq x \leq 0.4$ [5.160, 5.161]. The films were prepared by cosputtering and coevaporation onto unheated substrates followed by annealing up to 1200°C. The crystallization took place at about 690°C. The phases, which grow during annealing, correspond to the phase diagram of [5.55], but in the composition range of $0.35 \leq x \leq 0.4$ a metastable phase is formed before the final one. Single-phase films close to the Ir_3Si_5 composition were doped with iron. The best power factors were found in evaporated films with 1 at % of iron [5.160]. Assuming the thermal conductivity of the bulk material [5.54], the highest figure of merit of the iron-doped films reaches about 2/3 of that of SiGe.

5.5 Concluding Remarks

Charge carrier transport properties in semiconducting silicides discussed in this chapter are important for both extending our basic knowledge about the carrier behavior in semiconductors and the practical application of these new materials in electronics. Their general features can be summarized as follows.

Not intentionally doped semiconducting silicides are usually p-type. The nature of the defects responsible for the p-type character is not known. A conversion into n-type is possible by the addition of transition metals with a larger number of d electrons than the metal they replace. The activation energies of most n- and p-type dopants vary between 0.05 and 0.14 eV, which is in good agreement with typical values of activation energies of dopants in semiconductors.

The temperature dependence of the electrical conductivity indicates four ranges with different transport mechanisms: impurity band conduction, extrinsic range, saturation of band conduction with ionized impurities, and intrinsic range. At low temperatures a variable range hopping conduction has been observed in semiconducting iron disilicide β-FeSi$_2$. There is scatter in energy gap values estimated from the conductivity data in the intrinsic regime by different researchers. This is attributed to the quality of the samples investigated and to the presence of impurities in the silicides, which are known to form energy levels in the forbidden band of semiconductors as well as being scattering centers for free charge carriers.

The electrical parameters of semiconducting silicides strongly depend on their crystalline structure. The room temperature resistivity of polycrystalline films is higher by one order of magnitude than that of single crystals. It ranges from 0.5 mohm cm for CrSi$_2$ to 100 ohm cm for BaSi$_2$. The data for polycrystalline films often exhibit a large discrepancy, which is related to the structural peculiarities of the films. The carrier concentrations are relatively high even in epitaxial films and crystals. They usually vary between 10^{18} and 10^{20} cm^{-3} for polycrystalline films and appear to be one to two orders of magnitude lower in single crystals. The reliable room temperature mobility values range between 40 cm^2/V s (in β-FeSi$_2$) and 370 cm^2/V s (in ReSi$_{1.75}$). The electron-to-hole mobility ratio is much less than unity, i.e. the mobilities of electrons are much lower than those of holes. Electron and hole effective masses are $\geq 1m_0$ for all semiconducting silicides.

The anisotropic effects have been found in many single crystal silicides. A carrier transport anisotropy between the a and c axis directions is apparent in the single crystals of CrSi$_2$, MnSi$_{\sim 1.73}$, ReSi$_{1.75}$, and Ru$_2$Si$_3$.

The straightforward evaluation of the Hall-effect data has been found in β-FeSi$_2$ to have limited accuracy due to the occurrence of the anomalous Hall-effect at low temperatures. In thin films the silicon substrate exhibits a growing influence on Hall-effect measurements at high temperature. These phenomena define a rather narrow observation window for Hall-effect studies of free carriers in thin β-FeSi$_2$ films on insulating substrates. The effect has not been analyzed in other semiconducting silicides, but it may be rather important.

Using a reasonable set of material parameters, the temperature dependence of the mobility for semiconducting silicides has been explained by means of "classical" scattering mechanisms, namely by the acoustic, polar and nonpolar phonon modes as well as by the charged and neutral impurity scattering. At room temperature and above, acoustic phonon scattering seems to be a dominant carrier scattering mechanism in many semiconducting silicides, but other mechanisms might contribute as well. At low temperatures scattering by ionized impurities becomes dominant.

Semiconducting silicides look attractive for thermoelectrics. The numerous efforts to optimize their thermoelectric properties have been focused mainly on two problems: to find effective dopants for low ohmic material and to minimize the thermal lattice conductivity. However, the best values obtained until now for the power factor and figure of merit did not exceed the values of the established thermoelectric materials, the SiGe alloys and lead chalcogenides. Some of the promising estimations are not yet verified, especially those for Mg$_2$Si and Ru$_2$Si$_3$ based materials.

With respect to bulk material the current interest lies mainly in the development of sintering techniques for highly doped compounds, e.g. for β-FeSi$_2$ and Mg$_2$Si. There is no expectation to obtain bulk material with $ZT > 1$ but some of the semiconducting silicides are of further interest for thermoelectric applications in spite of their thermal stability and chemical resistance.

There is new interest in semiconducting silicides as thermoelectric thin film material. Compounds with energy gaps below 1 eV are promising candidates as thermoelectric components in multilayer systems with reduced dimensionality. Such systems are expected to have efficiencies much larger than the corresponding bulk material.

In conclusion, it should be noted that the charge carrier transport properties alone do not allow a clear understanding of all microscopic phenomena in semiconducting silicides. They should be implemented together with structural, fundamental electronic and optical properties.

References

Chapter 1

1.1 G.V. Samsonov, L.A. Dvorina, B.M. Rud: *Silicides* (Metallurgiya, Moscow, 1979) (in Russian); G.V. Samsonov, I.M. Vinitskii: *Handbook of Refractory Compounds* (IFI/Plenum, New York, 1980).

1.2 S.P. Murarka: *Silicides for VLSI Applications* (Academic Press, New York, 1983).

1.3 K. Maex, M. van Rossum (Eds.) *Properties of Metal Silicides* (INSPEC, IEE, London, 1995).

1.4 V.E. Borisenko, P.J. Hesketh: *Solid State Rapid Thermal Processing of Semiconductors*, (Plenum, New York, 1997).

1.5 *Powder Diffraction Data File* (JCPDS International Center for Diffraction Data, Swarthmore, PA, 1990), Card 35-773.

1.6 R.J. LaBotz, D.R. Mason, D.F. O'Kane: The thermoelectric properties of mixed crystals of $Mg_2Ge_xSi_{1-x}$, *J. Electrochem. Soc.* **110**(2), 127–134 (1963).

1.7 H. Schäfer, K.H. Janzon, A. Weiss: $BaSi_2$, a phase with discrete Si_4 tetrahedra, *Angew. Chem. Int. Ed. Engl.* **2**(7), 393–394 (1963).

1.8 M. Imai, T. Hirano: Electrical resistivity of metastable phases of $BaSi_2$ synthesized under high pressure and high temperature, *J. Alloys Comp.* **224**(1), 111–116 (1995).

1.9 R.A. McKee, F.J. Walker, J.R. Conner, R. Raj: $BaSi_2$ and thin film alkaline earth silicides on silicon, *Appl. Phys. Lett.* **63**(20), 2818–2820 (1993).

1.10 J. Evers, G. Oehlinger, A. Weiss: Solid solutions $M_{1-x} Sr_xSi_2$ (M= Ca, Eu, Ba) and $BaSi_{2-y}Ge_y$ with $SrSi_2$-type structure, *J. Less-Common Met.* **69**(2), 399–402 (1980).

1.11 A. Betz, H. Schäfer, A. Weiss, R. Wulf: Zur Kenntnis der digermanide des strontiums und bariums, $SrGe_2$ und $BaGe_2$, *Z. Naturforsch.* **23b**, 878 (1968).

1.12 I.J. Ohsugi, T. Kojima, I.A. Nishida: Temperature dependence of the magnetic susceptibility of a $CrSi_2$ single crystals, *Phys. Rev. B* **42**(16), 10761–10764 (1990).

1.13 K. Maex, M. Van Rossum, A. Reader: Crystal structure of TM silicides, in: *Properties of Metal Silicides*, edited by K. Maex, M. Van Rossum (INSPEC, IEE, London, 1995), pp. 3–14.

1.14 U. Gottlieb, B. Lambert-Andron, F. Nava, M. Affronte, O. Laborde, A. Rouault, R. Madar: Structural and electronic transport properties of $ReSi_{2-\delta}$ single crystals, *J. Appl. Phys.* **78**(6), 3902–3907 (1995).

1.15 M. Behar, H. Bernas, J. Desimoni, X.W. Lin, R.L. Maltez: Sequential phase formation by ion-induced epitaxy in Fe-implanted Si(001). Study of their properties and thermal behavior, *J. Appl. Phys.* **79**(2), 752–762 (1996).

1.16 Y. Dusausoy, J. Protas, R. Wandji, B. Roques: Structure cristalline du disiliciure de fer, $FeSi_2 \beta$, *Acta Cryst. B* **27**(6), 1209–1218 (1971).

1.17 D.J. Poutcharovsky, E. Parthé: The orthorhombic crystal structure of Ru_2Si_3, Ru_2Ge_3, Os_2Si_3 and Os_2Ge_3, *Acta Cryst. B* **30**, 2692–2696 (1974).

1.18 L. Schellenberg, H.F. Braun, J. Muller: The osmium-silicon phase diagram, *J. Less-Common Met.* **144**(2), 341–350 (1988).

1.19 L.N. Finnie: Structures and compositions of the silicides of ruthenium, osmium, rhodium and iridium, *J. Less-Common Met.* **4**(1), 24–34 (1962).

1.20 V.G. Weitz, L. Born, E. Hellner: Zur Structur des $OsGe_2$, Z. Metallkunde **51**(4), 238–244 (1960).

1.21 I. Engström, T. Lindsten, E. Zdansky: The crystal structure of the iridium silicide Ir_3Si_5, Acta Chem. Scand. A **41**(4), 237–242 (1987).

1.22 S. Nan, L. Yi-huan: X-ray measurement of the thermal expansion of germanium, silicon, indium antimonide and gallium arsenide, Acta Phys. Sinica, **20**(8), 699–704 (1964).

1.23 M. Östling, C. Zaring: Mechanical properties of TM silicides, in: *Properties of Metal Silicides*, edited by K. Maex, M. Van Rossum (INSPEC, IEE, London, 1995), pp. 15–30; M. Östling, C. Zaring, Thermal properties of TM silicides, ibid, pp. 31–44.

1.24 R.J. LaBotz, D.R. Mason: The thermal conductivities of Mg_2Si and Mg_2Ge, J. Electrochem. Soc. **110**(2), 121–26 (1963).

1.25 G.H. Li, H.S. Gill, R.A. Varin: Magnesium silicide intermetallic alloys, Metall. Trans. A **24**(11), 2383–2391 (1993).

1.26 S. Bose, H.N. Acharya, H.D. Banerjee: Electrical, thermal, thermoelectric and related properties of magnesium silicide semiconductor prepared from rice husk, J. Mater. Sci. **28**(20), 5461–5468 (1993).

1.27 V.S. Neshpor, V.L. Upko: Investigation of formation conditions and some physical properties of barium disilicide, J. Appl. Chem. **36**(5), 1139–1142 (1963) (in Russian).

1.28 G.G. Bentini, L. Correra: Analysis of thermal stresses induced in silicon during xenon arc lamp flash annealing, J. Appl. Phys. **154**(4), 2057–2062 (1983). Other approximations for temperature dependence of the silicon thermal conductivity may be found in [1.4].

1.29 P.M. Lee: Electronic structure of magnesium silicide and magnesium germanide, Phys. Rev. **135**(4A), A1110–A1114 (1964).

1.30 P. Canon, E.T. Conlin: Magnesium compounds: new dense phases, Science **145**(7), 487–489 (1964).

1.31 J. Evers: Transformation of three-connected silicon in barium disilicide, J. Solid State Chem. **32**(1), 77–86 (1980).

1.32 L.F. Mattheiss: Electronic structure of $CrSi_2$ and related refractory disilicides, Phys. Rev. B **43**(15), 12549–12555 (1991).

1.33 W.T. Lin, L.J. Chen: Localized epitaxial growth of molybdenum disilicide on silicon, J. Appl. Phys. **59**(5), 1518–1524 (1986).

1.34 W.T. Lin, L.J. Chen: Localized epitaxial growth of WSi_2 on silicon, J. Appl. Phys. **59**(10), 3481–3488 (1986).

1.35 H.W. Knott, M.H. Mueller, L. Heaton: The crystal structure of $Mn_{15}Si_{26}$, Acta Cryst. **23**(4), 549–555 (1967).

1.36 J.J. Chu, L.J. Chen, K.N. Tu: Localized epitaxial growth of $ReSi_2$ on (111) and (001) silicon, J. Appl. Phys. **62**(2), 461–465 (1987).

1.37 R.G. Long, M.C. Bost, J.E. Mahan: Optical and electrical properties of semiconducting rhenium disilicide thin films, Thin Solid Films **162**(1/2), 29–40 (1988).

1.38 T. Siegrist, F. Hulliger, G. Travaglini: The crystal structure and some properties of rhenium silicide ($ReSi_2$), J. Less-Common Met. **92**(1), 119–129 (1983).

1.39 N.E. Christensen: Electronic structure of β-$FeSi_2$, Phys. Rev. B **42**(11), 7148–7153 (1990).

1.40 I. Engström: The crystal structure of $OsSi_2$, Acta Chem. Scan. **24**(6), 2117–2125 (1970).

1.41 L. Miglio, F. Tavazza, G. Malegori: Stability hierarchy of the pseudomorphic $FeSi_2$ phases: α, γ, and defected CsCl, Appl. Phys. Lett. **67**(16), 2293–2295 (1995).

1.42 D.J. Poutcharovsky, K. Yvon, E. Parthé: Diffusionless phase transformations of Ru$_2$Si$_3$, Ru$_2$Ge$_3$ and Ru$_2$Sn$_3$, I. Crystal structure investigations, *J. Less-Common Met.* **40**(1), 139–144 (1975).

1.43 S. Petersson, J. Baglin, W. Hammer, F.M. d'Heurle, T.S. Kuan, I. Ohdomari, J. de Sousa Pires, P. Tove: Formation of iridium silicides from Ir thin films on Si substrates, *J. Appl. Phys.* **50**(5), 3357–3365 (1979).

1.44 J. de Sousa Pires, P. Ali, B. Crowder, F.M. d'Heurle, S. Petersson, L. Stolt, P. Tove: Measurements of the rectifying barrier heights of the various iridium silicides with n-Si, *Appl. Phys. Lett.* **35**(2), 202–204 (1979).

1.45 S. Petersson, J.A. Reimer, M.H. Brodsky, D.R. Campbell, F.M. d'Heurle, B. Karlsson, P.A. Tove: IrSi$_{1.75}$ a new semiconductor compound, *J. Appl. Phys.* **53**(4), 3342–3343 (1982).

1.46 I. Engström, F. Zackrisson: X-ray studies of silicon-rich iridium silicides, *Acta Chem. Scand.* **24**(6), 2109–2116 (1970).

1.47 M. Wittmer, P. Oelhafen, K.N. Tu: Electronic structure of iridium silicides, *Phys. Rev. B* **33**(8), 5391–5400 (1986).

1.48 C.E. Allevato, C.B. Vining: Phase diagram and electrical behavior of silicon-rich iridium silicide compounds, *J. Alloys Comp.* **200**(12), 99–105 (1993).

1.49 M. Setton: Ternary TM-TM-Si reactions, in: *Properties of Metal Silicides*, edited by K. Maex, M. Van Rossum (INSPEC, IEE, London, 1995), pp. 129–149.

1.50 E.I. Gladishevskii, V.I. Lakh, R.V. Skolozdra, B.I. Stadnik: Investigation of joint solubility of disilicides of transition metals from IV, V, and VI groups, *Powder Metallurgy* (**4**), 15–20 (1964) (in Russian).

1.51 Ya. Noda, H. Kon, Yo. Furukawa, N. Otsuka, I.A. Nishida, K. Masumoto: Preparation and thermoelectric properties of Mg$_2$Si$_{1-x}$Ge$_x$ (x=0.0–0.4) solid solution semiconductors, *Mater. Trans., JIM* **33**(9), 845–850 (1992).

1.52 J. Evers, A. Weiss: Electrical properties of alkaline earth disilicides and digermanides, *Mater. Res. Bull.* **9**(5), 549–554 (1974).

1.53 A.B. Filonov, I.E. Tralle, N.N. Dorozhkin, D.B. Migas, V.L. Shaposhnikov, G.V. Petrov, A.M. Anishchik, V.E. Borisenko: Semiconducting properties of hexagonal chromium, molybdenum, tungsten disilicides, *Phys. Stat. Sol. (b)* **186**(1), 209–215 (1994).

1.54 N. Boutarek, R. Madar: The influence of germanium substitution on the phase stability of 3d transition metal disilicides, *Appl. Surf. Sci.* **73**(1–4), 209–213 (1993).

1.55 N. Lundberg, M. Östling, F.M. d'Heurle: Chromium germanides: formation, structure and properties, *Appl. Surf. Sci.* **53**(1), 126–131 (1991).

1.56 A.W. Searcy, R.J. Peavler: The preparation and properties of molybdenum-germanium compounds, *J. Am. Chem. Soc.* **75**, 5657–5659 (1953).

1.57 O.Yu. Khyzhun, Ya.V. Zaulychny, E.A. Zhurakovsky: Electronic structure of tungsten and molybdenum germanides synthesized at high pressures, *J. Alloys Comp.* **244**(1–2), 107–112 (1996).

1.58 S.V. Popova, L.N. Fomicheva: Synthesis of tungsten germanides at high pressure, *Inorg. Mater.* **14**(4), 533–535 (1978).

1.59 M.C. Bost, J.E. Mahan: An optical determination on the bandgap of the most silicon-rich manganese silicide phase, *J. Electron. Mater.* **16**(6), 389–395 (1987).

1.60 T.A. Nguyen Tan, J.Y. Veuillen, P. Muret: Semiconducting rhenium silicide thin films on Si(111), *J. Appl. Phys.* **77**(6), 2514–2518 (1995).

1.61 V.I. Larchev, S.V. Popova: The new chimney-ladder phase Co$_2$Si$_3$ and Re$_4$Ge$_7$ formed by treatment at high temperatures and pressures, *J. Less-Common Met.* **84**(1), 87–91 (1982).

1.62 A.B. Filonov, D.B. Migas, V.L. Shaposhnikov, N.N. Dorozhkin, G.V. Petrov,

V.E. Borisenko, W. Henrion, H. Lange: Electronic and related properties of crystalline semiconducting iron disilicide, *J. Appl. Phys.* **79**(10), 7708–7712 (1996).

1.63 H. Chen, P. Han, X.D. Huang, L.Q. Hu, Y. Shi, Y.D. Zheng: Semiconducting Ge-Si-Fe alloy grown on Si(100) substrate by reactive deposition epitaxy, *Appl. Phys. Lett.* **69**(13), 1912–1914 (1996).

1.64 C.P. Susz, J. Muller, K. Yvon, E. Parthé: Diffusionless phase transformations of Ru_2Si_3, Ru_2Ge_3 and Ru_2Sn_3, II: Electrical and magnetic properties, *J. Less-Common Met.* **71**(1), P1–P8 (1980).

1.65 A.B. Filonov, D.B. Migas, V.L. Shaposhnikov, N.N. Dorozhkin, V.E. Borisenko, A. Heinrich, H. Lange: Electronic properties of isostructural ruthenium and osmium silicides and germanides, (1999) (to be published).

1.66 K. Mason, G. Müller-Vogt: Osmium disilicide: preparation, crystal growth, and physical properties of a new semiconducting compound, *J. Cryst. Growth* **63**(1), 34–38 (1983).

1.67 C.E. Allevato, C.B. Vining: Thermoelectric properties of semiconducting iridium silicides, in: *Proceedings of the 28th Intersociety Energy Conversion Engineering Conference* (Atlanta, Georgia, USA, 1993), vol. 1, pp. 1.239–1.243 (1993).

1.68 J. Schumann, D. Elefant, C. Gladun, A. Heinrich, W. Pitschke, H. Lange, W. Henrion, R. Grötzschel: Polycrystalline iridium silicide films. Phase formation, electrical and optical properties, *Phys. Stat. Sol. (a)* **145**(2), 429–439 (1994).

1.69 R.G. Long, J.E. Mahan: Two pseudobinary semiconducting silicides: $Re_xMo_{1-x}Si_2$ and $Cr_xV_{1-x}Si_2$, *Appl. Phys. Lett.* **56**(17), 1655–1657 (1990).

1.70 T. Tokushima, I. Nishida, K. Sakata, T. Sakata: The $CrSi_2$-$CoSi$ thermomodule and its applications, *J. Mater. Sci.* **4**(7), 978–984 (1969).

1.71 I.A. Saltykova, Kh.I. Gol'dberg, F.A. Sidorenko, P.V. Gel'd: About electronic structure of solid solutions of the quasibinary system $CrSi_2$-$MnSi_2$, *Powder Metallurgy* (6), 73–79 (1968) (in Russian).

1.72 I. Nishida, T. Sakata: Semiconducting properties of pure and Mn-doped chromium disilicides, *J. Phys. Chem. Solids* **39**(5), 499–505 (1978).

1.73 C.B. Vining, C.E. Allevato: Progress in doping of ruthenium silicide (Ru_2Si_3), in: *Proceedings of the 27th Intersociety Energy Conversion Engineering Conference* (San Diego, CA, 1992), vol. 3, pp. 3.489–3.492.

1.74 J.M. Harris, S.S. Lau, M.-A. Nicolet, R.S. Nowicki: Studies of the Ti-W metallization system on Si, *J. Electrochem. Soc.* **123**(1), 120–124 (1976).

1.75 F. Nava, C. Nobili, G. Ferla, G. Iannuzzi, G. Queirolo, G. Celotti: Ti-W alloy interaction with silicon in the presence of PtSi, *J. Appl. Phys.* **54**(5), 2434–2440 (1983).

1.76 P. Gas, F.J. Tardy, F.M. d'Heurle: Disilicide solid solutions, phase diagram, and resistivities. I. $TiSi_2$-WSi_2, *J. Appl. Phys.* **60**(1), 193–200 (1986).

1.77 S. Hong, C. Pirri, P. Wetzel, G. Gewinner: Synthesis of epitaxial ternary $Co_{1-x}Fe_xSi_2$ silicides with CsCl- and CaF_2-type cubic structures on Si(111) by codeposition techniques, *Phys. Rev. B* **55**(19), 13040–13050 (1997).

1.78 S. Teichert, R. Kilper, T. Franke, J. Erben, P. Häussler, W. Henrion, H. Lange, D. Panknin: Electrical and optical properties of thin $Fe_{1-x}Co_xSi_2$ films, *Appl. Surf. Sci.* **91**(1), 56–62 (1995).

1.79 I. Dézsi, Cs. Fetzer, I. Szűcs, G. Langouche, A. Vantomme: Phases of cobalt-iron ternary disilicides, *Appl. Phys. Lett.* **72**(22), 2826–2828 (1998).

1.80 D. Panknin, E. Wieser, W. Skorupa, W. Henrion, H. Lange: Buried $(Fe_{1-x}Co_x)Si_2$ layers with variable band gap formed by ion beam synthesis, *Appl. Phys. A* **62**(2), 155–162 (1996).

1.81   T. Kojima, M. Sakata, I. Nishida: Study on the formation of $Fe_{1-x}Mn_xSi_2$ from the sintered $FeSi-Fe_2Si_5$ eutectic alloy doped with manganese *J. Less-Common Met.* **162**(1), 39–49 (1990).

1.82   J. Tani, H. Kido: Mechanism of electrical conduction of Mn-doped β-$FeSi_2$, *J. Appl. Phys.* **86**(1), 464–467 (1999).

1.83   H. Takizawa, P.F. Mo, T. Endo, M. Shimada: Preparation and thermoelectric properties of β-$Fe_{1-x}Ru_xSi_2$, *J. Mater. Sci.* **30**(16), 4199–4203 (1995).

1.84   A.B. Filonov, D.B. Migas, V.L. Shapohnikov, N.N. Dorozhkin, V.E. Borisenko, H. Lange: Electronic properties of osmium disilicide, *Appl. Phys. Lett.* **70**(8), 976–977 (1997).

1.85   G.A. Golikov: *Manual for Physical Chemistry* (Visshaya Shkola, Moscow, 1988) (in Russian).

1.86   M.E. Schlesinger: Thermodynamics of solid transition-metal silicides, *Chem. Rev.* **90**(4), 607–628 (1990).

1.87   O. Knacke, O. Kubaschewski, K. Hesselmann (Ed.) *Thermochemical Properties of Inorganic Substances* (Springer, Berlin, 1991).

1.88   A.R. Miedema, P.F. de Chatel, F.R. de Boer: Cohesion in alloys - fundamentals of a semiempirical model, *Physica B+C* **100**(1), 1–28 (1980).

1.89   A.K. Niessen, F.R. de Boer: The enthalpy of formation of solid borides, carbides, nitrides, silicides and phosphides of transition and noble metals, *J. Less-Common Met.* **82**(1), 75–80 (1981).

1.90   T.B. Massalski, H. Okamoto, P.R. Subramanian, L. Kacprzak (Ed.) *Binary Alloy Phase Diagrams*, *Second Edition* (American Society for Metals, Metals Park, OH, 1990), vols. 1–2.

1.91   A.P. Zephirov (Ed.) *Thermodynamic Properties of inorganic matter* (Atomizdat, Moscow, 1965) (in Russian).

1.92   J. Evers: Semiconductor-metal transition in $BaSi_2$, *J. Less-Common Met.* **58**(1), 75–83 (1978).

1.93   A.B. Gokhale, G.J. Abbaschian: The Mo-Si (molybdenum-silicon) system, *J. Phase Equilibria* **12**(4), 493–498 (1991).

1.94   B.K. Yen: X-ray diffraction study of solid-state formation of metastable $MoSi_2$ and $TiSi_2$ during mechanical alloying, *J. Appl. Phys.* **81**(10), 7061–7063 (1997).

1.95   A.H. van Ommen, A.H. Reader, J.W.C. de Vries: Influence of microstructure on the resistivity of $MoSi_2$ thin films, *J. Appl. Phys.* **64**(7), 3574–3580 (1988).

1.96   W.T. Lin, L.J. Chen: Localized epitaxial growth of hexagonal and tetragonal $MoSi_2$ on (111) Si, *Appl. Phys. Lett.* **46**(11), 1061–1063 (1985).

1.97   C.M. Doland, R.J. Nemanich: Phase formation during reactive molybdenum-silicide formation, *J. Mater. Res.* **5**(12), 2854–2864 (1990).

1.98   F.M. d'Heurle, C.S. Petersson, M.Y. Tsai: Observations on the hexagonal form of $MoSi_2$ and WSi_2 films produced by ion implantation and on related snowplow effects, *J. Appl. Phys.* **51**(11), 5976–5980 (1980).

1.99   M.S. Chandrasekharaiah, J.L. Margrave, P.A.G. O'Hare: The disilicides of tungsten, molybdenum, tantalum, titanium, cobalt, and nickel, and platinum monosilicide: a survey of their thermodynamic properties, *J. Phys. Chem. Ref. Data* **22**(6), 1459–1468 (1993).

1.100 F. Nava, B.Z. Weiss, K.Y. Ahn, D.A. Smith, K.N. Tu: Thermal stability and electrical conduction behavior of coevaporated $WSi_{2\pm x}$ thin films, *J. Appl. Phys.* **64**(1), 354–364 (1988).

1.101 F. Nava, B.Z. Weiss, K. Ahn, K.N. Tu: Tungsten disilicide formation in codeposite amorphous WSi_x alloy thin films, *Le Vide, les Couches Minces* **42**(1), 225–228 (1987).

304 References

1.102 G.M. Lukashenko, V.R. Sidorko, B.Ya. Kotur: Phase constitution and thermodynamic properties of alloys of the Mn-Si system in the range from 25 to 37.5 at % Si, *Sov. Powder Metal Met. Ceram.* **20**(8), 571–574 (1981).

1.103 J.L. Jorda, M. Ishikawa, J. Muller: Phase relations and superconductivity in the binary Re-Si system, *J. Less-Common Met.* **85**(1), 27–35 (1982).

1.104 U. Gottlieb, M. Affronte, F. Nava, O. Laborde, A. Sulpice, R. Madar: Some physical properties of $ReSi_{1.75}$ single crystals, *Appl. Surf. Sci.* **91**(1), 82–86 (1995).

1.105 M.C. Bost, J.E. Mahan: Optical properties of semiconducting iron disilicide thin films, *J. Appl. Phys.* **58**(7), 2696–2703 (1985).

1.106 I. Yamauchi, A. Suganuma, T. Okamoto, I. Ohnaka: Effect of copper addition on the β-phase formation rate in $FeSi_2$ thermoelectric materials, *J. Mater. Sci.* **32**(17), 4603–4611 (1997).

1.107 I. Yamauchi, T. Okamoto, H. Ohata, I. Ohnaka: β-phase transformation and thermoelectric power in $FeSi_2$ and Fe_2Si_5 based alloys containing small amounts of Cu, *J. Alloys Comp.* **260**(1), 162–171 (1997).

1.108 C.E. Allevato, C.B. Vining: Phase diagram and electrical behavior of silicon-rich iridium silicide compounds, in: *Proceedings of the 27th Intersociety Energy Conversion Engineering Conference* (San Diego, CA, 1992), vol. 3, pp. 3.493–3.497 (1992).

1.109 M.C. Bost, J.E. Mahan: Summary abstract: Semiconducting silicides as potential materials for electro-optic very large scale integrated circuit interconnects, *J. Vac. Sci. Technol.* B **4**(6), 1336–1338 (1986).

1.110 R. Madar: Phase diagrams for TM-Si-Ge systems, in: *Properties of Metal Silicides*, edited by K. Maex, M. Van Rossum (INSPEC, IEE, London, 1995), pp. 124–125.

1.111 K.N. Tu, W.N. Hammer, J.O. Olowolafe: Shallow silicide contact, *J. Appl. Phys.* **51**(3), 1663–1668 (1980).

1.112 G. Ottaviani, K.N. Tu, J.W. Mayer, B.Y. Tsaur: Phase separation in alloy-Si interaction, *Appl. Phys. Lett.* **36**(4), 331–333 (1980).

1.113 C.J. Palmström, J. Gyulai, J.W. Mayer: Phase separation in interactions of tantalum-chromium alloy on Si, *J. Vac. Sci. Technol.* A **1**(2), 452–454 (1983).

1.114 D. Mangelinck, L. Wang, C. Lin, P. Gas, J. Grahn, M. Östling: Influence of the addition of Co and Ni on the formation of epitaxial semiconducting $β-FeSi_2$: Comparison of different evaporation methods, *J. Appl. Phys.* **83**(8), 4193–4201 (1998).

1.115 J.K. Burdett: Electronic influences on the crystal chemistry of transition metal-main group MX and MX_2 compounds, *J. Solid State Chem.* **45**(3), 399–410 (1982).

1.116 Y. Ohta, D.G. Pettifor: Size versus electronic factors in transition metal Laves phase stability, *J. Phys.: Condensed Matter* **2**(41), 8189–8194 (1990).

1.117 F.M. d'Heurle: Nucleation of a new phase from the interaction of two adjacent phases: Some silicides, *J. Mater. Res.* **3**(1), 167–195 (1988).

1.118 K. Maex: Silicides for integrated circuits: $TiSi_2$ and $CoSi_2$, *Mater. Sci. Eng.* **R11**(2–3), 53–153 (1993).

1.119 S.O. Hyatt, D.O. Northwood: A thermodynamic aspect to the sequence of phase formation in thin films: Application to Ti, V, Cr and Mn with Si, *Mater. Forum* **17**(3), 251–256 (1994).

1.120 R.M. Walser, R.W. Bené: First phase nucleation in silicon-transition-metal planar interfaces, *Appl. Phys. Lett.* **28**(10), 624–625 (1976).

1.121 R. Pretorius: Prediction of silicide first phase and phase sequence from heat of formation, in *Thin Films and Interfaces II*, edited by J.E.E. Baglin, D.R. Campbel, W.K. Chu (North-Holland, New York, 1984), pp. 15–20.

1.122 R.W. Bené: A kinetic model for solid-state silicide nucleation, *J. Appl. Phys.* **61**(5), 1826–1833 (1987).

1.123 R.M. Walser, R.W. Bené: First phase nucleation in silicon-transition-metal planar interfaces, *Appl. Phys. Lett.* **28**(10), 624–625 (1976).

1.124 R.W. Bené: A kinetic model for solid-state silicide nucleation, *J. Appl. Phys.* **61**(5), 1826–1833 (1987).

1.125 C.V. Thompson: On the role of diffusion in phase selection during reactions at interfaces, *J. Mater. Res.* **7**(2), 367–373 (1992).

1.126 P. Gas, F.M. d'Heurle: Kinetics of formation of TM silicide thin films: self-diffusion, in: *Properties of Metal Silicides*, edited by K. Maex, M. Van Rossum (INSPEC, IEE, London, 1995), pp. 279–292.

1.127 W.K. Chu, S.S. Lau, J.W. Mayer, H. Müller, K.N. Tu: Implanted noble gas atoms as diffusion markers in silicide formation, *Thin Solid Films* **25**(2), 393–402 (1975).

1.128 G. Ottaviani: Metallurgical aspects of the formation of silicides, *Thin Solid Films* **140**(1), 3–21 (1986).

1.129 E.G. Colgan, B.Y. Tsaur, J.W. Mayer: Phase formation in Cr-Si thin-film interactions, *Appl. Phys. Lett.* **37**(10), 938–940 (1980).

1.130 C.J. Bedeker, S. Kritzinger, J.C. Lombaard: Formation and microstructure of various thin film chromium silicide phases, *Thin Solid Films* **141**(1), 117–127 (1986).

1.131 C. Heck, M. Kusaka, M. Hirai, M. Iwami, H. Nakamura: Study of Cr silicide formation on Si(100) due to solid-phase reaction using soft X-ray emission spectroscopy, *Jpn. J. Appl. Phys.* **33**(12), 6667–6670 (1994).

1.132 A. Martinez, D. Esteve, A. Guivarc'h, P. Auvray, P. Henoc, G. Pelous: Metallurgical and electrical properties of chromium silicon interfaces, *Solid-State Electron.* **23**(1), 55–64 (1980).

1.133 J.M. Liang, L.J. Chen: Autocorrelation function analysis of phase formation in the initial stage of interfacial reactions of molybdenum thin films on (111)Si, *Appl. Phys. Lett.* **64**(10), 1224–1226 (1994).

1.134 M. Eizenberg, K.N. Tu: Formation and Schottky behavior of manganese silicides on *n*-type silicon, *J. Appl. Phys.* **53**(10), 6885–6890 (1982).

1.135 H.C. Cheng, T.R. Yew, L.J. Chen: Interfacial reactions of iron thin films on silicon, *J. Appl. Phys.* **57**(12), 5246–5250 (1985).

1.136 Q.G. Zhu, H. Iwaraki, E.D. Williams, R. Park: Formation of iron silicide thin films, *J. Appl. Phys.* **60**(7), 2629–2631 (1986).

1.137 M. Fanciulli, C. Rosenblad, G. Weyer, H. von Känel, N. Onda, V. Nevolin, A. Zenkevich: Phase transformations in layered Fe-Si structures, in: *Silicide Thin Films - Fabrication, Properties, and Applications*, edited by R.T. Tung, K. Maex, P.W. Pellegrini, L.H. Allen (MRS, Pittsburgh, Pennsylvania, 1996), pp. 319–324.

1.138 N.R. Baldwin, D.G. Ivey: Low temperature iron thin films-silicon reactions, *J. Mater. Sci.* **31**(1), 31–37 (1996).

1.139 J.M. Gallego, R. Miranda: The Fe/Si(100) interface, *J. Appl. Phys.* **69**(3), 1377–1383 (1991).

1.140 C.S. Petersson, J.E.E. Baglin, J.J. Dempsey, F.M. d'Heurle, S.J. La Placa: Silicides of ruthenium and osmium: thin film reactions, diffusion, nucleation, and stability, *J. Appl. Phys.* **53**(7), 4866–4883 (1982).

1.141 F.M. d'Heurle: Nucleation of a new phase from the interaction of two adjacent phases: Some silicides, *J. Mater. Res.* **3**(1), 167–195 (1988).

1.142 I. Ohdomari, T.S. Kuan, K.N. Tu: Microstructure and Schottky barrier height of iridium silicides formed on silicon, *J. Appl. Phys.* **50**(11), 7020–7029 (1979).

1.143 T. Rodriguez, H. Wolters, M. Fernandez, A. Almendra, M.F. da Silva, M. Clement, J.C. Soares, C. Ballesteros: Iridium silicides obtained by rapid thermal annealing, *Appl. Surf. Sci.* **73**, 182–185 (1993).

1.144   S.P. Murarka: Properties and applications of silicides, in: *Microelectronic Materials and Processes*, edited by R.A. Levy (Kluwer Academic Publishers, 1989), pp.275–323.

1.145   A. Vantomme,   S. Degroote,   J. Dekoster,   G. Langouche,   R. Pretorius: Concentration-controlled phase selection of silicide formation during reactive deposition, *Appl. Phys. Lett.* **74**(21), 3137–3139 (1999).

1.146   S.-L. Zhang, F.M. d'Heurle: Modellization of the growth of three intermediate phases, *Mater. Sci. Forum* **155/156**, 59–70 (1994).

1.147   J. Philibert: Reactive interdiffusion, *Mater. Sci. Forum* **155/156**, 15–30 (1994).

1.148   V.E. Borisenko, L.I. Ivanenko, E.A. Krushevski: Combined reaction and diffusion controlled kinetics of silicidation, *Thin Solid Films* **250**(1), 53–55 (1994).

1.149   K.N. Tu, J.W. Mayer: Silicide formation, in: *Thin Films. Interdiffusion and Reactions*, edited by J.M. Poate, K.N. Tu, J.W. Mayer (J. Wiley, New York, 1978), pp. 359–405.

1.150   P. Gas: Silicides thin films formed by metal/silicon reaction: Role of diffusion, *Mater. Sci. Forum* **155/156**, 39–54 (1994).

1.151   P.L. Janega, D. Landheer, J. McCaffrey, D. Mitchell, M. Buchanan, M. Denhoff: Contact resistivity of some magnesium/silicon and magnesium/silicide silicon structure, *Appl. Phys. Lett.* **53**(21), 2056–2058 (1988).

1.152   C.D. Lien, M.-A. Nicolet, S.S. Lau: Kinetics of silicides on Si (100) and evaporated silicon substrates, *Thin Solid Films* **143**(1), 63–72 (1986).

1.153   L.R. Zheng, E. Zingu, J.W. Mayer: Lateral silicide growth, in: *Thin Films and Interfaces II*, edited by J.E.E. Baglin, D.R. Campbel, W.K. Chu (North-Holland, Amsterdam, 1984), pp. 75–85.

1.154   J.O. Olowolafe, M.-A. Nicolet, J.W. Mayer: Formation kinetics of $CrSi_2$ films on Si substrates with and without interposed Pd_2Si layer, *J. Appl. Phys.* **47**(12), 5182–5186 (1976).

1.155   R.W. Bower, J.W. Mayer: Growth kinetics observed in the formation of metal silicides on silicon, *Appl. Phys. Lett.* **20**(9), 359–361 (1972).

1.156   A.P. Botha, R. Pretorius: Co_2Si, $CrSi_2$, $ZrSi_2$ and $TiSi_2$ formation studied by a radioactive ^{31}Si marker technique, *Thin Solid Films* **93**(1), 127–133 (1982).

1.157   N.I. Plusnin, N.G. Galkin, A.N. Kamenev, A.G. Lifshits, S.A. Lobachev: Atomic mixing at Si-Cr interface and initial stage of $CrSi_2$ epitaxy, *Poverkhnost* (9), 55–61 (1989) (in Russian).

1.158   A. Zalar, S. Hofmann, F. Pimentel, P. Panjan: Interfacial reactions and silicide formation in Si/Ni/Si and Si/Cr/Si sandwich layers, *Surf. Interface Anal.* **21**(8), 560–565 (1994).

1.159   W.Y. Hsieh, J.H. Lin, L.J. Chen: Simultaneous occurrence of multiphases in the interfacial reactions of ultrahigh vacuum deposited Hf and Cr thin films on (111)Si, *Appl. Phys. Lett.* **62**(10), 1088–1090 (1993).

1.160   A.P. Botha, R. Pretorius, S. Kritzinger: Determination of the diffusing species and mechanism of diffusion during $CrSi_2$ formation, using ^{31}Si as a marker, *Appl. Phys. Lett.* **40**(5), 412–414 (1982).

1.161   L.R. Zheng, L.S. Hung, S.H. Chen, J.W. Mayer: Effects of krypton on $CrSi_2$ formation, *J. Appl. Phys.* **59**(6), 1998–2001 (1986).

1.162   T.E. Sohlesinger, R.C. Cammarata, S.M. Prokes: Kinetics of silicide formation in chromium-amorphous silicon multilayered films, *J. Appl. Phys.* **59**(4), 449–451 (1991).

1.163   M. Natan, S.W. Duncan: Microstructure and growth kinetics of $CrSi_2$ on Si{100} studied using cross-sectional transmission electron microscopy, *Thin Solid Films* **123**(1), 69–85 (1985).

1.164   V.E. Borisenko, L.I. Ivanenko, S.Yu. Nikitin: Semiconducting properties of chromium disilicide, *Mikroelektronika* **21**(2), 61–77 (1992) (in Russian).

1.165   V.E. Borisenko, D.I. Zarovskii: Formation of chromium silicide during a second thermal treatment, *Zn. Tekn. Fiz.*, **55**(10), 2025–2026 (1985) (in Russian); V.E. Borisenko, D.I. Zarovskii: Formation of chromium silicides by annealing for times < 1 minute, *Sov. Phys. Tech. Phys.* **30**(10), 1188 (1985).

1.166   L.R. Zheng, L.R. Doolittle, J.W. Mayer: Silicon transport in lateral silicide growth of $CrSi_2$, *Mater. Res. Soc. Symp. Proc.* **71**, 287–295 (1986).

1.167   P. Gas, F.M. d'Heurle: Diffusion processes in silicides: A comparison between bulk and thin film phase formation, in: *Silicide Thin Films - Fabrication, Properties, and Applications,* edited by R.T. Tung, K. Maex, P.W. Pellegrini, L.H. Allen (MRS, Pittsburgh, Pennsylvania, 1996), pp. 39–50.

1.168   F. Nava, T. Tien, K.N. Tu: Temperature dependence of semiconducting and structural properties of Cr-Si thin films, *J. Appl. Phys.* **57**(6), 2018–2025 (1985).

1.169   V.E. Borisenko, D.I. Zarovskii, G.V. Litvinovich, V.A. Samuilov: Optical spectroscopy of chromium silicide formed by a flash heat treatment of chromium films on silicon, *Zurn. Prikl. Spectrosk.* **44**(2), 314–317 (1986) (in Russian).

1.170   H. Jiang, H.J. Whitlow, M. Ostling, E. Niemi, F.M. d'Heurle, C.S. Petersson: A quantitative study of oxygen behavior during $CrSi_2$ and $TiSi_2$ formation, *J. Appl. Phys.* **65**(2), 567–574 (1989).

1.171   A. Cros, R.A. Pollak, K.N. Tu: Interaction between chromium oxide and chromium silicide, *J. Appl. Phys.* **54**(1), 258–259 (1983).

1.172   A. Guivarc'h, P. Auvray, L. Berthou, M. Le Cun, J.P. Boulet, P. Henoc, G. Pelous: Reaction kinetics of molybdenum thin films on silicon (111) surface, *J. Appl. Phys.* **49**(1), 233–237 (1978).

1.173   P.R. Gage, R.W. Bartlett: Diffusion kinetics affecting formation of silicide coatings on molybdenum and tungsten, *Trans. Metall. Soc.* **233**, 832–834 (1965).

1.174   P. Urwank, E. Wieser, A. Hässner, Ch. Kaufmann, H. Lippmann, I. Melzer: Formation of $MoSi_2$ by light pulse irradiation, *Phys. Stat. Sol. (a)* **90**(2), 463–468 (1985).

1.175   S.A. Agamy, Y.Q. Ho, H.M. Naguib: Arsenic (1+) implantation and transient annealing of molybdenum disilicide ($MoSi_2$) thin films, *J. Vac. Sci. Technol.* **3**(3), 718–722 (1985).

1.176   R. Portius, D. Dietrich, A. Hässner, Th. Raschke, E. Wieser, W. Wolke: Formation of stable $MoSi_2$/poly-Si films by rapid thermal annealing, *Phys. Stat. Sol. (a)* **100**(1), 199–206 (1987).

1.177   V.Q. Ho, H.M. Naguib: Characterization of rapidly annealed Mo-polycide, *J. Vac. Sci. Technol. A* **3**(3), 896–899 (1985).

1.178   J. Baglin, J. Dampsey, Y. Hammer, F.M. d'Heurle, S. Petersson, C. Serrano: The formation of silicides in molybdenum-tungsten bilayer films on silicon substrates: a marker experiment, *J. Electron. Mater.* **8**(5), 641–661 (1979).

1.179   B.V. Cockeram, G. Wang, R.A. Rapp: Growth kinetics and pesting resistance of $MoSi_2$ and germanium-doped $MoSi_2$ diffusion coatings growth by the pack cementation method, *Materials and Corrosion* **46**(4), 207–217 (1995).

1.180   R.W. Bartlett, P.R. Gage, P.A. Larssen: Growth kinetics of intermediate silicides in the $MoSi_2$/Mo and WSi_2/W systems, *Trans. Metall. Soc.* **230**, 1528–1534 (1964).

1.181   D.L. Kwong, D.C. Meyers, N.S. Alvi, L.W. Li, E. Norbeck: Refractory metal silicide formation by ion-beam mixing and rapid thermal annealing, *Appl. Phys. Lett.* **47**(7), 688–691 (1985); D.L. Kwong, D.C. Meyers, N.S. Alvi: Molybdenum silicide formation by ion beam mixing and rapid thermal annealing, in: *VLSI Science and Technology,* edited by W.M. Bullis, S. Broydo (Electrochem. Soc., Inc., Pennington, 1985), pp. 195–202.

1.182 M.Y. Tsai, F.M. d'Heurle, C.S. Petersson, R.W. Johnson: Properties of tungsten silicide film on polycrystalline silicon, *J. Appl. Phys.* **52**(8), 5350–5355 (1981).

1.183 L.D. Locker, C.D. Capio: Reaction kinetics of tungsten thin films on silicon (100) surfaces, *J. Appl. Phys.* **44**(10), 4366–4369 (1973).

1.184 J. Lajzerowicz, J. Torres, G. Goltz, R. Pantel: Kinetics of WSi_2 formation at low and high temperatures, *Thin Solid Films* **140**(1), 23–28 (1986).

1.185 S-L. Zhang, R. Buchta, M. Östling: A study of silicide formation from LPCVD-tungsten films: Film texture and growth kinetics, *J. Mater. Res.* **6**(9), 1886–1891 (1991).

1.186 M.P. Siegal, W.R. Graham: The formation of thin film tungsten silicide annealed in ultra-high vacuum, *J. Appl. Phys.* **66**(12), 6073–6076 (1989).

1.187 T.O. Sedgwick, F.M. d'Heurle, S.A. Cohen: Short time annealing of coevaporated tungsten silicide films, *J. Electrochem. Soc.* **131**(10), 2446–2451 (1984).

1.188 T. Hara, H. Takahashi, S.-C. Chen: Ion implantation of arsenic in chemical vapor deposition tungsten silicide, *J. Vac. Sci. Technol. B* **3**(6), 1664–1667, (1985).

1.189 B.-Y. Tsaur, C.K. Chen, C.H. Anderson, D.L. Kwong: Selective tungsten silicide formation by ion-beam mixing and rapid thermal annealing, *J. Appl. Phys.* **57**(6), 1890–1894 (1985).

1.190 J. Kato, M. Asahina, H. Shimura, Y. Yamamoto: Rapid annealing of tungsten polycide films using halogen lamps, *J. Electrochem. Soc.* **133**(4), 794–798 (1986).

1.191 C. Nobili, M. Bosi, G. Ottaviani, G. Queirolo, L. Bacci: Rapid thermal annealing of WSi_x, *Appl. Surf. Sci.* **53**(1), 219–223 (1991).

1.192 M.P. Siegal, J.J. Santiago: Effects of rapid thermal processing on the formation of uniform tetragonal tungsten disilicide thin films on Si(100) substrates, *J. Appl. Phys.* **63**(2), 525–529 (1988).

1.193 T.S. Jaydev, A.Joshi: Conductivity changes in tungsten silicide films due to rapid thermal processing, *Electron. Lett.* **20**(14), 604–606 (1984).

1.194 L. Zhang, D.G. Ivey: Low temperature reactions of thin layers of Mn with Si, *J. Mater. Res.* **6**(7), 1518–1531 (1991).

1.195 Y.C. Lian, L.J. Chen: Localized epitaxial growth of $MnSi_{1.7}$ on silicon, *Appl. Phys. Lett.* **48**(5), 359–361 (1986).

1.196 J. Wang, M. Hirai, M. Kusaka, M. Iwami: Preparation of manganese silicide thin films by solid phase reaction, *Appl. Surf. Sci.* **113/114**, 53–56 (1997).

1.197 L. Zhang, D.G. Ivey: Reaction kinetics and opticals properties of semiconducting $MnSi_{1.73}$ grown on ⟨001⟩ oriented silicon, *J. Mater. Sci.* **2**(2), 116–123 (1991).

1.198 J. Alvarez, J.J. Hinarejos, E.G. Michel, R. Miranda: Determination of the Fe/Si(111) phase diagram by means of photoelectron spectroscopies, *Surf. Sci.* **287/288**, 490–494 (1993).

1.199 J. Alvarez, A.L. Vázquez de Parga, J.J. Hinarejos, J. de la Figuera, E.G. Michel, C. Ocal, R. Miranda: Geometric and electronic structure of epitaxial iron silicides, *J. Vac. Sci. Technol. A* **11**(4), 929–933 (1993).

1.200 M. Hasegawa, N. Kobayashi, N. Hayashi: Initial stages of reactions between monolayer Fe and Si(001) surfaces, in: *Silicide Thin Films - Fabrication, Properties, and Applications,* edited by R.T. Tung, K. Maex, P.W. Pellegrini, L.H. Allen (MRS, Pittsburgh, Pennsylvania, 1996), pp. 529–534.

1.201 S.S. Lau, J.S.-Y. Feng, J.O. Olowolafe, M.-A. Nicolet: Iron silicide film formation at low temperatures, *Thin Solid Films* **25**(2), 415–422 (1975).

1.202 K. Radermacher, S. Mantl, Ch. Dieker, H. Lüth, C. Freiburg: Growth kinetics of iron silicides fabricated by solid phase epitaxy or ion beam synthesis, *Thin Solid Films* **215**(1), 76–83 (1992).

1.203 U. Erlesand, M. Östling, K. Bodén: Formation of iron disilicide on amorphous silicon, *Appl. Surf. Sci.* **53**(1), 153–158 (1991).

1.204   P.H. Amesz, L.V. Jørgensen, M. Libezny, J. Poortmans, J. Nijs, A. van Veen, H. Schut, J.Th.M. de Hosson: Morphologies and growth modes of FeSi and β-FeSi$_2$ layers prepared by rapid thermal annealing, in: *Silicide Thin Films - Fabrication, Properties, and Applications,* edited by R.T. Tung, K. Maex, P.W. Pellegrini, L.H. Allen (MRS, Pittsburgh, Pennsylvania, 1996), pp. 373–378.

1.205   C.A. Dimitriadis, J.H. Werner: Growth mechanism and morphology of semiconducting FeSi$_2$ films, *J. Appl. Phys.* **68**(1), 93–96 (1990).

1.206   V.E. Borisenko, D.I. Zarovskii, L.I. Ivanenko: Structure and phase transitions in thin film Fe-Si system subjected to pulse thermal processing, *Mikroelektronika* **21**(6), 27–30 (1992) (in Russian).

1.207   D. Donoval, L. Stolt, H. Norde, J. de Sousa Pires, P.A. Tove, C.S. Petersson: Barrier heights to silicon, of ruthenium and its silicide, *J. Appl. Phys.* **53**(7), 5352–5353 (1982).

1.208   U. Gottlieb, O. Laborde, A. Rouault, R. Madar: Resistivity of Ru$_2$Si$_3$ single crystals, *Appl. Surf. Sci.* **73**, 243–245 (1993).

1.209   F.M. d'Heurle: Formation and oxidation mechanisms in two semiconducting silicides, *Thin Solid Films* **151**(1), 41–50 (1987).

1.210   Y.S. Chang, M.L. Chou: Solid-state epitaxy of osmium silicide on (111)Si under reducing atmosphere, *J. Appl. Phys.* **66**(7), 3011–3013 (1989).

1.211   R. Anderson, J. Baglin, J. Dempsey, W. Hammer, F.M. d'Heurle, S. Petersson: Nucleation-controlled thin-film interactions: Some silicides, *Appl. Phys. Lett.* **35**(3), 285–287 (1979).

1.212   T. Rodriguez, H. Wolters, A. Almendra, J. Sanz-Maudes, M.F. da Silva, J.C. Soares: Iridium silicide formation by rapid thermal annealing, in: *Infrared Detectors - Materials, Processing, and Devices,* edited by L.R. Dawson, A. Appelbaum (MRS, Pittsburgh, Pennsylvania, 1994), pp. 313–317.

1.213   T. Rodriguez, A. Almendra, M. Botella, M.F. da Silva, J.C. Soares, H. Wolters, C. Ballesteros: Iridium silicides formed by RTA in vacuum, in: *Silicide Thin Films - Fabrication, Properties, and Applications,* edited by R.T. Tung, K. Maex, P.W. Pellegrini, L.H. Allen (MRS, Pittsburgh, Pennsylvania, 1996), pp. 599–604.

1.214   U. Gösele, K.N. Tu: Growth kinetics of planar binary diffusion couples: "Thin-film case" versus "Bulk cases", *J. Appl. Phys.* **53**(4), 3252–3260 (1982).

1.215   O. Thomas, T.G. Finstad, F.M. d'Heurle: Respective mobilities of metal and silicon in disilicides: Bilayers of chromium with molybdenum or tungsten, *J. Appl. Phys.* **67**(5), 2410–2414 (1990).

1.216   M. Bartur: Thermal oxidation of transition metal silicides: the role of mass transport, *Thin Solid Films* **107**(1), 55–65 (1983).

1.217   W.J. Strydom, J.C. Lombaard, R. Pretorius: Thermal oxidation of the silicides CoSi$_2$, CrSi$_2$, NiSi$_2$, PtSi, TiSi$_2$ and ZrSi$_2$, *Thin Solid Films* **131**(3–4), 215–231 (1985).

1.218   F.M. d'Heurle: Diffusion, reaction, oxidation of silicides in electronics and elsewhere, *J. Phys.* **15**(11), 1707–1728 (1995).

1.219   O. Thomas, L. Stolt, P. Buaud, J.C. Poler, F.M. d'Heurle: Oxidation and formation mechanisms in disilicides: VSi$_2$ and CrSi$_2$, inert marker experiments and interpretation, *J. Appl. Phys.* **68**(12), 6213–6223 (1990).

1.220   R.D. Frampton, E.A. Irene, F.M. d'Heurle: A study of the oxidation of selected metal silicides, *J. Appl. Phys.* **62**(7), 2972–2980 (1987).

1.221   G.L.P. Berning: The room temperature oxidation of Cr, Si and CrSi$_2$, *Appl. Surf. Sci.* **40**(3), 209–212 (1989).

1.222   L. Stolt, O. Thomas, F.M. d'Heurle: Oxidation of titanium, manganese, iron, and niobium silicides: Marker experiments, *J. Appl. Phys.* **68**(10), 5133–5139 (1990).

1.223 L.J. Terminello, J.A. Yarmoff, F.M. d'Heurle, F.R. McFeely: Pressure dependence of oxide growth rate on silicon and metal silicides, *Surf. Sci.* **243**(1–3), 127–131 (1991).

1.224 F.M. d'Heurle, R.D. Frampton, E.A. Irene, H. Jiang, C.S. Petersson: Rate of formation of silicon dioxide; semiconducting ruthenium silicide, *Appl. Phys. Lett.* **47**(11), 1170–1172 (1985).

1.225 J.E.E. Baglin, F.M. d'Heurle, C.S. Petersson: Interface effects in the formation of silicon oxide on metal silicide layers over silicon substrates, *J. Appl. Phys.* **54**(4), 1849–1854 (1983).

1.226 T. Mochizuki, M. Kashiwagi: Characterization of thin film molybdenum silicide oxide, *J. Electrochem. Soc.* **127**(5), 1128–1135 (1980).

1.227 S. Zirinsky, W. Hammer, F.M. d'Heurle, J. Baglin: Oxidation mechanisms in WSi₂ thin films, *Appl. Phys. Lett.* **33**(1), 76–78 (1983).

1.228 M. Bartur, M.-A. Nicolet: Marker experiments for diffusion in the silicide during oxidation of PdSi, Pd₂Si, CoSi₂, and NiSi₂ films on <Si>, *J. Appl. Phys.* **54**(9), 5404–5415 (1983).

1.229 B.E. Deal, A.S. Grove: General relationship for the thermal oxidation of silicon, *J. Appl. Phys.* **36**(12), 3770–3778 (1965).

1.230 L.N. Lie, W.A. Tiller, K.C. Saraswat: Thermal oxidation of silicides, *J. Appl. Phys.* **56**(7), 2127–2132 (1984).

1.231 S.-L. Zhang, F.M. d'Heurle: The kinetics of oxidation: refinements for silicides, silicon and other materials, *Philos. Mag. A* **66**(3), 415–424 (1992).

1.232 S.-L. Zhang, R. Ghez, F.M. d'Heurle: A mathematical model for silicide oxidation, *J. Electrochem. Soc.* **137**(2), 659–663 (1990).

1.233 H.Z. Massoud, J.D. Plummer, E.A. Irene: Thermal oxidation of silicon in dry oxygen. Accurate determination of the kinetic rate constants, *J. Electrochem. Soc.* **132**(7), 1745–1753 (1985).

1.234 E.A. Irene: Silicon oxidation studies: A revised model for thermal oxidation, *J. Appl. Phys.* **54**(9), 5416–5420 (1983).

1.235 E.A. Irene, D.W. Dong: Silicon oxidation study: The oxidation of heavily B- and P-doped single crystal silicon, *J. Electrochem. Soc.* **125**(7), 1146–1151 (1978).

1.236 A.S. Wakita, T.W. Sigmon, J.F. Gibbons: Oxidation kinetics of laser formed MoSi₂ on polycrystalline silicon, *J. Appl. Phys.* **54**(5), 2711–2715 (1983).

1.237 W.A. Pliskin: Separation of the linear and parabolic terms in the steam oxidation of silicon, *IBM J. Res. Dev.* **10**(3), 198–205 (1966).

1.238 A. Cros: Silicon surfaces: Metallic character, oxidation and adhesion, *J. Phys.* **44**(6), 707–711 (1983).

1.239 H. Jiang, C.S. Petersson, M.-A. Nicolet: Thermal oxidation of transition metal silicides, *Thin Solid Films* **140**(1), 115–129 (1986).

Chapter 2

2.1 V.E. Borisenko, P.J. Hesketh: *Solid State Rapid Thermal Processing of Semiconductors*, (Plenum, New York, 1997).

2.2 A. Zur, T.C. McGill: Lattice match: An application to heteroepitaxy, *J. Appl. Phys.* **55**(2), 378–386 (1984).

2.3 A. Zur, T.C. McGill, M.-A. Nicolet: Transition-metal silicides lattice-matched to silicon, *J. Appl. Phys.* **57**(2), 600–603 (1985).

2.4 K.S. An, R.J. Park, J.S. Kim, S.B. Lee, T. Abukawa, S. Kono, T. Kinoshita, A. Kakisaki, T. Ishii: Initial interface formation study of the Mg/Si(111) system, *J. Appl. Phys.* **78**(2), 1151–1155 (1995).

2.5 R.A. McKee, F.J. Walker, J.R. Conner, R. Raj: $BaSi_2$ and thin film alkaline earth silicides on silicon, *Appl. Phys. Lett.* **63**(20), 2818–2820 (1993).

2.6 A. Zur, T.C. McGill, M.-A. Nicolet: Tables of Lattice Matches between Transition-Metal Silicides and Silicon, Physics Auxiliary Publication Service, American Institute of Physics, Doc. No. JAPIA-57-0600-76 (1984).

2.7 F.Y. Shiau, H.C. Cheng, L.J. Chen: Localized epitaxial growth of $CrSi_2$ on silicon, *J. Appl. Phys.* **59**(8), 2784–2787 (1986).

2.8 R.W. Fathauer, P.J. Grunthaner, T.L. Lin, K.T. Chang, J.H. Mazur: Nucleation and growth of $CrSi_2$ on Si(111), in: *Heteroepitaxy in Silicon: Fundamentals, Structures, Devices*, edited by H.K. Choi, H. Ishiwara, R. Hull, R.J. Nemanich (MRS, Pittsburgh, PA, 1988), pp. 453–458.

2.9 R.G. Long, J.P. Becker, J.E. Mahan, A. Vantomme, M.-A. Nicolet: Heteroepitaxial relationships for $CrSi_2$ thin films on Si(111), *J. Appl. Phys.* **77**(7), 3088–3094 (1995).

2.10 J.E. Mahan, K.M. Geib, G.Y. Robinson, G.Bai, M.-A. Nicolet: Reflection high-energy electron diffraction patterns of $CrSi_2$ films on (111) silicon, *J. Vac. Sci. Technol. B* **9**(1), 64–68 (1991).

2.11 W.T. Lin, L.J. Chen: Localized epitaxial growth of WSi_2 on silicon, *J. Appl. Phys.* **59**(10), 3481–3488 (1986).

2.12 J.E. Mahan, K.M. Geib, J.Y. Robinson, R.G. Long, Y. Xinghua, G. Bai, M.-A. Nicolet, M. Nathan: Epitaxial tendencies of $ReSi_2$ on (001) silicon, *Appl. Phys. Lett.* **56**(24), 2439–2441 (1990).

2.13 J.E. Mahan, K.M. Geib, G.Y. Robinson, R.G. Long, Y. Xinghua, G. Bai, M.-A. Nicolet, M. Nathan: Epitaxial films of semiconducting $FeSi_2$ on (001) silicon, *Appl. Phys. Lett.* **56**(21), 2126–2128 (1990).

2.14 K.M. Geib, J.E. Mahan, R.G. Long, M. Nathan, G. Bay: Epitaxial orientation and morphology of β-$FeSi_2$ on (001) silicon, *J. Appl. Phys.* **70**(3), 1730–1736 (1991).

2.15 J.E. Mahan, V. Le Thanh, J. Chevrier, I. Berbezier, J. Derrien, R.G. Long: Surface electron-diffraction patterns of β-$FeSi_2$ films epitaxially grown on silicon, *J. Appl. Phys.* **74**(3), 1747–1761 (1993).

2.16 N. Cherief, C.D.'Anterroches, R.C. Cinti, T.A. Nguyen Tan, J. Derrien: Semiconducting silicide-silicon heterojunction elaboration by solid phase epitaxy, *Appl. Phys. Lett.* **55**(16), 1671–1673 (1989).

2.17 L.A. Clevenger, R.W. Mann: Formation of epitaxial TM silicides, in: *Properties of Metal Silicides*, edited by K. Maex, M. Van Rossum (INSPEC, IEE, London, 1995), pp. 61–70.

2.18 I. Markov, A. Milchev: The effect of anharmonicity in epitaxial interfaces. II. Equilibrium structure of thin epitaxial films, *Surf. Sci.* **136**(2–3), 519–531 (1984).

2.19 I. Markov, S. Stoyanov: Mechanisms of epitaxial growth, *Contemporary Physics* **28**(3), 267–320 (1987).

2.20 C. Wigren, J.N. Andersen, R. Nyholm: Epitaxial silicide formation in the Mg/Si(111) system, *Surf. Sci.* **289**(2), 290–296 (1993).

2.21 R.W.Fathauer, P.J. Grunthaner, T.L. Lin, K.T. Chang, J.H. Mazur, D.N. Jamieson: Molecular-beam epitaxy of $CrSi_2$ on Si(111), *J. Vac. Sci. Technol. B* **6**(2), 708–712 (1988).

2.22 A.M. Rocher, A. Oustry, M.J. David, M. Caumont: Role of the surface steps on the growth of $CrSi_2$ on {111} silicon, in: *Silicide Thin Films - Fabrication, Properties, and Applications,* edited by R.T. Tung, K. Maex, P.W. Pellegrini, L.H. Allen (MRS, Pittsburgh, Pennsylvania, 1996), pp. 535–540.

2.23 W.T. Lin, L.J. Chen: Localized epitaxial growth of hexagonal and tetragonal MoSi$_2$ on (111) Si, *Appl. Phys. Lett.* **46**(11), 1061–1063 (1985).

2.24 J.J. Chu, L.J. Chen, K.N. Tu: Localized epitaxial growth of ReSi$_2$ on (111) and (001) silicon, *J. Appl. Phys* **62**(2), 461–465 (1987).

2.25 G. Bai, M.-A. Nicolet, J.E. Mahan, K.M. Geib: Channeling of MeV ions in polyatomic epitaxial films: ReSi$_2$ on Si(100), *Phys. Rev. B* **41**(13), 8603–8607 (1990).

2.26 J.E. Mahan, G. Bai, M.-A. Nicolet, R.G. Long, K.M. Geib: Microstructure and morphology of some epitaxial ReSi$_2$ films on (001) silicon, *Thin Solid Films* **207**(1–2), 223–230 (1992).

2.27 L. Schellenberg, H.F. Braun, J. Muller: The osmium-silicon phase diagram, *J. Less-Common Met.* **144**(2), 341–350 (1988).

2.28 I. Engström: The crystal structure of OsSi$_2$, *Acta Chem. Scan.* **24**(6), 2117–2125 (1970).

2.29 A. Ishizaka, Y. Shiraki: Low temperature surface cleaning of silicon and its application to silicon MBE, *J. Electrochem. Soc.* **133**(4), 666–671 (1986).

2.30 A. Kugimiya, Y. Hirofuji, N. Matsuo: Si-beam radiation cleaning in molecular-beam epitaxy, *Jpn. J. Appl. Phys.* **24**(5), 564–567 (1985).

2.31 A.K. Niessen, F.R. de Boer: The enthalpy of formation of solid borides, carbides, nitrides, silicides and phosphides of transition and noble metals, *J. Less-Common Met.* **82**(1), 75–80 (1981).

2.32 J.P. Becker, R.J. Long, J.E. Mahan: Reflection high-energy electron diffraction patterns of carbide-contaminated silicon surfaces, *J. Vac. Sci. Technol. A* **12**(1), 174–178 (1994).

2.33 H. von Känel, R. Stalder, H. Sirringhaus, N. Onda, J. Henz: Epitaxial silicides with the fluorite structure, *Appl. Surf. Sci.* **53**, 196–205 (1991).

2.34 A. Vantomme, J.E. Mahan, G. Langouche, J.P. Becker, M.V. Bael, K. Temst, Ch.V. Haesendonck: Thin film growth of semiconducting Mg$_2$Si by codeposition, *Appl. Phys. Lett.* **70**(9), 1086–88 (1997).

2.35 D. Vandré, L. Incoccia, G. Kaindl: Structural studies of the Mg/Si(111) interface formation, *Surf. Sci.* **225**(3), 233–241 (1990).

2.36 M. Östling, C. Zaring: Mechanical properties of TM silicides, in: *Properties of Metal Silicides*, edited by K. Maex, M. Van Rossum (INSPEC, IEE, London, 1995), pp. 15–30; M. Östling, C. Zaring: Thermal properties of TM silicides, ibid, pp. 31–44.

2.37 I. Ohdomari, T.S. Kuan, K.N. Tu: Microstructure and Schottky barrier height of iridium silicides formed on silicon, *J. Appl. Phys.* **50**(11), 7020–7029 (1979).

2.38 C. Krontiras, L. Grönberg, I. Suni, F.M. d'Heurle, J. Tersoff, I. Engström, B. Karlsson, C.S. Petersson: Some properties of ReSi$_2$, *Thin Solid Films* **161**(1/2), 197–206 (1988).

2.39 Y.S. Chang, M.L. Chou: Solid-state epitaxy of osmium silicide on (111)Si under reducing atmosphere, *J. Appl. Phys.* **66**(7), 3011–3013 (1989).

2.40 S. Petersson, J. Baglin, W. Hammer, F.M. d'Heurle, T.S. Kuan, I. Ohdomari, J. de Sousa Pires, P. Tove: Formation of iridium silicides from Ir thin films on Si substrates, *J. Appl. Phys.* **50**(5), 3357–3365 (1979).

2.41 W. Henrion, M. Rebien, V.N. Antonov, O. Jepsen, H. Lange: Optical characterization of Ru$_2$Si$_3$ by spectroscopic ellipsometry, UV-VIS-NIR spectroscopy and band structure calculations, *Thin Solid Films* **313/314**, 218–221 (1998).

2.42 T. Rodriguez, H. Wolters, M. Fernandez, A. Almendra, M.F. da Silva, M. Clement, J.C. Soares, C. Ballesteros: Iridium silicides obtained by rapid thermal annealing, *Appl. Surf. Sci.* **73**, 182–185 (1993).

2.43   P.L. Janega, D. Landheer, J. McCaffrey, D. Mitchell, M. Buchanan, M. Denhoff: Contact resistivity of some magnesium/silicon and magnesium/silicide silicon structure, *Appl. Phys. Lett.* **53**(21), 2056–2058 (1988).

2.44   Y.C. Lian, L.J. Chen: Localized epitaxial growth of $MnSi_{1.7}$ on silicon, *Appl. Phys. Lett.* **48**(5), 359–361 (1986).

2.45   R.G. Long, M.C. Bost, J.E. Mahan: Optical and electrical properties of semiconducting rhenium disilicide thin films, *Thin Solid Films* **162**(1/2), 29–40 (1988).

2.46   H.C. Cheng, T.R. Yew, L.J. Chen: Interfacial reactions of iron thin films on silicon, *J. Appl. Phys.* **57**(12), 5246–5250 (1985).

2.47   S. Luby, G. Leggieri, A. Luches, M. Jergel, G. Majni, E. Majkova, M. Ožvold: Interfacial reactions of thin iron films on silicon under amorphous silicon and SiO_x capping, *Thin Solid Films* **245**(1), 55–59 (1994).

2.48   N.R. Baldwin, D.G. Ivey: Low temperature iron thin films-silicon reactions, *J. Mater. Sci.* **31**(1), 31–37 (1996).

2.49   K. Radermacher, S. Mantl, Ch. Dieker, H. Lüth, C. Freiburg: Growth kinetics of iron silicides fabricated by solid phase epitaxy or ion beam synthesis, *Thin Solid Films* **215**(1), 76–83 (1992).

2.50   W.K. Chu, S.S. Lau, J.W. Mayer, H. Müller, K.N. Tu: Implanted noble gas atoms as diffusion markers in silicide formation, *Thin Solid Films* **25**(2), 393–402 (1975).

2.51   A. Martinez, D. Esteve, A. Guivarc'h, P. Auvray, P. Henoc, G. Pelous: Metallurgical and electrical properties of chromium silicon interfaces, *Solid-State Electron.* **23**(1), 55–64 (1980).

2.52   L.R. Zheng, L.S. Hung, S.H. Chen, J.W. Mayer: Effects of krypton on $CrSi_2$ formation, *J. Appl. Phys.* **59**(6), 1998–2001 (1986).

2.53   H. Lange, M. Giehler, W. Henrion, F. Fenske, I. Sieber, G. Oertel: Growth and optical characterization of $CrSi_2$ thin films, *Phys. Stat. Sol. (b)* **171**, 63–76 (1992).

2.54   F. Nava, T. Tien, K.N. Tu: Temperature dependence of semiconducting and structural properties of Cr-Si thin films, *J. Appl. Phys.* **57**(6), 2018–2025 (1985).

2.55   C. Heck, M. Kusaka, M. Hirai, M. Iwami, Y. Yokota: Thin film silicon compound growth mechanisms: $CrSi_2/Si(001)$, *Thin Solid Films* **281/282**, 94–97 (1996).

2.56   F.Y. Shiau, H.C. Cheng, L.J. Chen: Epitaxial growth of $CrSi_2$ on (111)Si, *Appl. Phys. Lett.* **45**(5), 524–526 (1984).

2.57   R.G. Long, J.E. Mahan: Two pseudobinary semiconducting silicides: $Re_xMo_{1-x}Si_2$ and $Cr_xV_{1-x}Si_2$, *Appl. Phys. Lett.* **56**(17), 1655–1657 (1990).

2.58   C.M. Doland, R.J. Nemanich: Phase formation during reactive molybdenum-silicide formation, *J. Mater. Res.* **5**(12), 2854–2864 (1990).

2.59   J.M. Harris, S.S. Lau, M.-A. Nicolet, R.S. Nowicki: Studies of the Ti-W metallization system on Si, *J. Electrochem. Soc.* **123**(1), 120–124 (1976).

2.60   F. Nava, C. Nobili, G. Ferla, G. Iannuzzi, G. Queirolo, G. Celotti: Ti-W alloy interaction with silicon in the presence of PtSi, *J. Appl. Phys.* **54**(5), 2434–2440 (1983).

2.61   M. Eizenberg, K.N. Tu: Formation and Schottky behavior of manganese silicides on n-type silicon, *J. Appl. Phys.* **53**(10), 6885–6890 (1982).

2.62   M.C. Bost, J.E. Mahan: An optical determination on the bandgap of the most silicon-rich manganese silicide phase, *J. Electron. Mater.* **16**(6), 389–395 (1987).

2.63   L. Zhang, D.G. Ivey: Reaction kinetics and opticals properties of semiconducting $MnSi_{1.73}$ grown on ⟨001⟩ oriented silicon, *J. Mater. Sci.* **2**(2), 116–123 (1991).

2.64   Ch. Krontiras, K. Pomoni, M. Roilos: Resistivity and the Hall effect for thin $MnSi_{1.73}$ films, *J. Phys. D: Appl. Phys.* **21**(3), 509–512 (1988).

2.65   T.A. Nguyen Tan, J.Y. Veuillen, P. Muret: Semiconducting rhenium silicide thin films on Si(111), *J. Appl. Phys.* **77**(6), 2514–2518 (1995).

2.66   C.A. Dimitriadis, J.H. Werner: Growth mechanism and morphology of semiconducting FeSi$_2$ films, *J. Appl. Phys.* **68**(1), 93–96 (1990).

2.67   U. Erlesand, M. Östling, K. Bodén: Formation of iron disilicide on amorphous silicon, *Appl. Surf. Sci.* **53**(1), 153–158 (1991).

2.68   C.A. Dimitriadis: Electrical properties of β-FeSi$_2$/Si heterojunctions, *J. Appl. Phys.* **70**(10), 5423–5426 (1991).

2.69   S. Kennou, N. Cherief, R.C. Cinti, T.A. Nguyen Tan: The influence of steps of epitaxial growth of iron-silicide on Si (001), *Surf. Sci.* **211**(1–3), 685–691 (1989).

2.70   Ch. Stuhlmann, J. Schmidt, H. Ibach: Semiconducting iron disilicide films on silicon (111): A high-resolution electron energy loss spectroscopy study, *J. Appl. Phys.* **72**(12), 5905–5911 (1992).

2.71   K. Konuma, J. Vrijmoeth, P.M. Zagwijn, J.W.M. Frenken, E. Vlieg, J.F. van der Veen: Formation of epitaxial β-FeSi$_2$ films on Si (001) as studied by medium energy ion scattering, *J. Appl. Phys.* **73**(3), 1104–1109 (1992).

2.72   D.R. Peale, R. Haight, J. Ott: Heteroepitaxy of β-FeSi$_2$ on unstrained and strained Si (100) surfaces, *Appl. Phys. Lett.* **62**(12), 1402–1404 (1993).

2.73   D. Mangelinck, L. Wang, C. Lin, P. Gas, J. Grahn, M. Östling: Influence of the addition of Co and Ni on the formation of epitaxial semiconducting β-FeSi$_2$: Comparison of different evaporation methods, *J. Appl. Phys.* **83**(8), 4193–4201 (1998).

2.74   D. Donoval, L. Stolt, H. Norde, J. de Sousa Pires, P.A. Tove, C.S. Petersson: Barrier heights to silicon, of ruthenium and its silicide, *J. Appl. Phys.* **53**(7), 5352–5353 (1982).

2.75   C.S. Petersson, J.E.E. Baglin, J.J. Dempsey, F.M. d'Heurle, S.J. La Placa: Silicides of ruthenium and osmium: thin film reactions, diffusion, nucleation, and stability, *J. Appl. Phys.* **53**(7), 4866–4883 (1982).

2.76   Y.S. Chang, J.J. Chu: The structure identification of epitaxial ruthenium silicide (Ru$_2$Si$_3$) on (111) Si, *Mater. Lett.* **5**(3), 67–71 (1987).

2.77   Y.S. Chang, M.L. Chou: Formation and structure of epitaxial ruthenium silicides on (111) Si, *J. Appl. Phys.* **68**(5), 2411–2414 (1990).

2.78   S. Petersson, J.A. Reimer, M.H. Brodsky, D.R. Campbell, F.M. d'Heurle, B. Karlsson, P.A. Tove: IrSi$_{1.75}$ a new semiconductor compound, *J. Appl. Phys.* **53**(4), 3342–3343 (1982).

2.79   T. Rodriguez, H. Wolters, A. Almendra, J. Sanz-Maudes, M.F. da Silva, J.C. Soares, in: *Infrared Detectors - Materials, Processing, and Devices*, edited by L.R. Dawson, A. Appelbaum (MRS, Pittsburgh, Pennsylvania, 1994), pp. 313–317.

2.80   T. Rodriguez, A. Almendra, M. Botella, M.F. da Silva, J.C. Soares, H. Wolters, C. Ballesteros: Iridium silicides formed by RTA in vacuum, in: *Silicide Thin Films - Fabrication, Properties, and Applications,* edited by R.T. Tung, K. Maex, P.W. Pellegrini, L.H. Allen (MRS, Pittsburgh, Pennsylvania, 1996), pp. 599–604.

2.81   M. Wittmer, W. Lüthy, M. von Allmen: Laser induced reaction of magnesium with silicon, *Phys. Lett.* **75A**(1–2), 127–130 (1979).

2.82   L.R. Zheng, J.W. Mayer: Suppression of lateral diffusion in the Cr-Si system by ion irradiation, *Appl. Phys. Lett.* **45**(6), 636–638 (1984).

2.83   E. D'Anna, G. Leggieri, A. Luches, G. Majni, G. Ottaviani: Chromium silicide formation under pulsed heat flow, *Thin Solid Films* **136**(1), 93–104 (1986).

2.84   E. D'Anna, G. Leggieri, A. Luches, G. Majni: Chromium silicide formation with multiple electron beam pulses, *Thin Solid Films* **140**(1), 163–166 (1986).

2.85   E. D'Anna, A.V. Drigo, G. Leggieri, A. Luches, G. Majni, P. Mengucci: Synthesis of chromium silicide with laser pulses, *Appl. Phys. A*, **50**(4), 411–415 (1990).

2.86   J. Wang, M. Hirai, M. Kusaka, M. Iwami: Preparation of manganese silicide thin films by solid phase reaction, *Appl. Surf. Sci.* **113/114**, 53–56 (1997).

2.87   V.E. Borisenko, D.I. Zarovskii, L.I. Ivanenko: Structure and phase transitions in thin film Fe-Si system subjected to pulse thermal processing, *Mikroelektronika* **21**(6), 27–30 (1992) (in Russian).

2.88   M. Sauvage-Simkin, N. Jedrecy, A. Waldhauer, R. Pinchaux: Structural approach of the Fe-Si(111) annealed interfaces, *Physica B* **198**(1), 48–54 (1994).

2.89   J. Derrien, J. Chevrier, A. Younsi, V. Le Thanh, J.P. Dussaunlcy, N. Cherief: Structure and electronic properties of epitaxially grown silicides, *Phys. Scr.* **35**, 251–260 (1991).

2.90   W. Raunau, H. Niehus, T. Schilling, G. Comsa: Scanning tunneling microscopy and spectroscopy of iron silicide epitaxially grown on Si (111), *Surf. Sci.* **286**(3), 203–211 (1993).

2.91   H.C. Cheng, T.R. Yew, L.J. Chen: Epitaxial growth of $FeSi_2$ in Fe thin films on Si with a thin interposing Ni layer, *Appl. Phys. Lett.* **47**(2), 128–130 (1985).

2.92   J. de Sousa Pires, P. Ali, B. Crowder, F.M. d'Heurle, S. Petersson, L. Stolt, P. Tove: Measurements of the rectifying barrier heights of the various iridium silicides with n-Si, *Appl. Phys. Lett.* **35**(2), 202–204 (1979).

2.93   S.P. Murarka: Properties and applications of silicides, in: *Microelectronic Materials and Processes,* edited by R.A. Levy (Kluwer Academic Publishers, 1989), pp.275–323.

2.94   M.Y. Tsai, F.M. d'Heurle, C.S. Petersson, R.W. Johnson: Properties of tungsten silicide film on polycrystalline silicon, *J. Appl. Phys.* **52**(8), 5350–5355 (1981).

2.95   S.F. Gong, X.-H. Li, H.T.G. Hentzell, J. Strandberg: Electrical and structural properties of thin film of sputtered $CrSi_2$, *Thin Solid Films* **208**(1), 91–95 (1992).

2.96   A.H. van Ommen, A.H. Reader, J.W.C. de Vries: Influence of microstructure on the resistivity of $MoSi_2$ thin films, *J. Appl. Phys.* **64**(7), 3574–3580 (1988).

2.97   F. Nava, B.Z. Weiss, K.Y. Ahn, D.A. Smith, K.N. Tu: Thermal stability and electrical conduction behavior of coevaporated $WSi_{2\pm x}$ thin films, *J. Appl. Phys.* **64**(1), 354–364 (1988).

2.98   F. Nava, B.Z. Weiss, K. Ahn, K.N. Tu: Tungsten disilicide formation in codeposite amorphous WSi_x alloy thin films, *Le Vide, les Couches Minces* **42**(1), 225–228 (1987).

2.99   W. Pitschke, A. Heinrich, J. Schumann: Rhenium silicide thin films: structural analysis of the $ReSi_{2-\delta}$-phase, in: *Proceedings of the 16th International Conference on Thermoelectrics* (IEEE, Inc., Piscataway, NJ, 1997), pp. 299–302.

2.100   M. Powalla, K. Herz: Co-evaporated thin films of semiconducting β-$FeSi_2$, *Appl. Surf. Sci.* **65/66**, 482–488 (1993).

2.101   K. Herz, M. Powalla: Electrical and optical properties of thin β-$FeSi_2$ films on Al_2O_3 substrates, *Appl. Surf. Sci.* **91**(1), 87–92 (1995).

2.102   M. Libezny, J. Poortmans, J. Dekoster, S. Degroote, A. Vantomme, B.G.M. De Lange, G. Langouche, J. Nijs: RTA-preparation of β-$FeSi_2$ layers from MBE-grown Fe-Si films deposited on Si and relaxed SiGe (100) substrates, in: *Rapid Thermal and Integrated Processing IV,* edited by J.C. Sturm, J.C. Gelpey, S.R.J. Brueck, A. Kermani, J.L. Regolini (MRS, Pittsburgh, Pennsylvania, 1995), pp.407–412.

2.103   D.H. Tassis, C.A. Dimitriadis, S. Boultadakis, J. Arvanitidis, S. Ves, S. Kokkou, S. Logothetidis, O. Valassiades, P. Poulopoulos, N.K. Flevaris: Influence of conventional furnace and rapid thermal annealing on the quality of polycrystalline β-$FeSi_2$ thin films grown from vapor-deposited Fe/Si multilayers, *Thin Solid Films* **310**, 115–122 (1997).

2.104   D.H. Tassis,   C.L. Mitsas,   T.T. Zorba,   M. Angelakeris,   C.A. Dimitriadis, O. Valassiades, D.I. Siapkas, G. Kiriakidis: Optical and electrical characterization of high quality β-FeSi$_2$ thin films grown by solid phase epitaxy, *Appl. Surf. Sci.* **102**, 178–183 (1996).

2.105   D.H. Tassis,   C.L. Mitsas,   T.T. Zorba,   C.A. Dimitriadis,   O. Valassiades, D.I. Siapkas, M. Angelakeris, P. Poulopoulos, N.K. Flevaris, G. Kiriakidis: Infrared spectroscopic and electronic transport properties of polycrystalline semiconducting FeSi$_2$ thin films, *J. Appl. Phys.* **80**(2), 962–968 (1996).

2.106   A. Valentini, E. Chimenti, A. Cola, G. Leo, F. Quaranta, L. Vasanelli: Triple ion beam sputtering deposition of β-FeSi$_2$. Advances in crystal growth, *Mater. Sci. Forum* **203**, 173–178 (1996).

2.107   D.R. Peale, R. Haight, F.K. LeGoues: Strain relaxation in ultrathin epitaxial films of β-FeSi$_2$ on unstrained and strained Si (100) surfaces, *Thin Solid Films* **264**(1), 28–39 (1995).

2.108   H. Sirringhaus, N. Onda, E. Müller-Gubler, P. Müller, R. Stalder, H. von Känel: Phase transition from pseudomorphic FeSi$_2$ to β-FeSi$_2$/Si(111) studied by *in situ* scanning tunneling microscopy, *Phys. Rev. B* **47**(16), 10567–10577 (1993).

2.109   M. Döscher, B. Selle, M. Pauli, F. Kothe, J. Szymanski, J. Müller, H. Lange: Influence of stoichiometric variations and rapid thermal processing of β-FeSi$_2$ thin films on their electrical and microstructural properties, in: *Silicide Thin Films - Fabrication, Properties, and Applications*, edited by R.T. Tung, K. Maex, P.W. Pellegrini, L.H. Allen (MRS, Pittsburgh, Pennsylvania, 1996), pp. 325–330.

2.110   S. Hong,   C. Pirri,   P. Wetzel,   G. Gewinner: Synthesis of epitaxial ternary Co$_{1-x}$Fe$_x$Si$_2$ silicides with CsCl- and CaF$_2$-type cubic structures on Si(111) by codeposition techniques, *Phys. Rev. B* **55**(19), 13040–13050 (1997).

2.111   T. Takada, H. Katsumata, Y. Makita, N. Kobayashi, M. Hasegawa, S. Uekusa: Fabrication of p-β-Fe$_{1-x}$Mn$_x$Si$_2$/n-Si heterostructure diode and their electrical and optical properties, in: *Thermoelectric Materials - New Directions and Approaches*, edited by T.M. Tritt, M.G. Kanatzidis, H.B. Lyon, G.D. Mahan (MRS, Pittsburgh, Pennsylvania, 1997), pp. 267–272.

2.112   M. Wittmer, P. Oelhafen, K.N. Tu: Electronic structure of iridium silicides, *Phys. Rev. B* **33**(8), 5391–5400 (1986).

2.113   J. Schumann,   D. Elefant,   C. Gladun,   A. Heinrich,   W. Pitschke,   H. Lange, W. Henrion, R. Grötzschel: Polycrystalline iridium silicide films. Phase formation, electrical and optical properties, *Phys. Stat. Sol.(a)* **145**(2), 429–439 (1994).

2.114   W. Pitschke, R. Kurt, A. Heinrich, J. Schumann, J. Thomas, M. Mäder: Structure and phase formation in amorphous Ir$_x$Si$_{1-x}$ thin films at high temperatures, in: *Proceedings of the 15th International Conference on Thermoelectrics* (IEEE, Inc., Piscataway, NJ, 1996), pp. 499–503.

2.115   U. Gottlieb, B. Lambert-Andron, F. Nava, M. Affronte, O. Laborde, A. Rouault, R. Madar: Structural and electronic transport properties of ReSi$_{2-\delta}$ single crystals, *J. Appl. Phys.***78**(6), 3902–3907 (1995).

2.116   S. Mantl: Ion beam synthesis of epitaxial silicides: Fabrication, characterization and application, *Mater. Sci. Rep.* **8**, 1–95 (1992).

2.117   W.-Z. Li, H. Kheyrandish, Z. Al-Tamimi, W.A. Grant: Cr$^+$ ion irradiation and thermal annealing of chromium films on silicon for formation of silicides, *Nucl. Instrum. Methods Phys. Res. B* **19/20**, 723–730 (1987).

2.118   F.M. d'Heurle, C.S. Petersson, M.Y. Tsai: Observations on the hexagonal form of MoSi$_2$ and WSi$_2$ films produced by ion implantation and on related snowplow effects, *J. Appl. Phys.* **51**(11), 5976–5980 (1980).

2.119   A.E. White, K.T. Short, D.J. Eaglesham: Electrical and structural properties of Si/CrSi$_2$/Si heterostructures fabricated using ion implantation, *Appl. Phys. Lett.* **56**(13), 1260–1262 (1990).

2.120   N. Kobayashi, H. Katsumata, H.L. Shen, M. Hasegawa, Y. Makita, H. Shibata, S. Kimura, A. Obara, S. Uekusa, T. Hatano: Structural and optical characterization of β-FeSi$_2$ layers on Si formed by ion beam synthesis, *Thin Solid Films* **270**(1–2), 406–410 (1995).

2.121   H. Katsumata, Y. Makita, N. Kobayashi, M. Hasegawa, H. Shibata, S. Uekusa: Synthesis of β-FeSi$_2$ for optical applications by Fe triple-energy ion implantation into Si(100) and Si(111) substrates, *Thin Solid Films* **281/282**, 252–255 (1996).

2.122   H. Katsumata, Y. Makita, N. Kobayashi, H. Shibata, M. Hasegawa, I. Aksenov, S. Kimura, A. Obara, S. Uekusa: Optical absorption and photoluminescence studies of β-FeSi$_2$ prepared by heavy implantation of Fe$^+$ ions into Si, *J. Appl. Phys.* **80**(10), 5955–5962 (1996).

2.123   H. Katsumata, Y. Makita, N. Kobayashi, H. Shibata, M. Hasegawa, S. Uekusa: Effect of multiple-step annealing on the formation of semiconducting β-FeSi$_2$ and metallic α-Fe$_2$Si$_5$ on Si (100) by ion beam synthesis, *Jpn. J. Appl. Phys.* **36**(5), 2802–2812 (1997).

2.124   Z. Yang, K.P. Homewood, M.S. Finney, M.A. Harry, K.J. Reeson: Optical absorption study of ion beam synthesized polycrystalline semiconducting FeSi$_2$, *J. Appl. Phys.* **78**(3), 1958–1963 (1995).

2.125   Z. Yang, G. Shao, K.P. Homewood, K.J. Reeson, M.S. Finney, M. Harry: Order domain boundaries in ion beam synthesized semiconducting FeSi$_2$ layers, *Appl. Phys. Lett.* **67**(5), 667–669 (1995).

2.126   K. Radermacher, S. Mantl, R. Apetz, Ch. Dieker, H. Lüth: Ion beam synthesis of buried epitaxial FeSi$_2$, *Mater. Sci. Eng. B* **12**(1–2), 115–118 (1992).

2.127   S. Eisebitt, J.-E. Rubensson, M. Nicodemus, T. Böske, S. Blügel, W. Eberhardt, K. Radermacher, S. Mantl, G. Bihlmayer: Electronic structure of buried α-FeSi$_2$ and β-FeSi$_2$ layers: Soft-x-ray-emission and -absorption studies compared to band-structure calculations, *Phys. Rev. B* **50**(24), 18330–18340 (1994).

2.128   D. Gerthsen, K. Radermacher, Ch. Dieker, S. Mantl: Structural properties of ion-beam-synthesized β-FeSi$_2$ in Si(111), *J. Appl. Phys.* **71**(8), 3788–3794 (1992).

2.129   K. Radermacher, S. Mantl, Ch. Dieker, H. Lüth: Ion beam synthesis of buried α-FeSi$_2$ and β-FeSi$_2$ layers, *Appl. Phys. Lett.* **59**(17), 2145–2147 (1996).

2.130   D.J. Oostra, C.W.T. Bulle-Lieuwma, D.E.W. Vandenhoudt, F. Felten, J.C. Jans: β-FeSi$_2$ in (111) Si and in (001) Si formed by ion-beam synthesis, *J. Appl. Phys.* **74**(7), 4347–4353 (1993).

2.131   D.J. Oostra, D.E.W. Vandenhoudt, C.W.T. Bulle-Lieuwma, E.P. Naburgh: Ion-beam synthesis of a Si/β-FeSi$_2$/Si heterostructure, *Appl. Phys. Lett.* **59**(14), 1737–1739 (1991).

2.132   M. Behar, H. Bernas, J. Desimoni, X.W. Lin, R.L. Maltez: Sequential phase formation by ion-induced epitaxy in Fe-implanted Si(001). Study of their properties and thermal behavior, *J. Appl. Phys.* **79**(2), 752–762 (1996).

2.133   B.X. Liu, D.H. Zhu, H.B. Lu, F. Pan, K. Tao: Synthesis of β- and α-FeSi$_2$ phases by Fe ion implantation into Si using metal vapor vacuum arc ion source, *J. Appl. Phys.* **75**(8), 3847–3854 (1994).

2.134   T.D. Hunt, K.J. Reeson, R.M. Gwilliam, K.P. Homewood, R.J. Wilson, R.S. Spraggs, B.J. Sealy, C.D. Meekison, G.R. Booker, P. Oberschachtsiek: Determination of the optical and materials properties of β-FeSi$_2$ layers fabricated using ion beam synthesis, *Mater. Res. Soc. Symp. Proc.* **260**, 239–244 (1992).

2.135   S. Jin, X.N. Li, Z. Zhang, C. Dong, Z.X. Gong, H. Bender, T.C. Ma: Ion beam syntheses and microstructure studies of a new FeSi$_2$ phase. *J. Appl. Phys.* **80**(6), 3306–3309 (1996).

2.136 D. Panknin, E. Wieser, W. Skorupa, W. Henrion, H. Lange: Buried $(Fe_{1-x}Co_x)Si_2$ layers with variable band gap formed by ion beam synthesis, *Appl. Phys. A* **62**(2), 155–162 (1996).

2.137 E. Wieser, D. Panknin, W. Skorupa, G. Querner, W. Henrion, J. Albrecht: Ion beam synthesys of ternary $(Fe_{1-x}Co_x)Si_2$, *Nucl. Instrum. Methods Phys. Res. B* **80/81** (Pt. 2), 867–871 (1993).

2.138 M.S. Finney, Z. Yang, M.A. Harry, K.J. Reeson, K.P. Homewood, R.M. Gwilliam, B.J. Sealy: Effects of annealing and cobalt implantation on the optical properties of β-$FeSi_2$, in: *Silicides, Germanides, and Their Interfaces*, edited by R.W. Fathauer, S. Mantl, L.J. Schowalter, K.N. Tu (MRS, Pittsburgh, Pennsylvania, 1994), pp. 173–178.

2.139 H. Reuther, M. Dobler: Implantation and growth of large β-$FeSi_2$ precipitates and α-$FeSi_2$ network structures in silicon, *Appl. Phys. Lett.* **69**(21), 3176–3178 (1996).

2.140 S. Kruijer, W. Keune, M. Dobler, H. Reuther: Depth analysis of phase formation in Si after high-dose Fe ion implantation by depth-selective conversion-electron Mössbauer spectroscopy, *Appl. Phys. Lett.* **70**(20), 2696–2698 (1997).

2.141 J. Desimoni, H. Bernas, M. Behar, X.W. Lin, J. Washburn, Z. Liliental-Weber: Ion beam synthesis of cubic $FeSi_2$, *Appl. Phys. Lett.* **62**(3), 306–308 (1993).

2.142 Z. Yang, G. Shao, K.P. Homewood: Metastable γ phase in ion beam synthesized $FeSi_2$, *Appl. Phys. Lett.* **68**(13), 1784–1786 (1996).

2.143 M.A. Harry, G. Curello, M.S. Finney, K.J. Reeson, B.J. Sealy: Structural properties of ion beam synthesized iron-cobalt silicide, *J. Phys. D: Appl. Phys.* **29**(7), 1822–1830 (1996).

2.144 Zh. Tan, F. Namavar, S.M. Heald, J.I. Budnick: Sequential-ion-implantation synthesys of ternary metal silicides, *Appl. Phys. Lett.* **63**(6), 791–793 (1993).

2.145 S. Sivaram: *Chemical Vapor Deposition* (Van Nostrand Reinhold, New York, 1995).

2.146 S. Motojima, K. Sugiyama: Chemical vapor growth of Cr_5Si_3 whiskers and hollow crystals, *J. Cryst. Growth* **55**(3), 611–613 (1981).

2.147 E. Blanquet, C. Vahlas, C. Bernard, R. Madar, J. Palleau, J. Torres: Chemical vapor deposition of $TaSi_2$ and WSi_2 at atmospheric pressure from *in situ* prepared metal chlorides, *J. de Physique* **50**(5), 557–563 (1989).

2.148 R. Madar, C. Bernard: Chemical vapour deposition of metal silicides in silicon microelectronics, *Appl. Surf. Sci.* **53**(1), 1–10 (1991).

2.149 A.-M. Dutron, E. Blanquet, N. Bourhila, R. Madar, C. Bernard: A thermodinamic and experimental approach to $ReSi_2$ LPCVD, *Thin Solid Films* **259**(1), 25–31 (1995).

2.150 J.L. Regolini, F. Trincat, I. Berbezier, Y. Shapira: Selective and epitaxial deposition of β-$FeSi_2$ on silicon by rapid thermal processing-chemical vapor deposition using a solid iron source, *Appl. Phys. Lett.* **60**(8), 956–958 (1992).

2.151 Y. Morand, E. Blanquet, N. Bourhila, N. Thomas, C. Bernard, R. Madar: Thermodynamic and experimental study of β-$FeSi_2$ LPCVD, in: *Silicides, Germanides, and Their Interfaces*, edited by R.W. Fathauer, S. Mantl, L.J. Schowalter, K.N. Tu (MRS, Pittsburgh, Pennsylvania, 1994), pp. 91–96.

2.152 E. Hengge, H. Keller-Rudek, D. Koschel, U. Krürke, P. Merlet: *Gmelin Handbook of Inorganic Chemistry, 8th Edition, Silicon, V.B1* (Spinger, Berlin, 1982).

2.153 D.R. Lide (Ed.) *CRC Handbook of Chemistry and Physics*, 75th edition (CRC Press, Boca Raton, Florida, 1994).

2.154 R.L. Moon, Y.-M. Houng: Organometallic vapor phase epitaxy of III-V materials, in: *Chemical Vapor Deposition. Principles and Applications*, edited by M.L. Hitchman, K.F. Jensen (Academic Press, London, 1993), pp. 245–384.

2.155 E.K. Broadbent, W.T. Stacy: Selective tungsten processing by low pressure CVD, *Solid State Technology* **2**(12), 51–59(1985).

2.156 K. Othmer: *Encyclopedia of Chemical Technology,* 3rd edition (J. Wiley, New York, 1980).

2.157 T.P. Chow, D.M. Brown, A.J. Steckl, M. Garfinkel: Silane silicidation of Mo thin films, *J. Appl. Phys.* **51**(11), 5981–5985 (1980).

2.158 D. Dobkin, L. Bartholomew, G. McDaniel, J. DeDontney: Atmospheric pressure chemical vapor deposition of tungsten silicide, *J. Electrochem. Soc.* **137**(5), 1623–1626 (1990).

2.159 C.S. Ozkan, M. Moinpour: Tungsten silicide films based on dichlorosilane chemistry for sub 0.5 micron applications, in: *Silicide Thin Films - Fabrication, Properties, and Applications,* edited by R.T. Tung, K. Maex, P.W. Pellegrini, L.H. Allen (MRS, Pittsburgh, Pennsylvania, 1996), pp. 301–306.

2.160 J.P. André, H. Alaoui, A. Deswarte, Y. Zheng, J.F. Pétroff, X. Wallart, J.P. Nys: Iron silicide growth on Si(111) substrate using the metalorganic vapor phase epitaxy process, *J. Cryst. Growth* **144**(1), 29–40 (1994).

2.161 J.S. Byun, B.H. Lee, J.-S. Park, J.J. Kim: Characterization of the dopand effect on dichlorosilane-based tungsten silicide deposition, *J. Electrochem. Soc.* **144**(10), 3572–3582 (1997).

2.162 S. Inoue, N. Toyokura, T. Nakamura, M. Maeda, M. Takagi: Properties of molybdenum silicide film deposited by chemical vapor deposition, *J. Electrochem. Soc.* **130**(7), 1603–1607 (1983).

2.163 K. Akitmoto, K. Watanabe: Formation of W_xSi_{1-x} by plasma chemical vapor deposition, *Appl. Phys. Lett.* **39**(5), 445–447 (1981).

2.164 C.C. Tang, J.K. Chu, D.W. Hess: Plasma-Enhanced Deposition of Tungsten, Molybdenum, and Tungsten Silicide Films, *Solid State Technology* **26**(3), 125–128 (1983).

2.165 J.L. Regolini, F. Trincat, I. Berbezier, J. Palleau, J. Mercier, D. Bensahel: Selective and epitaxial deposition of β-FeSi$_2$ onto silicon by RPT-CVD, *J. Phys. III France* **2**, 1445–1452 (1992).

2.166 I. Berbezier, J.L. Regolini, C. d'Anterroches: Epitaxial orientation of β-FeSi$_2$/Si heterojunctions obtained by RPT chemical vapor deposition, *Microsc. Microanal. Microstruct.* **4**, 5–21 (1993).

2.167 J.L. Regolini, F. Trincat, I. Sagnes, Y. Shapira, G. Brémond, D. Bensahel: Characterization of semiconducting iron disilicide obtained by LRP/CVD, *IEEE Trans. Electron Devices* **39**(1), 200–201 (1992).

2.168 A. Rizzi, B.N.E. Rösen, D. Freundt, Ch. Dieker, H. Lüth, D. Gerthsen: Heteroepitaxy of β-FeSi$_2$ on Si by gas-source MBE, *Phys. Rev. B* **51**(24), 17780–17794 (1995).

2.169 B. Rösen, H.Ch. Schäfer, Ch. Dieker, H. Lüth, A. Rizzi, D. Gerthsen: Electron energy loss spectroscopy on FeSi$_2$/Si(111) heterostructures grown by gas source molecular-beam epitaxy, *J. Vac. Sci. Technol. B* **11**(4), 1407–1412 (1993).

2.170 H.Ch. Schäfer, B. Rösen, H. Moritz, A. Rizzi, B. Lengeler, H. Lüth, D. Gerthsen: Gas source molecular-beam epitaxy of FeSi$_2$/Si(111) heterostructures, *Appl. Phys. Lett.* **62**(18), 2271–2273 (1993).

2.171 B.N.E. Rösen, D. Freundt, Ch. Dieker, D. Gerthsen, A. Rizzi, R. Carius, H. Lüth: Characterization of β-FeSi$_2$/Si heterostructures grown by gas- source-MBE, in: *Silicides, Germanides, and Their Interfaces,* edited by R.W. Fathauer, S. Mantl, L.J. Schowalter, K.N. Tu (MRS, Pittsburgh, Pennsylvania, 1994), pp.139–144.

2.172 J. Derrien, J. Chevrier, V. Le Thanh, T.E. Crumbaker, J.Y. Natoli, I. Berbezier: Silicide epilayers: Recent development and prospects for a Si-compatible technology, *Appl. Surf. Sci.* **70/71**, 546–558 (1993).

2.173   J.Y. Natoli, I. Berbezier, J. Derrien: Growth of β-FeSi₂ on Si(111) by chemical beam epitaxy, *Appl. Phys. Lett.* **65**(11), 1439–1441 (1994).

2.174   J. Derrien, I. Berbezier, J. Chevrier, A. Ronda, J.Y. Natoli: Interface phase transition as observed in ultra thin FeSi₂ epilayers, *Appl. Surf. Sci.* **92**(1–4), 311–320 (1996).

2.175   see for example: V.E. Borisenko, A.B. Filonov, S.V. Gaponenko, V.S. Gurin (Eds.) *Physics, Chemistry and Application of Nanostructures* (World Scientific, Singapore, 1997, 1999).

2.176   E. Fogarassy, S. Lazare (Eds.) *Laser Ablation of Electronic Materials* (Elsevier Science Publishers B.V., 1992).

2.177   A. Mele, A. Giardini, R. Teghil: Thin film epitaxial growth by laser ablation, in: *Frontiers in Nanoscale Science of Micro/Submicron Devices*, edited by A.-P. Jauho, E.V. Buzaneva (Kluwer Academic Publisher, Dordrecht, 1996), pp. 67–83.

2.178   C.H. Olk, O.P. Karpenko, S.M. Yalisove, G.L. Doll, J.F. Mansfield: Growth of epitaxial β-FeSi₂ thin films by pulsed laser deposition on silicon (111), *J. Mater. Res.* **9**(11), 2733–2736 (1994).

2.179   S. Teichert, R. Kilper, T. Franke, J. Erben, P. Häussler, W. Henrion, H. Lange, D. Panknin: Electrical and optical properties of thin Fe₁₋ₓCoₓSi₂ films, *Appl. Surf. Sci.* **91**(1), 56–62 (1995).

2.180   T. Tsunoda, M. Mukaida, A. Watanabe, Y. Imai: Composition dependence of morphology, structure, and thermoelectric properties of FeSi₂ films prepared by sputtering deposition, *J. Mater. Res.* **11**(8), 2062–2070 (1996).

2.181   O.P. Karpenko, C.H.Olk, G.L. Doll, J.F. Mansfield, S.M. Yalisove: Structural analysis of pulsed laser deposited FeSi₂ films, in: *Silicides, Germanides, and Their Interfaces*, edited by R.W. Fathauer, S. Mantl, L.J. Schowalter, K.N. Tu (MRS, Pittsburgh, Pennsylvania, 1994), pp. 103–108.

2.182   H. Kakemoto, Y. Makita, A. Obara, Y. Tsai, S. Sakuragi, S. Ando, T. Tsukamoto: Structural and optical properties of β-FeSi₂/Si(100) prepared by laser ablation method, in: *Thermoelectric Materials - New Directions and Approaches*, edited by T.M. Tritt, M.G. Kanatzidis, H.B. Lyon, G.D. Mahan (MRS, Pittsburgh, Pennsylvania, 1997), pp. 273–278.

2.183   A. Vantomme, S. Degroote, J. Dekoster, G. Langouche, R. Pretorius: Concentration-controlled phase selection of silicide formation during reactive deposition, *Appl. Phys. Lett.* **74**(21), 3137–3139 (1999).

2.184   A. Vantomme, M.-A. Nicolet, R.G. Long, J.E. Mahan: Reactive deposition epitaxy of CrSi₂, *Appl. Surf. Sci.* **73**, 146–152 (1993).

2.185   K.H. Kim, G. Bai, M.-A. Nicolet, J.E. Mahan, K.M. Geib: Amorphization and recrystallization of epitaxial ReSi₂ films grown on Si(100), *Appl. Phys. Lett.* **58**(17), 1884–1886 (1991).

2.186   I. Ali, P. Muret, T.A. Nguyen Tan: Properties of semiconducting rhenium silicide thin films grown epitaxially on silicon (111), *Appl. Surf. Sci.* **102**, 147–150 (1996).

2.187   A. Vantomme, M.-A. Nicolet, R.G. Long, J.E. Mahan: Epitaxial ternary ReₓMo₁₋ₓSi₂ thin films on Si(100), *J. Appl. Phys.***75**(8), 3924–3927 (1994).

2.188   V.V. Kleshkovskaya, T.S. Kamilov, S.I. Adasheva, S.S. Khudaiberdiev, V.I. Muratova: Crystal structure of the films of highest manganese silicide on silicon, *Crystallography Reports* **39**(5), 815–819 (1994).

2.189   J.M. Gallego, J. Alvarez, J.J. Hinarejos, E.G. Michel, R. Miranda: The growth and characterization of iron silicides on Si (100), *Surf. Sci.* **251/252**, 59–63 (1991).

2.190   M. Tanaka, Y. Kumagai, T. Suemasu, F. Hasegawa: Formation of β-FeSi₂ layers on Si(001) substrates, *Jpn. J. Appl. Phys.* **36**(6A), 3620–3624 (1997).

2.191   L. Wang, C. Lin, Q. Chen, X. Lin, R. Ni, S. Zou: Reactive deposition epitaxial growth of β-FeSi$_2$ film Si(111): In situ observation by reflective high energy electron diffraction, *Appl. Phys. Lett.* **66**(25), 3453–3455 (1995).

2.192   Z. Liu, M. Okoshi, M. Hanabusa: Formation of β-FeSi$_2$ films by pulsed laser deposition using iron target, *J. Vac. Sci. Techol. A* **17**(2), 619–623 (1999).

2.193   Le Thanh Vinh, J. Chevrier, J. Derrien: Epitaxial growth of Fe-Si compounds on the silicon (111) face, *Phys. Rev. B* **46**(24), 15946–15954 (1992).

2.194   T. Suemasu, M. Tanaka, T. Fujii, S. Hashimoto, Y. Kumagai, F. Hasegawa: Aggregation of monocrystalline β-FeSi$_2$ by annealing and by Si overlayer growth, *Jpn. J. Appl. Phys.* **36**(9A/B), L1225–L1228 (1997).

2.195   A. Terrasi, S. Ravesi, M.G. Grimaldi, C. Spinella: Ion beam assisted growth of β-FeSi$_2$, *J. Vac. Sci. Technol. A* **12** (2), 289–294 (1994).

2.196   N.P. Barradas, D. Panknin, E. Wieser, B. Schmidt, M. Betzl, A. Mücklich, W. Skorupa: Influence of the ion irradiation on the properties of β-FeSi$_2$ layers prepared by ion beam assisted deposition, *Nucl. Instrum. Methods Phys. Rep. B* **127/128**, 316–320 (1997).

2.197   D. Sander, A. Enders, J. Kirschner: Stress evolution during the growth of ultrathin layers of iron and iron silicide on Si(111), *Appl. Phys. Lett.* **67**(13), 1833–1835 (1995).

2.198   H. Chen, P. Han, X.D. Huang, L.Q. Hu, Y. Shi, Y.D. Zheng: Semiconducting Ge-Si-Fe alloy grown on Si(100) substrate by reactive deposition epitaxy, *Appl. Phys. Lett.* **69**(13), 1912–1914 (1996).

2.199   J.E. Mahan, A. Vantomme, G. Langouche, J.P. Becker: Semiconducting Mg$_2$Si thin films prepared by molecular-beam epitaxy, *Phys. Rev. B* **54**(23), 16965–16971 (1996).

2.200   L. Haderbache, P. Wetzel, C. Pirri, J.C. Peruchetti, D. Bolmont, G. Gewinner: Molecular beam epitaxy of monotype CrSi$_2$ on Si(111), *Surf. Sci.* **209**(3), L139–L143 (1989).

2.201   P. Wetzel, C. Pirri, J.C. Peruchetti, D. Bolmont, G. Gewinner: Epitaxial growth of CrSi and CrSi$_2$ on Si(111), *Solid State Commun.* **65**(10), 1217–1220 (1988).

2.202   N.I. Plusnin, N.G. Galkin, V.G. Lifshits, S.A. Lobachev: Formation of interfaces and templates in the Si(111)-Cr system, *Surf. Rev. Lett.* **2**(4), 439–449 (1995).

2.203   N.G. Galkin, T.V. Velitchko, S.V. Skripka, A.B. Khrustalev: Semiconducting and structural properties of CrSi$_2$ A-type epitaxial films on Si(111), *Thin Solid Films* **280**(1–2), 211–220 (1996).

2.204   J. Chevrier, P. Stocker, Le Thanh Vinh, J.M. Gay, J. Derrien: Epitaxial growth of α-FeSi$_2$ on Si(111) at low temperature, *Europhys. Lett.* **22**(6), 449–454 (1993).

2.205   H.-U. Nissen, E. Müller, H.R. Deller, H. von Känel: TEM investigation of iron disilicide films on Si(001) grown by molecular beam epitaxy, *Phys. Stat. Sol. (a)* **150**(2), 395–406 (1995).

2.206   H. von Känel, U. Kafader, P. Sutter, N. Onda, H. Sirringhaus, E. Müller, U. Kroll, C. Schwarz, S. Goncalves-Conto: Epitaxial semiconducting and metallic iron silicides, in: *Silicides, Germanides, and Their Interfaces*, edited by R.W. Fathauer, S. Mantl, L.J. Schowalter, K.N. Tu (MRS, Pittsburgh, Pennsylvania, 1994), pp. 73–84.

2.207   H. Lange, W. Henrion, B. Selle, G.-U. Reinsperger, G. Oertel, H. von Känel: Optical properties of β-FeSi$_2$ films grown on Si substrates with different degree of structural perfection, *Appl. Surf. Sci.* **102**(1), 169–172 (1996).

2.208   C. Lin, L. Wang, X. Chen, L.F. Chen, L.M. Wang: Structural characterization of codeposition growth β-FeSi$_2$ film, *Jpn. J. Appl. Phys.* **37**(2), 622–625 (1998).

2.209   L.N. Finnie: Structures and compositions of the silicides of ruthenium, osmium, rhodium and iridium, *J. Less-Common Met.* **4**(1), 24–34 (1962).

2.210 S. Mantl: Molecular beam allotaxy: a new approach to epitaxial heterostructures, *J. Phys. D: Appl. Phys.* **31**(1), 1–17 (1998).

Chapter 3

3.1 K.-Th. Wilke, J. Bohm: *Kristallzüchtung* (Deutscher Verlag der Wissenschaften, Berlin, 1988).

3.2 *Handbook of crystal growth*, edited by D.T.J. Hurle (North-Holland, Amsterdam, 1994).

3.3 H. Tabata, T. Hirano: Growth of $MoSi_2$ single crystals by the floating zone method, *J. Jpn. Inst. Metals* **52**(11), 1154–1158 (1988).

3.4 K. Mason, G. Müller-Vogt: Osmium disilicide preparation, crystal growth, and physical properties of a new semiconducting compound, *J. Cryst. Growth* **63**(1), 34–38 (1983).

3.5 G. Krabbes: Vapour growth, in: *Science and Technology of Crystal Growth*, edited by J.P. van der Erden, O.S.L. Bruisma (Kluwer Academic Publishers, Amsterdam, 1995), pp. 123–136.

3.6 R. Krausze, G. Krabbes, M. Khristov: Estimation of the chemical transport behaviour of transition metal silicides $MeSi_x$ (Me = Cr, Ti, V, Mo, Fe, Mn) with X_2 (X = Cl, Br, I), *Cryst. Res. Technol.* **26**(2), 179–185 (1991).

3.7 T. Kojima, I. Nishida: Crystal growth of $MnSi_{1.73}$ by chemical transport method, *Jpn. J. Appl. Phys.* **14**(1), 141–142 (1975).

3.8 G. Krabbes, H. Oppermann, E. Wolf: Application of thermodynamic models to chemical transport reactions with systems containing several coexisting solid phases or phases with a homogeneity range, *J. Cryst. Growth* **64**, 353–366 (1983).

3.9 H. Schäfer, Der chemische transport und die "Löslichkeit des bodenkörpers in der gasphase", *Z. Anorg. Allg. Chem.* **400**, 242–252 (1973).

3.10 R. Krausze, M. Khristov, P. Peshev, G. Krabbes: Crystal growth of chromium silicides by chemical vapour transport with halogens - 1. Growth of chromium disilicide single crystals, *Z. Anorg. Allg. Chem.* **579**, 231–239 (1989).

3.11 W.G. Moffatt: *The Handbook of Binary Phase Diagrams*, edited by (General Electric, Schenectady, N.Y. USA, 1984).

3.12 *Binary Alloy Phase Diagrams, Second Edition*, edited by T.B. Massalski, H. Okamoto, P.R. Subramanian, L. Kacprzak (American Society for Metals, Metals Park, OH, 1990), vols. 1–2.

3.13 H.J. Scheel, D. Elwell: Stable growth rates and temperature programming in flux growth, *J. Cryst. Growth* **12**, 153–161 (1972).

3.14 V.N. Gurin, M.M. Korsukova: Crystal growth of refractory compounds from solutions in metallic melts, *Prog. Crystal Growth Charact.* **6**, 59–101 (1983).

3.15 J.W. Rutter, B. Chalmers: A prismatic substructure formed during solidification of metals, *Can. J. Phys.* **31**, 15-39 (1953).

3.16 W.A. Tiller, J.W. Rutter, K.A. Jackson, B. Chalmers: The redistribution of solute atoms during the solidification of metals, *Acta Met.* **1**, 428–437 (1953).

3.17 I. Nishida: The crystal growth and thermoelectric properties of chromium disilicide, *J. Mater. Sci.* **7**, 1119–1124 (1972).

3.18 O. Thomas, J.P. Senateur, R. Madar, O. Laborde, E. Rosencher: Molybdenum disilicide: crystal growth, thermal expansion and resistivity, *Solid State Commun.* **55**, 629–632 (1985).

3.19 U. Gottlieb, O. Laborde, A. Rouault, M. Madar: Resistivity of Ru_2Si_3 single crystals, *Appl. Surf. Sci.* **73**, 243–245 (1993).

3.20 U. Gottlieb, B. Lambert-Andron, F. Nava, M. Affronte, O. Laborde, A. Rouault, M. Madar: Structural and electronic transport properties of ReSi$_{2-\delta}$ single crystals, *J. Appl. Phys.* **78**(6), 3902–3907 (1995).

3.21 G. Behr: Untersuchungen zur Züchtung und Perfektion von Einkristallen intermetallischer Verbindungen im System V-Cr-Si, *Dissertation,* ZFW der AdW der DDR, Dresden, 1982.

3.22 G. Behr, K. Bartsch, M. Jurisch, M. Schönherr, E. Wolf: Preparation of high purity V$_3$Si single crystals using a siliciothermic reaction, *Cryst. Res. Technol.* **20**(9), K93–K95 (1985).

3.23 G. Behr, J. Werner, G. Weise, A. Heinrich, A. Burkov, C. Gladun: Preparation and properties of high purity β-FeSi$_2$ single crystals, *Phys. Stat. Sol. (a)* **160**(2), 549–556 (1997).

3.24 J.J. Nickl, J.D. Koukoussas: Transportreaktionen und Einkristallzüchtung von Siliciden der IVA-VIA Metalle, *J. Less-Common Met.* **23**(1), 73–81 (1971).

3.25 T. Kojima, I. Nishida: Crystal growth of Mn$_{15}$Si$_{26}$, *J. Cryst. Growth* **47**, 589–592 (1979).

3.26 M. Khristov, G. Gyurov, P. Peschev: Preparation of rhenium disilicide crystals, *Cryst. Res. Technol.* **24**(2), K22–K25 (1989).

3.27 R. Wandji, Y, Dusausoy, J. Protas, B. Roques, *C.R. Acad. Sci. Paris* **267**, 1587 (1968).

3.28 J. Ouvrard, R. Wandji, B. Roques: Contribution des réactions chimiques de transport à l'étude structurale des siliciures de fer, *J. Cryst. Growth* **13/14**, 406–409 (1972).

3.29 Ch. Kloc, E. Arushanov, M. Wendl, H. Hohl, U. Malang, E. Bucher: Preparation and properties of FeSi, α-FeSi$_2$ and β-FeSi$_2$ single crystals, *J. Alloys Comp.* **219**, 93–96 (1995).

3.30 Y. Tomm, L. Ivanenko, K. Irmscher, St. Brehme, W. Henrion, I. Sieber, H. Lange: Effects of doping on the electronic properties of semiconducting iron disilicide, *Mater. Sci. Eng. B* **37**, 215–218 (1996).

3.31 T. Brehme, L. Ivanenko, Y. Tomm, G.-U. Reinsperger, P. Strauß, H. Lange: Hall effect investigation of doped and undoped β-FeSi$_2$, in: *Silicide Thin Films - Fabrication, Properties, and Applications,* edited by R.T. Tung, K. Maex, P.W. Pellegrini, L.H. Allen (MRS, Pittsburgh, Pennsylvania, 1996), pp. 355–360.

3.32 H. Reuther, G. Behr, M. Dobler, A. Teresiak: Angle dependent Mössbauer spectroscopy on β-FeSi$_2$ single crystals, *Hyperfine Interactions (C)* **3**, 385–388 (1998).

3.33 E. Arushanov, Ch. Kloc, H. Hohl, E. Bucher: The Hall effect in β-FeSi$_2$ single crystals, *J. Appl. Phys.* **75**(10), 5106–5109 (1994).

3.34 E. Arushanov, Ch. Kloc, E. Bucher: Impurity band in p-type β-FeSi$_2$, *Phys. Rev. B* **50**(4), 2653–2656 (1994).

3.35 A. Heinrich, C. Gladun, A. Burkov, Y. Tomm, S. Brehme, H. Lange: Thermopower and electrical resistivity of β-FeSi$_2$ single crystals doped with Cr, Co and Mn, in: *Proceedings of the 14th International Conference on Thermoelectrics* (Ioffe-Institute, St. Petersburg, 1995), pp.259–263.

3.36 G. Weise, G. Owsian: Preparation of highly pure iron by vapour deposition, *Kristall und Technik* **11**(7), 729–737 (1976).

3.37 K. Friedrich, J. Barthel, J. Kunze: Über den einfluß der oberflächengase auf die analytisch ermittelten gasgehalte von molybdän- und wolframeinkristallen, *J. Less-Common Met.* **14**, 55–68 (1968).

3.38 W.N. Gurin, A.P. Obukhov, M.M. Korsukowa, Z.P. Terentjewa, I.R. Kozlowa: Synthese hochschmelzender verbindungen der übergangsmetalle aus deren Lösung in metallschmelzen, *Planseeberichte für Pulvermetallourgie* **19**, 86–90 (1971).

3.39   V.N. Gurin,   A.P. Obukhov,   T.I. Mazina,   Z.P. Terentjeva,   I.R. Kozlova, M.M. Korsukova: Formation of monocrystals of refractory compounds from the solution in the zinc melt, *Izv. Akad. Nauk SSSR, Neorg. Mater.* **5**(11), 1995–1998 (1969) (in Russian).

3.40   V.N. Gurin,   A.P. Obukhov,   T.I. Mazina,   M.M. Korsukova,   Z.P. Terentjeva: Spontaneus   crystallization   of   silicides   from   the   zinck   melt   solutions *Izv. Akad. Nauk SSSR, Neorg. Mater.* **6**(9), 1607–1612 (1970) (in Russian).

3.41   V.N. Gurin,   Z.P. Terentjeva,   I.R. Kozlova,   A.P. Obukhov: Morphology and regimes of spontaneous crystallization of transition-metal silicides from solutions in metal melts, *Izv. Akad. Nauk SSSR, Neorg. Mater.* **8**(11), 1917–1920 (1972) (in Russian).

3.42   S. Okada, T. Atoda: Preparation of Cr_5Si_3, $CrSi$ and $CrSi_2$ single crystals in molten tin, *Nippon Kagaku Kaishi* **5**, 746–748 (1983).

3.43   P. Peshev, M. Khristov, G. Gyurov: The growth of titanium, chromium and molybdenum disilicide crystals from high-temperature solutions, *J. Less-Common Met.* **153**(1), 15–22 (1989).

3.44   T. Siegrist, F. Hulliger, G. Travaglini: The crystal structure and some properties of $ReSi_2$, *J. Less-Common Met.* **92**(1), 119–129 (1983).

3.45   N.Kh. Abrikosov, L.I. Petrova: Preparation of crystals of the low-temperature modification of $FeSi_2$, *Inorg. Mater.* **11**, 184–186 (1975).

3.46   R.G. Morris, R.D. Redin, G.C. Danielson: Semiconducting properties of Mg_2Si single crystals, *Phys. Rev.* **109**(6), 1909–1915 (1958).

3.47   R.J. LaBotz, D.R. Mason: The thermal conductivities of Mg_2Si and Mg_2Ge, *J. Electrochem. Soc.* **110**(2), 121–126 (1963).

3.48   J. Evers, A. Weiss, *Mat. Res. Bull.* **9**, 549–554 (1974).

3.49   D. Shinoda, S. Atanabe, Y. Sasaki: Semiconducting properties of chromium disilicide, *J. Phys. Soc. Jpn.* **19**(3), 269–272 (1964).

3.50   B.K. Voronov, L.D. Dudkin, N.N. Trusova: Anisotropy of thermoelectric properties of chromium disilicide and highest manganese silicide, *Kristallografija* **12**, 519–521 (1967) (in Russian).

3.51   I.J. Ohsugi, T. Kojima, I.A. Nishida: Temperature dependence of the magnetic susceptibility of a $CrSi_2$ single crystal, *Phys. Rev. B* **42**(16), 10761–10764 (1990).

3.52   T. Hirano, M. Kaise: Electrical resistivities of single-crystalline transition-metal disilicides, *J. Appl. Phys.* **68**(2), 627–633 (1990).

3.53   H.W. Knott, M.H. Mueller, L. Heaton: The crystal structure of $Mn_{15}Si_{26}$, *Acta Cryst.* **23**(4), 549–555 (1967).

3.54   L.M. Levinson, *G.E. Technical Report*, No. 72CRD111, March 1972.

3.55   L.M. Levinson: Investigation of the defect manganese silicide Mn_nSi_{2n-m}, *J. Solid State Chem.* **6**, 126–135 (1973).

3.56   I. Kawasumi, I. Nishida, K. Masumoto, M. Sakata: On the striations in $MnSi_{2-x}$ crystals, *Jpn. J. Appl. Phys.* **15**(7), 1405–1406 (1976).

3.57   I. Kawasumi, M. Sakata, I. Nishida, K. Masumoto: Crystal growth of manganese silicide, $MnSi_{~1.73}$ and semiconducting properties of $Mn_{15}Si_{26}$, *J. Mater. Sci.* **16**, 355–366 (1981).

3.58   Y. Fujino, D. Shinoda, S. Asanabe, Y. Sasaki: Phase diagram of the partial system of MnSi-Si, *Jpn. J. Appl. Phys.* **3**(8), 431–435 (1964).

3.59   M.A. Morokhovets, E.I. Elagina, N.Kh. Abrikosov: Phase diagram of Mn-Si system in the region of the higest manganese silicide, *Izv. Acad. Nauk SSSR, Neorgan. Mater.* **2**(4), 650–656 (1966) (in Russian).

3.60 L.D. Ivanova, N.Kh. Abrikosov, E.I. Elagina, V.D. Khvostikova: Formation and investigation of the higest manganese silicide, *Izv.Akad. Nauk SSSR, Neorg. Materialy* **5**(11), 1933–1937 (1969) (in Russian).

3.61 G. Zwilling, H. Novotny: Die kristallstruktur der mangansilicide im bereich von MnSi₋₁.₇, *Monatsh. Chem.* **102**(3), 672–677 (1971).

3.62 N.Kh. Abrikosov, L.D. Ivanova V.G. Muravov: Formation and investigation of single crystals of the highest manganese silicide with Ge and CrSi₂, *Izv. Akad. Nauk SSSR, Neorg. Materialy* **8**(7), 1194–1200 (1972) (in Russian).

3.63 N.Kh. Abrikosov, L.D. Ivanova: Investigation of single crystals of solid solution of the highest manganese silicide and FeSi₂ *Izv. Akad. Nauk SSSR, Neorg. Materialy* **10**(6), 1016–1022 (1974).

3.64 H. Shibata, Y. Makita, H. Kakemoto, Y. Tsai, S. Sakuragi, H. Katsumata, A. Obara, N. Kobayashi, S. Uekusa, T. Tsukamoto, T. Tsunoda, Y. Imai: Electrical properties of β-FeSi₂ bulk crystal grown by horizontal gradient freeze method, in: *Proceedings of the 15ᵗʰ International Conference on Thermoelectrics*, edited by J.-P. Fleurial (IEEE, Inc., Piscataway, NJ, 1996), pp. 62–66.

3.65 C.B. Vining, C.E. Allevato: Intrinsic thermoelectric properties of single crystal Ru₂Si₃, in: *Proceedings of the 10ᵗʰ International Conference on Thermoelectrics*, edited by D.M. Rowe (Babrow Press, Cardiff, 1991), pp. 167–173.

3.66 I. Engström, F. Zackrisson: X-ray studies of silicon-rich iridium silicides, *Acta Chem. Scand.* **24**(6), 2109–2116 (1970).

3.67 I. Engström, T. Lindstein, E. Zdansky: The crystal structure of the iridium silicide Ir₃Si₅, *Acta Chem. Scand. A* **41**, 237–242 (1987).

3.68 C.E. Allevato, C.B. Vining: Thermoelectric Properties of semiconducting iridium silicides, in: *Proceedings of the 28ᵗʰ Intersociety Energy Conversion Engineering Conference* (Atlanta, Georgia, USA, 1993), vol. 1, pp. 1.239–1.243.

3.69 C.R. Whitsett, G.C. Danielson: Electrical properties of the semiconductors Mg₂Si and Mg₂Ge, *Phys. Rev.* **100**, 1261–1262 (1955).

3.70 R.J. LaBotz, D.R. Mason, D.F. O'Kane: The thermoelectric properties of mixed crystals of Mg₂GeₓSi₁₋ₓ, *J. Electrochem. Soc.* **110**(2), 127–134 (1963).

3.71 J. Evers, A. Weiss, E. Kaldis, J. Muheim: Purification of calcium and barium by reactive distillation, *J. Less-Common Met* **30**(1), 83–95 (1973).

3.72 J. Evers, E. Kaldis, J. Muheim, A. Weiss, *J. Less-Common Met.* **30**, 169 (1973).

3.73 E. Kaldis: High-temperature chemical transport, *J. Cryst. Growth* **9**, 281–294 (1971).

3.74 M. Hansen: *Constitution of Binary Alloys* (McGraw-Hill Co., New York, 1958), p. 561.

3.75 M. Jurisch, G. Behr: Growth and perfection of Cr₃Si single crystals, *Acta. Phys. Hung. (HU)* **47**(1–3), 201–203 (1979).

3.76 G. Behr, M. Jurisch, J. Barthel: Crystal growth of V₃Si and (VₓCr₁₋ₓ)₃Si by zone floating, *Crystal Properties and Preparation* **36–38**, 544–552 (1991).

3.77 R.L. Coble: Sintering crystalline solids. II. Experimental test of diffusion models in powder compacts, *J. Appl. Phys.* **32**(5), 793–799 (1961).

3.78 B. Borén, *Arkiv Kemi Mineral Geol.* **11A**, 11 (1933).

3.79 G. Zwilling, *Dissertation*, Universität Wien, 1970.

3.80 U. Birkholz, J. Schelm: Mechanism of electrical conduction in β-FeSi₂, *Phys. Stat. Sol.* **27**, 413–425 (1968).

3.81 D.J. Poutcharovsky, E. Parthé: The orthorhombic crystal structure of Ru₂Si₃, Ru₂Ge₃, Os₂Si₃ and Os₂Ge₃, *Acta Cryst. B* **30**, 2692–2696 (1974).

3.82 F. Nava, K.N. Tu, O. Thomas, J.P. Senateur, R. Madar, A. Borghesi, G. Guizzetti, U. Gottlieb, O. Laborde, O. Bisi: Electrical and optical properties of silicide single crystals and thin films, *Mater. Sci. Rep.* **9**, 141–200 (1993).

3.83 C.E. Allevato, C.B. Vining: Phase diagram and electrical behavior of silicon-rich iridium silicide compounds, *J. Alloys Comp.* **200**(12), 99–105 (1993).

Chapter 4

4.1 B.K. Ridley: *Quantum Processes in Semiconductors*, Second Edition (Clarendon Press, Oxford, 1988).

4.2 N. Peyghambarian, S. W. Koch, A. Mysyrowicz: *Introduction to Semiconductor Optics* (Prentice Hall, Englewood Cliffs, New Jersey, 1993).

4.3 C.F. Klingshirn: *Semiconductor Optics* (Springer, Berlin, 1995).

4.4 P.Y. Yu, M. Cardona: *Fundamentals of Semiconductors* (Springer, Berlin, 1996).

4.5 *Density Functionals: Theory and Applications*, edited by D. Joubert (Springer, Berlin, 1998).

4.6 F. Meloni, E. Mooser, A. Baldereschi: Bonding nature of conduction states in electron-deficient semiconductors: Mg_2Si, *Physica B + C (Amsterdam)* **117–118** (1), 72–74 (1983).

4.7 D.M. Wood, A. Zunger: Electronic structure of generic semiconductors: Antifluorite silicide and III-V compounds, *Phys. Rev. B* **34**(6), 4105–4120 (1986).

4.8 N.O. Folland: Self-consistent calculations of the energy band structure of Mg_2Si, *Phys. Rev.* **158**(3), 764–775 (1967).

4.9 P.M. Lee: Electronic structure of magnesium silicide and magnesium germanide, *Phys. Rev.* **135**(4A), A1110–A1114 (1964).

4.10 F. Aymerich, G. Mula: Pseudopotential band structure of Mg_2Si, Mg_2Ge, Mg_2Sn, and of the solid solution $Mg_2(Ge, Sn)$, *Phys. Stat. Sol.* **42**(2), 697–704 (1970).

4.11 V.K. Bashenov, A.M. Mutal, V.V. Timofeenko: Valence-band density of states for Mg_2Si from pseudopotential calculation, *Phys. Stat. Sol. (b)* **87**(1), K77–K79 (1978).

4.12 M.Y. Au-Yang, M.L. Cohen: Electronic structure and optical properties of Mg_2Si, Mg_2Ge, and Mg_2Sn, *Phys. Rev.* **178**(3), 1358–1364 (1969).

4.13 A.J. Bevolo, H.R. Shanks: Valence band study of Mg_2Si by Auger spectroscopy, *J. Vac. Sci. Technol. A* **1**(2), 574–577 (1983).

4.14 V.M. Glazov, V.B. Koltsov, V.A. Kurbatov: Investigation of the Hall-effect in Mg_2Si, Mg_2Ge, Mg_2Sn compounds in solid and liquid state, *Soviet Physics Semiconductors* **20**(5), 527–530 (1986) (in Russian).

4.15 R.G. Morris, R.D. Redin, G.C. Danielson: Semiconducting properties of Mg_2Si single crystals, *Phys. Rev.* **109**(6), 1909–1915 (1958).

4.16 R.J. LaBotz, D.R. Mason, D.F. O'Kane: The thermoelectric properties of mixed crystals of $Mg_2Ge_xSi_{1-x}$, *J. Electrochem. Soc.* **110**(2), 127–34 (1963).

4.17 J.E. Mahan, A. Vantomme, G. Langouche, J.P. Becker: Semiconducting Mg_2Si thin films prepared by molecular-beam epitaxy, *Phys. Rev. B* **54**(23), 16965–16971 (1996).

4.18 V.M. Glazov, L.M.Pavlova, K.B. Poyarkov: Thermodynamics of semiconducting compounds Mg_2B^{IV} (B^{IV} - Si, Ge, Sn, Pb), *Obzory po electr. technike,* Ser. 6, Mater. **9**(917), 1–44 (1982) (in Russian).

4.19 J. Tejeda, M. Cardona: Valence bands of the Mg_2X (X = Si, Ge, Sn) semiconducting compounds, *Phys. Rev. B* **14**(6), 2559–2568 (1976).

4.20 F. Vazquez, R.A. Forman, M. Cardona: Electroreflectance measurements on Mg_2Si, Mg_2Ge, and Mg_2Sn, *Phys. Rev.* **176**(3), 905–908 (1968).

4.21 S.G. Kroitoru, V.V. Sobolev: Reflectance spectra of Mg_2Si and Mg_2Sn, *Opt. Spectr. (USSR)* **21**(1), 48–50 (1966).

4.22 A. Stella, D.W. Lynch: Photoconductivity in Mg_2Si and Mg_2Ge, *J. Phys. Chem. Solids* **25**(12), 1253–1259 (1964).

4.23 A. Stella, A.D. Brothers, R.H. Hopkins, D.W. Lynch: Pressure coefficient of the band gap in Mg_2Si, Mg_2Ge, and Mg_2Sn, *Phys. Stat. Sol.* **23**(2), 697–702 (1967).

4.24 P. Koenig, D.W. Lynch, G.C. Danielson: Infrared absorption in magnesium silicide and magnesium germanide, *J. Phys. Chem. Solids* **20**(1/2), 122–126 (1961).

4.25 D. Panke, E. Wölfel: Die Verteilung der Valenzelektronen im Mg_2Si, *Z. Kristallogr.* **129**(1), 9–28 (1969).

4.26 W.B. Whitten, P.L. Chung, G.C.Danielson: Elastic constants and lattice vibration frequencies of Mg_2Si, *J. Phys. Chem. Solids* **26**(1), 49–56 (1965).

4.27 J. Evers: Semiconductor-metal transition in $BaSi_2$, *J. Less-Common Met.* **58**(1), 75–83 (1978).

4.28 J. Evers, A. Weiss: Electrical properties of alkaline earth disilicides and digermanides, *Mater. Res. Bull.* **9**(5), 549–554 (1974).

4.29 V.S. Neshpor, V.L. Upko: Investigation of formation conditions and some physical properties of barium disilicide, *J. Appl. Chem.* **36**(5), 1139–1142 (1963) (in Russian).

4.30 A.B. Filonov, I.E. Tralle, N.N. Dorozhkin, D.B. Migas, V.L. Shaposhnikov, G.V. Petrov, A.M. Anishchik, V.E. Borisenko: Semiconducting properties of hexagonal chromium, molybdenum, tungsten disilicides, *Phys. Stat. Sol. (b)* **186**(1), 209–215 (1994).

4.31 M.P.C.M. Krijn, R. Eppenga: First-principles electronic structure and optical properties of $CrSi_2$, *Phys. Rev. B* **44**(16), 9042–9044 (1991).

4.32 M.C. Bost, J.E. Mahan: An investigation of the optical constants and band gap of chromium disilicide, *J. Appl. Phys.* **63**(3), 839–844 (1988).

4.33 F. Nava, T. Tien, K.N. Tu: Temperature dependence of semiconducting and structural properties of Cr-Si thin films, *J. Appl. Phys.* **57**(6), 2018–2025 (1985).

4.34 V.E. Borisenko, L.I. Ivanenko, S.Yu. Nikitin: Semiconducting properties of chromium disilicide, *Mikroelektronika* **21**(2), 61–77 (1992) (in Russian).

4.35 V.E. Borisenko, D.I. Zarovskii, G.V. Litvinovich, V.A. Samuilov: Optical spectroscopy of chromium silicide formed by a flash heat treatment of chromium films on silicon, *Zurn. Prikl. Spectrosk.* **44**(2), 314–317 (1986) (in Russian).

4.36 E.N. Nikitin, V.I. Tarasov, V.K. Zaitsev: Electrical properties of some solid solutions of 3d-transition metal silicides, *Fiz. Tverdogo Tela* **15**(4), 1254–1256 (1973) (in Russian).

4.37 S. Halilov, E. Kulatov: Electron and optical spectra in the indirect-gap semiconductor $CrSi_2$, *Semicond. Sci. Technol.* **7**(3), 368–372 (1992).

4.38 I. Nishida, T. Sakata: Semiconducting properties of pure and Mn-doped chromium disilicides, *J. Phys. Chem. Solids* **39**(5), 499–505 (1978).

4.39 W. Henrion, H. Lange, E. Jahne, M. Giehler: Optical properties of chromium and iron disilicide layers, *Appl. Surf. Sci.* **70/71**, 569–572 (1993).

4.40 L.F. Mattheiss: Structural effects on the calculated semiconductor gap of $CrSi_2$, *Phys. Rev. B* **43**(2), 1863–1866 (1991).

4.41 H. Lange, M. Giehler, W. Henrion, F. Fenske, I. Sieber, G. Oertel: Growth and optical characterization of $CrSi_2$ thin films, *Phys. Stat. Sol. (b)* **171**(1), 63–76 (1992).

4.42 D. Shinoda, S. Asanabe, Y. Sasaki: Semiconducting properties of chromium disilicide, *J. Phys. Soc. Jap.* **19**(3), 269–272 (1964).

4.43 V. Bellani, G. Guizzetti, F. Marabelli, A. Piaggi, A. Borghesi, F. Nava, V.N. Antonov, Vl.N. Antonov, O. Jepsen, O.K. Andersen, V.V. Nemoshkalenko: Theory and experiment on the optical properties of $CrSi_2$, *Phys. Rev. B* **46**(15), 9380–9389 (1992).

4.44 N.G. Galkin, T.V. Velitchko, S.V. Skripka, A.B. Khrustalev: Semiconducting and structural properties of $CrSi_2$ A-type epitaxial films on Si(111), *Thin Solid Films* **280**(1–2), 211–220 (1996).

4.45 L.F. Mattheiss: Electronic structure of $CrSi_2$ and related refractory disilicides, *Phys. Rev. B* **43**(15), 12549–12555 (1991).

4.46 S. Valeri, L. Grandi, C. Calandra: Two hole spectra and valence states in chromium and chromium silicides, *J. Electron Spectrosc. Relat. Phenom.* **43**(2), 121–130 (1987).

4.47 A. Franciosi, J.H. Weaver, D.G. O'Neill, F.A. Schmidt, O. Bisi, C. Calandra: Electronic structure of Cr silicides and Si-Cr interface reactions, *Phys. Rev. B* **28**(12), 7000–7008 (1983).

4.48 J.H. Weaver, A. Franciosi, V.L. Moruzzi: Bonding in metal disilicides $CaSi_2$ through $NiSi_2$: experiment and theory, *Phys. Rev. B* **29**(6), 3293–3302 (1984).

4.49 W. Speier, E.v. Leuken, J.C. Fuggle, D.D. Sarma, L. Kumar, B. Dauth, K.N.J. Buschow: Photoemission and inverse photoemission of transition-metal silicides, *Phys. Rev. B* **39**(9), 6008–6016 (1989).

4.50 P.J.W. Weijs, H. van Leuken, R.A. de Groot, J.C. Fuggle, S. Reiter, G. Wiech, K.H.J. Buschow: X-ray-emission studies of chemical bonding in transition-metal silicides, *Phys. Rev. B* **44**(15), 8195–8202 (1991).

4.51 C. Mariani, U. Del Pennino, S. Valeri: Electron energy loss spectroscopic investigation of Cr-$L_{2,3}$ core levels in Cr and chromium silicides, *Solid State Commun.* **61**(1), 5–7 (1987).

4.52 C. Mariani, M.G. Betti, U. del Pennino, A.M. Fiorello, S. Nannarone: Photoabsorption spectroscopy of $CrSi_2$: an investigation of unoccupied states, *Europhys. Lett.* **5**(3), 283–286 (1988).

4.53 F. Nava, B.Z. Weiss, K. Ahn, K.N. Tu: Tungsten disilicide formation in codeposite amorphous WSi_x alloy thin films, *Le Vide, les Couches Minces* **42**(1), 225–228 (1987).

4.54 F. Nava, B.Z. Weiss, K.Y. Ahn, D.A. Smith, K.N. Tu: Thermal stability and electrical conduction behavior of coevaporated $WSi_{2\pm x}$ thin films, *J. Appl. Phys.* **64**(1), 354–364 (1988).

4.55 F.M. d'Heurle, F.K. LeGoues, R. Joshi, I. Suni: Stacking faults in WSi_2: Resistivity effects, *Appl. Phys. Lett.* **48**(5), 332–334 (1986).

4.56 C. Krontiras, I. Suni, F.M. d'Heurle, R. Joshi, F.K. Le Goues: Electronic transport properties of thin films of WSi_2 and $MoSi_2$, *J. Phys. F: Met. Phys.* **17**(9), 1953–1961 (1987).

4.57 I. Nishida: Semiconducting properties of nonstoichiometric manganese silicides, *J. Mater. Sci.* **7**, 435–440 (1972).

4.58 M.C. Bost, J.E. Mahan: An optical determination on the bandgap of the most silicon-rich manganese silicide phase, *J. Electron. Mater.* **16**(6), 389–395 (1987).

4.59 L. Zhang, D.G. Ivey: Reaction kinetics and opticals properties of semiconducting $MnSi_{1.73}$ grown on ⟨001⟩ oriented silicon, *J. Mater. Sci.* **2**(2), 116–123 (1991).

4.60 Ch. Krontiras, K. Pomoni, M. Roilos: Resistivity and the Hall effect for thin $MnSi_{1.73}$ films, *J. Phys. D: Appl. Phys.* **21**(3), 509–512 (1988).

4.61 I. Kawasumi, M. Sakata, I. Nishida, K. Masumoto: Crystal growth of manganese silicide, $MnSi_{1.73}$ and semiconducting properties of $Mn_{15}Si_{26}$, *J. Mater. Sci.* **16**(2), 355–366 (1981).

4.62   V.S. Neshpor, G.V. Samsonov: Investigation of electrical conductance of transition metal silicides, *Sov. Phys. - Solid State*, **2**(9), 1966–1970 (1960).

4.63   T. Siegrist, F. Hulliger, G. Travaglini: The crystal structure and some properties of rhenium silicide (ReSi₂), *J. Less-Common Met.* **92**(1), 119–129 (1983).

4.64   R.G. Long, M.C. Bost, J.E. Mahan: Optical and electrical properties of semiconducting rhenium disilicide thin films, *Thin Solid Films* **162**(1/2), 29–40 (1988).

4.65   C. Krontiras, L. Grönberg, I. Suni, F.M. d'Heurle, J. Tersoff, I. Engström, B. Karlsson, C.S. Petersson: Some properties of ReSi₂, *Thin Solid Films* **161**(1/2), 197–206 (1988).

4.66   T.A. Nguyen Tan, J.Y. Veuillen, P. Muret: Semiconducting rhenium silicide thin films on Si(111), *J. Appl. Phys.***77**(6), 2514–2518 (1995).

4.67   B.K. Bhattacharyya, D.M. Bylander, L. Kleinman: Fully relativistic energy bands and cohesive energy of ReSi₂, *Phys. Rev. B* **33**(6), 3947–3951 (1986).

4.68   S. Itoh: Electronic structure of refractory metal disilicides in C11b structure, *Mater. Sci. Eng.* **6**(1), 37–41 (1990).

4.69   U. Gottlieb, B. Lambert-Andron, F. Nava, M. Affronte, O. Laborde, A. Rouault, R. Madar: Structural and electronic transport properties of ReSi₂₋δ single crystals, *J. Appl. Phys.***78**(6), 3902–3907 (1995).

4.70   U. Gottlieb, M. Affronte, F. Nava, O. Laborde, A. Sulpice, R. Madar: Some physical properties of ReSi₁.₇₅ single crystals, *Appl. Surf. Sci.* **91**(1), 82–86 (1995).

4.71   A.B. Filonov, D.B. Migas, V.L. Shaposhnikov, N.N. Dorozhkin, V.E. Borisenko, H. Lange, A. Heinrich: Electronic properties of semiconducting rhenium silicide, *Europhys. Lett.* **46**(3), 376–381 (1999).

4.72   I. Ali, P. Muret, T.A. Nguyen Tan: Properties of semiconducting rhenium silicide thin films grown epitaxially on silicon (111), *Appl. Surf. Sci.* **102**, 147–150 (1996).

4.73   A.B. Filonov, D.B. Migas, V.L. Shaposhnikov, N.N. Dorozhkin, G.V. Petrov, V.E. Borisenko, W. Henrion, H. Lange: Electronic and related properties of crystalline semiconducting iron disilicide, *J. Appl. Phys.* **79**(10), 7708–7712 (1996).

4.74   V.N. Antonov, O. Jepsen, W. Henrion, M. Rebien, P. Stauß, H. Lange: Electronic structure and optical properties of β-FeSi₂, *Phys. Rev. B* **57**(15), 8934–8938 (1998).

4.75   H. Katsumata, Y. Makita, H. Takahashi, H. Shibata, N. Kobayashi, M. Hasegawa, S. Kimura, A. Obara, J. Tanabe, S. Uekusa: Optical, electrical and structural properties of polycrystalline β-FeSi₂ thin films fabricated by electron beam evaporation of ferrosilicon, in: *Proceedings of the 15ᵗʰ International Conference on Thermoelectrics* (IEEE, Inc., Piscataway, NJ, 1996), pp. 479–483.

4.76   J. van Ek, P.E.A. Turchi, P.A. Sterne: Fe, Ru, and Os disilicides: Electronic structure of ordered compounds, *Phys. Rev. B* **54**(11), 7897–7908 (1996).

4.77   C.H. Olk, S.M. Yalisove, G.L. Doll: Defect-induced absorption-band-edge values in β-FeSi₂, *Phys. Rev. B* **52** (3), 1692–1697 (1995).

4.78   R. Eppenga: *Ab initio* band-structure calculation of the semiconductor β-FeSi₂, *J. Appl. Phys.* **68**(6), 3027–3029 (1990).

4.79   H. Lange, W. Henrion, B. Selle, G.-U. Reinsperger, G. Oertel, H. von Känel: Optical properties of β-FeSi₂ films grown on Si substrates with different degree of structural perfection, *Appl. Surf. Sci.* **102**(1), 169–172 (1996).

4.80   L. Miglio, G. Malegori: Origin and nature of the band gap in β-FeSi₂, *Phys. Rev. B* **52**(3), 1448–1451 (1995).

4.81   H. von Känel, U. Kafader, P. Sutter, N. Onda, H. Sirringhaus, E. Müller, U. Kroll, C. Schwarz, S. Goncalves-Conto: Epitaxial semiconducting and metallic iron silicides, in: *Silicides, Germanides, and Their Interfaces*, edited by R.W. Fathauer, S. Mantl, L.J. Schowalter, K.N. Tu (MRS, Pittsburgh, Pennsylvania, 1994), pp. 73–84.

4.82 S.J. Clark, H.M. Al-Allak, S. Brand, R.A. Abram: Structure and electronic properties of FeSi$_2$, *Phys. Rev. B* **58**(16), 10389–10393 (1998).

4.83 K. Radermacher, O. Skeide, R. Carius, J. Klomfaß, S. Mantl: Electrical and optical properties of FeSi$_2$ layers, in: *Silicides, Germanides, and Their Interfaces*, edited by R.W. Fathauer, S. Mantl, L.J. Schowalter, K.N. Tu (MRS, Pittsburgh, Pennsylvania, 1994), pp. 115–120.

4.84 K. Radermacher, R. Carius, S. Mantl: Optical and electrical properties of buried semiconducting β-FeSi$_2$, *Nucl. Instrum. Met. Phys. Res. B* **84**(1), 163–167 (1994).

4.85 J.L. Regolini, F. Trincat, I. Sagnes, Y. Shapira, G. Brémond, D. Bensahel: Characterization of semiconducting iron disilicide obtained by LRP/CVD, *IEEE Trans. Electron Devices* **39**(1), 200–201 (1992).

4.86 N.E. Christensen: Electronic structure of β-FeSi$_2$, *Phys. Rev. B* **42**(11), 7148–7153 (1990).

4.87 D.H. Tassis, C.L. Mitsas, T.T. Zorba, M. Angelakeris, C.A. Dimitriadis, O. Valassiades, D.I. Siapkas, G. Kiriakidis: Optical and electrical characterization of high quality β-FeSi$_2$ thin films grown by solid phase epitaxy, *Appl. Surf. Sci.* **102**, 178–183 (1996).

4.88 D.H. Tassis, C.L. Mitsas, T.T. Zorba, C.A. Dimitriadis, O. Valassiades, D.I. Siapkas, M. Angelakeris, P. Poulopoulos, N.K. Flevaris, G. Kiriakidis: Infrared spectroscopic and electronic transport properties of polycrystalline semiconducting FeSi$_2$ thin films, *J. Appl. Phys.* **80**(2), 962–968 (1996).

4.89 B.N.E. Rösen, D. Freundt, Ch. Dieker, D. Gerthsen, A. Rizzi, R. Carius, H. Lüth: Characterization of β-FeSi$_2$/Si heterostructures grown by gas-source-MBE, in: *Silicides, Germanides, and Their Interfaces*, edited by R.W. Fathauer, S. Mantl, L.J. Schowalter, K.N. Tu (MRS, Pittsburgh, Pennsylvania, 1994), pp.139–144.

4.90 H. Kakemoto, Y. Makita, A. Obara, Y. Tsai, S. Sakuragi, S. Ando, T. Tsukamoto: Structural and optical properties of β-FeSi$_2$/Si(100) prepared by laser ablation method, in: *Thermoelectric Materials - New Directions and Approaches*, edited by T.M. Tritt, M.G. Kanatzidis, H.B. Lyon, G.D. Mahan (MRS, Pittsburgh, Pennsylvania, 1997), pp. 273–278.

4.91 H. Katsumata, Y. Makita, N. Kobayashi, M. Hasegawa, H. Shibata, S. Uekusa: Synthesis of β-FeSi$_2$ for optical applications by Fe triple-energy ion implantation into Si(100) and Si(111) substrates, *Thin Solid Films* **281–282**, 252–255 (1996).

4.92 S. Eisebitt, J.-E. Rubensson, M. Nicodemus, T. Böske, S. Blügel, W. Eberhardt, K. Radermacher, S. Mantl, G. Bihlmayer: Electronic structure of buried α-FeSi$_2$ and β-FeSi$_2$ layers: Soft-x-ray-emission and -absorption studies compared to band-structure calculations, *Phys. Rev. B* **50**(24), 18330–18340 (1994).

4.93 K. Lefki, P. Muret: Photoelectric study of β-FeSi$_2$ on silicon: Optical threshold as a function of temperature, *J. Appl. Phys.* **74**(2), 1138–1142 (1993).

4.94 L. Miglio, V. Meregalli: Theory of FeSi$_2$ direct gap semiconductor on Si(100), *J. Vac. Sci. Technol. B* **16**(3), 1604–1609 (1998).

4.95 H. Katsumata, Y. Makita, N. Kobayashi, H. Shibata, M. Hasegawa, I. Aksenov, S. Kimura, A. Obara, S. Uekusa: Optical absorption and photoluminescence studies of β-FeSi$_2$ prepared by heavy implantation of Fe$^+$ ions into Si, *J. Appl. Phys.* **80**(10), 5955–5962 (1996).

4.96 H. Katsumata, Y. Makita, N. Kobayashi, H. Shibata, M. Hasegawa, S. Uekusa: Effect of multiple-step annealing on the formation of semiconducting β-FeSi$_2$ and metallic α-Fe$_2$Si$_5$ on Si(100) by ion beam synthesis, *Jpn. J. Appl. Phys.* **36**(5), 2802–2812 (1997).

4.97 A. Valentini, E. Chimenti, A. Cola, G. Leo, F. Quaranta, L. Vasanelli: Triple ion beam sputtering deposition of β-FeSi$_2$, Advances in crystal growth, *Mater. Sci. Forum* **203**, 173–178 (1996).

4.98 L. Wang, L. Qin, Y. Zheng, W. Shen, X. Chen, X. Lin, C. Lin, S. Zou: Optical transition properties of β-FeSi$_2$ film, *Appl. Phys. Lett.* **65**(24), 3105–3107 (1994).

4.99 D. Panknin, W. Henrion, E. Wieser, M. Voelskow, W. Skorupa, H. Vöhse: Electrical and optical properties of Co alloyed β-FeSi$_2$ formed by ion beam synthesis, *4th Int. Conf. Form. Sem.*, 14-18 June, (1993).

4.100 Z. Yang, K.P. Homewood, M.S. Finney, M.A. Harry, K.J. Reeson: Optical absorption study of ion beam synthesized polycrystalline semiconducting FeSi$_2$, *J. Appl. Phys.* **78**(3), 1958–1963 (1995).

4.101 W. Raunau, H. Niehus, T. Schilling, G. Comsa: Scanning tunneling microscopy and spectroscopy of iron silicide epitaxially grown on Si (111), *Surf. Sci.* **286**(3), 203–211 (1993).

4.102 C.A. Dimitriadis, J.H. Werner, S. Logothetidis, M. Stutzmann, J. Weber, R. Nesper: Electronic properties of semiconducting FeSi$_2$ films, *J. Appl. Phys.* **68**(4), 1726–1734 (1990).

4.103 K. Lefki, P. Muret, N. Cherief, R.C. Cinti: Optical and electrical characterization of β-FeSi$_2$ epitaxial thin films on silicon substrates, *J. Appl. Phys.* **69**(1), 352–357 (1991).

4.104 M. Powalla, K. Herz: Co-evaporated thin films of semiconducting β-FeSi$_2$, *Appl. Surf. Sci.* **65/66**, 482–488 (1993).

4.105 L. Wang, M. Östling, K. Yang, L. Qin, C. Lin, X. Chen, S. Zou, Y. Zheng, Y. Qian: Optical transition in β-FeSi$_2$ films, *Phys. Rev. B* **54**(16), R11126–R11128 (1996).

4.106 M.S. Finney, Z. Yang, M.A. Harry, K.J. Reeson, K.P. Homewood, R.M. Gwilliam, B.J. Sealy: Effects of annealing and cobalt implantation on the optical properties of β-FeSi$_2$, in: *Silicides, Germanides, and Their Interfaces*, edited by R.W. Fathauer, S. Mantl, L.J. Schowalter, K.N. Tu (MRS, Pittsburgh, Pennsylvania, 1994), pp. 173–178.

4.107 M. Ožvold, V. Boháč, V. Gašparík, G. Leggieri, Š. Luby, A. Luches, E. Majková, P. Mrafko: The optical band gap of semiconducting iron disilicide thin films, *Thin Solid Films* **263**(1), 92–98 (1995).

4.108 L. Wang, C. Lin, X. Chen, S. Zou, L. Qin, H. Shi, W.Z. Shen, M. Östling: A clarification of optical transition of β-FeSi$_2$ thin film, *Solid State Commun.* **97**(5), 385–388 (1996).

4.109 M.C. Bost, J.E. Mahan: Optical properties of semiconducting iron disilicide thin films, *J. Appl. Phys.* **58**(7), 2696–2703 (1985).

4.110 N. Kobayashi, H. Katsumata, H.L. Shen, M. Hasegawa, Y. Makita, H. Shibata, S. Kimura, A. Obara, S. Uekusa, T. Hatano: Structural and optical characterization of β-FeSi$_2$ layers on Si formed by ion beam synthesis, *Thin Solid Films* **270**(1–2), 406–410 (1995).

4.111 D.J. Oostra, C.W.T. Bulle-Lieuwma, D.E.W. Vandenhoudt, F. Felten, J.C. Jans: β-FeSi$_2$ in (111) Si and in (001) Si formed by ion-beam synthesis, *J. Appl. Phys.* **74**(7), 4347–4353 (1993).

4.112 K. Lefki, P. Muret: Internal photoemission in metal/ β-FeSi$_2$/Si heterojunctions, *Appl. Surf. Sci.* **65/66**, 772–776 (1993).

4.113 E. Arushanov, E. Bucher, Ch. Kloc, O. Kulikova, L. Kulyuk, A. Siminel: Photoconductivity in *n*-type β-FeSi$_2$ single crystals, *Phys. Rev. B* **52**(1), 20–23 (1995).

4.114 M.C. Bost, J.E. Mahan: A clarification of the index of refraction of beta-iron disilicide, *J. Appl. Phys.* **64**(4), 2034–2037 (1988).

4.115   M. De Crescenzi, G. Gaggiotti, N. Motta, F. Patella, A. Balzarotti, G. Mattogno, J. Derrien: Electronic structure of epitaxial β-FeSi$_2$ on Si (111), *Surf. Sci.* **251/252**,175–179 (1991).

4.116   M. Ožvold, V. Gašparík, M. Dubnička: The temperature dependence of the direct gap of β-FeSi$_2$ films, *The Solid Films* **295**(1–2), 147–150 (1997).

4.117   C. Giannini, S. Lagomarsino, F. Scarinci, P. Castrucci: Nature of the band gap of polycrystalline β-FeSi$_2$ films, *Phys. Rev. B* **45**(15), 8822–8824 (1992).

4.118   Y. Isoda, M.A. Okamoto, T. Ohkoshi, I.A. Nishida: Semiconducting properties and thermal shock resistance of boron doped iron disilicide, in: *Proceedings of the 12th International Conference on Thermoelectrics* (Yokohama, 1993), pp.192–196.

4.119   R.M. Ware, D.J. McNeill: Iron disilicide as a thermoelectric generator material, *Proc. IEE.* **111**(1), 178–182 (1964).

4.120   T. Kojima: Semiconducting and thermoelectric properties of sintered iron disilicide, *Phys. Stat. Sol. (a)* **111**(1), 233–242 (1989).

4.121   L. Miglio, V. Meregalli, O. Jepsen: Strain dependent gap nature of epitaxial β-FeSi$_2$ in silicon by first principles calculations, *Appl. Phys. Lett.* **75**(3), 385–387 (1999).

4.122   H. Chen, P. Han, X.D. Huang, L.Q. Hu, Y. Shi, Y.D. Zheng: Semiconducting Ge–Si–Fe alloy grown on Si(100) substrate by reactive deposition epitaxy, *Appl. Phys. Lett.* **69**(13), 1912–1914 (1996).

4.123   P. Pecheur, G. Toussaint: Electronic structure of Ru$_2$Si$_3$, *Mater. Res. Soc. Symp. Proc.* **234**, 157–164 (1991).

4.124   W.B. Pearson: Phases with Nowotny chimney-ladder structures considered as 'electron' phases, *Acta Cryst. B* **26**, 1044–1046 (1970).

4.125   H. Voellenkle: Die kristallstructur von Ru$_2$Ge$_3$, *Monatsh. Chem.* **105**, 1217–1227 (1974).

4.126   C.P. Susz, J. Muller, K. Yvon, E. Parthé: Diffusionless phase transformations of Ru$_2$Si$_3$, Ru$_2$Ge$_3$ and Ru$_2$Sn$_3$, II: Electrical and magnetic properties, *J. Less-Common Met.* **71**(1), P1–P8 (1980).

4.127   W. Wolf, G. Bihlmayer, S. Blügel: Electronic structure of the Nowotny chimney-ladder silicide Ru$_2$Si$_3$, *Phys. Rev. B* **55**(11), 6918–6926 (1997).

4.128   W. Henrion, M. Rebien, V.N. Antonov, O. Jepsen, H. Lange: Optical characterization of Ru$_2$Si$_3$ by spectroscopic ellipsometry, UV-VIS-NIR spectroscopy and band structure calculations, *Thin Solid Films* **313/314**, 218–221 (1998).

4.129   A.B. Filonov, D.B. Migas, V.L. Shaposhnikov, N.N. Dorozhkin, V.E. Borisenko, A. Heinrich, H. Lange: Electronic properties of isostructural ruthenium and osmium silicides and germanides, (1999) (to be published).

4.130   C.B. Vining, C.E. Allevato: Intrinsic thermoelectric properties of single crystal Ru$_2$Si$_3$, in: *Proceedings of the 10th International Conference on Thermoelectrics*, edited by D.M. Rowe (Babrow Press, Cardiff, 1991), pp. 167–173.

4.131   T. Ohta, C.B. Vining, C.E. Allevato: Characteristics of a promising new thermoelectric material: ruthenium silicide, in: *Proceedings of the 26th Intersociety Energy Conversion Engineering Conference* (American Nuclear Society, La Grange Park, IL, 1991), vol. 3, pp. 196–201.

4.132   L. Schellenberg, H.F. Braun, J. Muller: The osmium-silicon phase diagram, *J. Less-Common Met.* **144**(2), 341–350 (1988).

4.133   P. Pecheur, G. Toussaint: Electronic structure and bonding of the Nowotny chimney-ladder compound Ru$_2$Si$_3$, *Phys. Lett. A* **160**(2), 193–196 (1991).

4.134   K. Mason, G. Müller-Vogt: Osmium disilicide: preparation, crystal growth, and physical properties of a new semiconducting compound, *J. Cryst. Growth* **63**(1), 34–38 (1983).

4.135   A.B. Filonov, D.B. Migas, V.L. Shapohnikov, N.N. Dorozhkin, V.E. Borisenko, H. Lange: Electronic properties of osmium disilicide,   *Appl. Phys. Lett.* **70**(8), 976–977 (1997).

4.136   C.E. Allevato, C.B. Vining: Phase diagram and electrical behavior of silicon-rich iridium silicide compounds, *J. Alloys Comp.* **200**(12), 99–105 (1993).

4.137   C.E. Allevato, C.B. Vining: Phase diagram and electrical behavior of silicon-rich iridium silicide compounds, in: *Proceedings of the 27th Intersociety Energy Conversion Engineering Conference* (San Diego, CA, 1992), vol. 3, pp. 3.493–3.497.

4.138   W. Pitschke, R. Kurt, A. Heinrich, J. Schumann, J. Thomas, M. Mäder: Structure and phase formation in amorphous Ir_xSi_{1-x} thin films at high temperatures, in: *Proceedings of the 15th International Conference on Thermoelectrics* (IEEE, Inc., Piscataway, NJ, 1996), pp. 499–503.

4.139   S. Petersson,   J.A. Reimer,   M.H. Brodsky,   D.R. Campbell,   F.M. d'Heurle, B. Karlsson,   P.A. Tove: $IrSi_{1.75}$ a   new   semiconductor   compound, *J. Appl. Phys.* **53**(4), 3342–3343 (1982).

4.140   C.E. Allevato, C.B. Vining: Thermoelectric properties of semiconducting iridium silicides, in: *Proceedings of the 28th Intersociety Energy Conversion Engineering Conference* (Atlanta, Georgia, USA, 1993), vol. 1, pp. 1.239–1.243 (1993).

4.141   J. Schumann,   D. Elefant,   C. Gladun,   A. Heinrich,   W. Pitschke,   H. Lange, W. Henrion, R. Grötzschel: Polycrystalline iridium silicide films. Phase formation, electrical and optical properties, *Phys. Stat. Sol.(a)* **145**(2), 429–439 (1994).

4.142   M. Wittmer, P. Oelhafen, K.N. Tu: Electronic structure of iridium silicides, *Phys. Rev. B* **33**(8), 5391–5400 (1986).

4.143   Y.M. Yarmoshenko, S.N. Shamin, L.V. Elokhina, V.E. Dolgih, E.Z. Kurmaev, S. Bartkowski, M. Neumann, D.L. Ederer, K. Göransson, B. Noläng, I. Engström: Valence band spectra of 4d and 5d silicides, *J. Phys.* **9**(43), 9403–9414 (1997).

4.144   S.R. Nishitani, S. Fujii, M. Mizuno, I.Tanaka, H. Adachi: Chemical bonding of 3d transition-metal disilicides, *Phys. Rev. B* **58**(15), 9741–9745 (1998).

4.145   W.J. Scouler: Optical properties of Mg_2Si, Mg_2Ge, and Mg_2Sn from 0.6 to 11.0 eV at 77K, *Phys. Rev.* **178**(3), 1353–1357 (1969).

4.146   W. Henrion, H. Lange, E. Jahne, M. Giehler: Optical properties of chromium and iron disilicide layers, *Appl. Surf. Sci.* **70/71**, 569–572 (1993).

4.147   H. Lange, W. Henrion, F. Fenske, T. Zettler, J. Schumann, S. Teichert: Optical interband properties of some semiconducting silicides, *Phys. Stat. Sol. (b)* **194**(1), 231–240 (1996).

4.148   H. Lange, E. Jahne, O. Günther, W. Henrion, M. Giehler, J. Schumann: Optical properties of semiconducting silicides, in: *Silicides, Germanides, and Their Interfaces*, edited by R.W. Fathauer, S. Mantl, L.J. Schowalter, K.N. Tu (MRS, Pittsburgh, Pennsylvania, 1994), pp. 479–484.

4.149   M. Tanaka, S. Kurita, M. Fujisawa, F. Levy: Dielectric properties of single-crystal $TiSi_2$ from 0.6 to 20 eV, *Phys. Rev. B* **43** (11), 9133–9137 (1991).

4.150   M. Amiotti,   A. Borghesi,   F. Marabelli,   A. Piaggi,   G. Guizzetti,   F. Nava: Optical and   electronic   properties   of   5th-column   transition   metal   disilicides, *Appl. Surf. Sci.* **53**, 230–236 (1991).

4.151   L.F. Mattheiss, J.C. Hensel: Electronic structure of $TiSi_2$, *Phys. Rev. B* **39**(11), 7754–7759 (1989).

4.152   A. Borghesi, A. Piaggi, G. Guizzetti, F. Nava, M.Bacchetta: Optical properties of vanadium silicide polycrystalline films, *Phys. Rev. B* **40**(5), 3249–3253 (1989).

4.153   J.C. Lasjaunias, U. Gottlieb, O. Laborde, O. Thomas, R. Madar: Transport and low temperature specific heat measurements of $CrSi_2$ single crystals, in: *Silicide Thin Films - Fabrication, Properties, and Applications,* edited by R.T. Tung, K. Maex, P.W. Pellegrini, L.H. Allen (MRS, Pittsburgh, Pennsylvania, 1996), pp. 343–348.

4.154   W. Henrion,   St. Brehme,   I. Sieber,   H. von Känel,   Y. Tomm,   H. Lange: Optical and electrical properties of iron disilicide with different degree of structural perfection, *Solid State Phenom.* **51/52**, 341–346 (1996).

4.155   M. Fanciulli,   C. Rosenblad,   G. Weyer,   H. von Känel,   N. Onda: Conversion electron Mössbauer spectroscopy study of iron disilicide films grown by MBE, *Thin Solid Films* **275**(1), 8–11 (1996).

4.156   A.B. Filonov,   D.B. Migas,   V.L. Shaposhnikov,   V.E. Borisenko,   W. Henrion, M. Rebien, H. Lange, G. Behr: Theoretical and experimental study of interband optical transitions in semiconducting iron disilicides, *J. Appl. Phys.* **83**(8), 4410–4414 (1998).

4.157   O.K. Andersen: Linear methods in band theory, *Phys. Rev. B* **12**(8), 3060–3083 (1975).

4.158   U. von Barth, L. Hedin: A local exchange-correlation potential for the spin polarized case, *J. Phys. C* **5**(13) 1629–1642 (1972).

4.159   W. Henrion, M. Rebien, V.N. Antonov, O. Jepsen: Optical characterization of Ru_2Si_3 and Ru_2Ge_3 by various spectroscopic methods and by band structure calculations, *Solid State Phenom.* **67/68**, 471–477 (1999).

4.160   G. Guizzetti,   F. Marabelli,   M. Patrini,   P. Pellegrino,   B. Pivac,   L. Miglio, V. Meregalli, H. Lange, W. Hemrion, V. Tomm: Measurement and simulation of anisotropy in the infrared and Raman spectra of β-$FeSi_2$ single crystals, *Phys. Rev. B* **55**(21), 14290–14297 (1997).

4.161   M. Rebien, W. Henrion, U. Müller, S. Gramlich: Exiton absorption in β-$FeSi_2$ epitaxial films, *Appl. Phys. Lett.* **74**(7) 970–972 (1999).

4.162   D. Panknin, E. Wieser, W. Skorupa, W. Henrion, H. Lange: Buried $(Fe_{1-x}Co_x)Si_2$ layers with variable band gap formed by ion beam synthesis, *Appl. Phys. A* **62**(2), 155–162 (1996).

4.163   N.G. Galkin, A.M. Maslov, A.V. Konchenko: Optical and photospectral properties of $CrSi_2$ A-type epitaxial films on Si(111), *Thin Solid Films* **311**(1–2), 230–238 (1997).

4.164   V.K. Zaitsev, S.V. Ordin, V.I. Tarassov, M.I. Fedorov: Optical properties of high manganese silicides, *Sov. Phys. - Solid State* **21**(8), 1454–1455 (1979).

4.165   D. Leong, M. Harry, K.J. Reeson, K.P. Homewood: A silicon/iron-disilicide light-emitting diode operating at a wavelength of 1.5 μm, *Nature* **387**(12), 686–688 (1997).

4.166   D.N. Leong, M.A. Harry, K.J. Reeson, K.P. Homewood: On the origin of the 1.5 μm luminescence in ion beam synthesized β-$FeSi_2$, *Appl. Phys. Lett.* **68**(12), 1649–1650 (1996).

4.167   H. Lange: Electronic properties of semiconducting silicides, *Phys. Stat. Sol. (b)* **201**(1), 3–65 (1997).

4.168   T. Ruf: *Phonon Raman Scattering in Semiconductors, Quantum Wells and Superlattices,* (Springer, Berlin, 1998).

4.169   G. Guizzetti, F. Marabelli, M. Patrini, Y. Mo, N. Ondo, H. von Känel: Phonon spectra dependence on growth and sample treatment in β-$FeSi_2$, in: *Silicides, Germanides, and Their Interfaces*, edited by R.W. Fathauer, S. Mantl, L.J. Schowalter, K.N. Tu (MRS, Pittsburgh, Pennsylvania, 1994), pp. 127–132.

4.170   A. Borghesi, A. Piaggi, A. Franchini, G. Guizzetti, F. Nava, G. Santoro: Far-infrared vibrational spectroscopy in $CrSi_2$, *Europhys. Lett.* **11**(1), 61–65 (1990).

4.171   F. Fenske, H. Lange, G. Oertel, G.-U. Reinsperger, J. Schumann, B. Selle: Characterization of semiconducting silicide films by infrared vibrational spectroscopy, *Mater. Chem. Phys.* **43**(3), 238–242 (1996).

4.172   U. Birkholz, H. Finkenrath, J. Naegele, N. Uhle: Infrared reflectivity of semiconducting $FeSi_2$, *Phys. Stat. Sol.* **30**(1), K81–K85 (1968).

4.173 K. Lefki, P. Muret, E. Bustarret, N. Boutarek, R. Madar, J. Chevrier, J. Derrien: Infrared and Raman characterization of beta iron silicide, *Solid State Commun.* **80**(10), 791–795 (1991).
4.174 O. Chaix-Pluchery, G. Lucazeau: Vibrational study of transition metal disilicides, MnSi$_2$ (M = Nb, Ta, V, Cr), *J. Raman Spectrosc.* **29**(2), 159–164 (1998).
4.175 L. Hedin: New method for calculating the one-particle Green's function with application to the electron-gas problem, *Phys. Rev.* **139**(3A), A796–A823 (1965).
4.176 M. Rohlfing, P. Krüger, J. Pollmann: Quasiparticle band-structure calculations for C, Si, Ge, GaAs, and SiC using Gaussian-orbital basis sets, *Phys. Rev. B* **48**(24), 17791–17805 (1993).
4.177 J.P. Perdew, J.A. Chevary, S.H. Vosko, K.A. Jackson, M.R. Pederson, D.J. Singh, C. Fiolhais: Atoms, molecules, solids and surfaces: Applications of the generalized gradient approximation for exchange and correlation, *Phys. Rev. B* **46**(11), 6671–6687 (1992).
4.178 J.P. Perdew, K. Burke, M. Ernzerhof: Generalized gradient approximation made simple, *Phys. Rev. Lett.* **77** (18), 3865–3868 (1996).

Chapter 5

5.1 J.S. Blakemore: *Solid State Physics* (Cambridge University Press, Cambridge, 1985).
5.2 R.A. Smith: *Semiconductors* (Cambridge University Press, Cambridge, 1978).
5.3 B.K. Ridley: *Quantum Processes in Semiconductors* (Clarendon Press, Oxford 1982).
5.4 V.L. Bonch-Bruevich, S.G. Kalashnikov: *Semiconductor Physics* (Nauka, Moscow, 1990) (in Russian).
5.5 P.Y. Yu, M. Cardona: *Fundamentals of Semiconductors* (Springer, Berlin, 1995).
5.6 L.N. Guseva, B.I. Ovechkin: Thermoelectric properties of chromium silicides, *Dokl. Akad. Nauk SSSR* **112**(4), 681–683 (1957) (in Russian).
5.7 P.V. Gel'd: Thermal and thermoelectric properties of transition metal-silicon alloys, *Zurn. Tekn. Fiz.* **27**(1), 113–118 (1957) (in Russian).
5.8 V.S. Neshpor, G.V. Samsonov: Investigation of electrical conductance of transition metal silicides, *Sov. Phys. – Solid State*, **2**(9), 1966–1970 (1960).
5.9 R.G. Morris, R.D. Redin, G.C. Danielson: Semiconducting properties of Mg$_2$Si single crystals, *Phys. Rev.* **109**(6), 1909–1915 (1958).
5.10 S.P. Murarka, M.H. Read, C.J. Doherty, D.B. Fraser: Resistivity of thin film transition metal silicides, *J. Electrochem. Soc.* **129**(2), 293–301 (1982).
5.11 C.P. Susz, J. Muller, K. Yvon, E. Parthe: Diffusionless phase transformations of Ru$_2$Si$_3$, Ru$_2$Ge$_3$ and Ru$_2$Sn$_3$, II: Electrical and magnetic properties, *J. Less-Common Met.* **71**(1), P1–P8 (1980).
5.12 S. Petersson, J.A. Reimer, M.H. Brodsky, D.R. Campbell, F.M. d'Heurle, B. Karlsson, P.A. Tove: IrSi$_{1.75}$ a new semiconductor compound, *J. Appl. Phys.* **53**(4), 3342–3343 (1982).
5.13 K. Mason, G. Müller-Vogt: Osmium disilicide: preparation, crystal growth, and physical properties of a new semiconducting compound, *J. Cryst. Growth* **63**(1), 34–38 (1983).
5.14 L. Schellenberg, H.F. Braun, J. Muller: The osmium-silicon phase diagram, *J. Less-Common Met.* **144**(2), 341–350 (1988).

5.15 F. Nava, B.Z. Weiss, K.Y. Ahn, D.A. Smith, K.N. Tu: Thermal stability and electrical conduction behavior of coevaporated $WSi_{2\pm x}$ thin films, *J. Appl. Phys.* **64**(1), 354–364 (1988).

5.16 O. Bisi, L. Braicovich, C. Carbone, I. Lindau, A. Iandelli, G.L. Olcese, A. Palenzona: Chemical bond and electronic state in calcium silicides: Theory and comparison with synchrotron-radiation photoemission, *Phys. Rev. B* **40**(15), 10194–10209 (1989).

5.17 C.R.Whitsett, G.C.Danielson: Electrical properties of the semiconductors Mg_2Si and Mg_2Ge, *Phys. Rev.* **100**(4), 1261–1262, (1955).

5.18 R.J. LaBotz, D.R. Mason, D.F. O'Kane: The thermoelectric properties of mixed crystals of $Mg_2Ge_xSi_{1-x}$, *J. Electrochem. Soc.* **110**(2), 127–134 (1963).

5.19 U. Winkler: Die elektrischen Eigenschaften der intermetallischen Verbindungen Mg_2Si, Mg_2Ge, Mg_2Sn und Mg_2Pb, *Helv. Phys. Acta.* **28**(7), 633–666 (1955).

5.20 S. Bose, H.N. Acharya, H.D. Banerjee: Electrical, thermal, thermoelectric and related properties of magnesium silicide semiconductor prepared from rice husk, *J. Mater. Sci.* **28**(20), 5461–5468 (1993).

5.21 V.S. Neshpor, V.L. Upko: Investigation of formation conditions and some physical properties of barium disilicide, *J. Appl. Chem.* **36**(5), 1139–1142 (1963) (in Russian).

5.22 J. Evers, A. Weiss: Electrical properties of alkaline earth disilicides and digermanides, *Mater. Res. Bull.* **9**(5), 549–554 (1974).

5.23 M. Imai, T. Hirano: Electrical resistivity of metastable phases of $BaSi_2$ synthesized under high pressure and high temperature, *J. Alloys Comp.* **224**(1), 111–116 (1995).

5.24 I. Nishida: The crystal growth and thermoelectric properties of chromium disilicide, *J. Mater. Sci.* **7**(10), 1119–1124 (1972).

5.25 T. Hirano, M. Kaise: Electrical resistivities of single-crystalline transition-metal disilicides, *J. Appl. Phys.* **68**(2), 627–633 (1990).

5.26 D. Shinoda, S. Asanabe, Y. Sasaki: Semiconducting properties of chromium disilicide, *J. Phys. Soc. Jap.* **19**(3), 269–272 (1964).

5.27 S.F. Gong, X.-H. Li, H.T.G. Hentzell, J. Strandberg: Electrical and structural properties of thin film of sputtered $CrSi_2$, *Thin Solid Films* **208**(1), 91–95 (1992).

5.28 F. Nava, T. Tien, K.N. Tu: Temperature dependence of semiconducting and structural properties of Cr-Si thin films, *J. Appl. Phys.* **57**(6), 2018–2025 (1985).

5.29 V.E. Borisenko, L.I. Ivanenko, S.Yu. Nikitin: Semiconducting properties of chromium disilicide, *Mikroelektronika* **21**(2), 61–77 (1992) (in Russian).

5.30 M.C. Bost, J.E. Mahan: An investigation of the optical constants and band gap of chromium disilicide, *J. Appl. Phys.* **63**(3), 839–844 (1988).

5.31 N.G. Galkin, T.V. Velitchko, S.V. Skripka, A.B. Khrustalev: Semiconducting and structural properties of $CrSi_2$ A-type epitaxial films on Si(111), *Thin Solid Films* **280**(1–2), 211–220 (1996).

5.32 I. Kawasumi, M. Sakata, I. Nishida, K. Masumoto: Crystal growth of manganese silicide, $MnSi_{1.73}$ and semiconducting properties of $Mn_{15}Si_{26}$, *J. Mater. Sci.* **16**(2), 355–366 (1981).

5.33 I. Nishida: Semiconducting properties of nonstoichiometric manganese silicides, *J. Mater. Sci.* **7**, 435–440 (1972).

5.34 M. Eizenberg, K.N. Tu: Formation and Schottky behavior of manganese silicides on n-type silicon, *J. Appl. Phys.* **53**(10), 6885–6890 (1982).

5.35 S. Teichert, R. Kilper, J. Erben, D. Franke, B. Gebhard, T. Franke, P. Häussler, W. Henrion, H. Lange: Preparation and properties of thin polycrystalline $MnSi_{1.73}$ films, *Appl. Surf. Sci.* **104/105**, 679–684 (1996).

5.36 Ch. Krontiras, K. Pomoni, M. Roilos: Resistivity and the Hall effect for thin MnSi$_{1.73}$ films, *J. Phys. D: Appl. Phys.* **21**(3), 509–512 (1988).

5.37 U. Gottlieb, B. Lambert-Andron, F. Nava, M. Affronte, O. Laborde, A. Rouault, R. Madar: Structural and electronic transport properties of ReSi$_{2-\delta}$ single crystals, *J. Appl. Phys.* **78**(6), 3902–3907 (1995).

5.38 C. Krontiras, L. Grönberg, I. Suni, F.M. d'Heurle, J. Tersoff, I. Engström, B. Karlsson, C.S. Petersson: Some properties of ReSi$_2$, *Thin Solid Films* **161**(1/2), 197–206 (1988).

5.39 R.G. Long, M.C. Bost, J.E. Mahan: Optical and electrical properties of semiconducting rhenium disilicide thin films, *Thin Solid Films* **162**(1/2), 29–40 (1988).

5.40 I. Ali, P. Muret, T.A. Nguyen Tan: Properties of semiconducting rhenium silicide thin films grown epitaxially on silicon (111), *Appl. Surf. Sci.* **102**, 147–150 (1996).

5.41 E. Arushanov, Ch. Kloc, E. Bucher: Impurity band in p-type β-FeSi$_2$, *Phys. Rev. B* **50**(4), 2653–2656 (1994).

5.42 S. Brehme, L. Ivanenko, Y. Tomm, G.-U. Reinsperger, H. Lange: Hall effect investigation of doped and undoped β-FeSi$_2$, in: *Silicide Thin Films - Fabrication, Properties, and Applications,* edited by R.T. Tung, K. Maex, P.W. Pellegrini, L.H. Allen (MRS, Pittsburgh, Pennsylvania, 1996), pp. 355–360.

5.43 G. Behr, J. Werner, G. Weise, A. Heinrich, A. Burkov, C. Gladun: Preparation and properties of high-purity β-FeSi$_2$ single crystals, *Phys. Stat. Sol. (a)* **160**(2), 549–556 (1997).

5.44 O. Valassiades, C.A. Dimitriadis, J.H. Werner: Galvanomagnetic behavior of semiconducting FeSi$_2$ films, *J. Appl. Phys.* **70**(2), 890–893 (1991).

5.45 C.A. Dimitriadis, J.H. Werner, S. Logothetidis, M. Stutzmann, J. Weber, R. Nesper: Electronic properties of semiconducting FeSi$_2$ films, *J. Appl. Phys.* **68**(4), 1726–1734 (1990).

5.46 D.H. Tassis, C.L. Mitsas, T.T. Zorba, C.A. Dimitriadis, O. Valassiades, D.I. Siapkas, M. Angelakeris, P. Poulopoulos, N.K. Flevaris, G. Kiriakidis: Infrared spectroscopic and electronic transport properties of polycrystalline semiconducting FeSi$_2$ thin films, *J. Appl. Phys.* **80**(2), 962–968 (1996).

5.47 A. Rizzi, B.N.E. Rösen, D. Freundt, Ch. Dieker, H. Lüth, D. Gerthsen: Heteroepitaxy of β-FeSi$_2$ on Si by gas-source MBE, *Phys. Rev. B* **51**(24), 17780–17794 (1995).

5.48 K. Radermacher, R. Carius, S. Mantl: Optical and electrical properties of buried semiconducting β-FeSi$_2$, *Nucl. Instrum. Methods Phys. Res. B* **84**(1), 163–167 (1994).

5.49 U. Gottlieb, O. Laborde, A. Rouault, R. Madar: Resistivity of Ru$_2$Si$_3$ single crystals, *Appl. Surf. Sci.* **73**, 243–245 (1993).

5.50 T. Ohta, C.B. Vining, C.E. Allevato: Characteristics of a promising new thermoelectric material: ruthenium silicide, in: *Proceedings of the 26th Intersociety Energy Conversion Engineering Conference*, vol. **3** (American Nuclear Society, La Grange Park, IL, 1991), pp. 196–201.

5.51 C.B. Vining, C.E. Allevato: Intrinsic thermoelectric properties of single crystal Ru$_2$Si$_3$, in: *Proceedings of the 10th International Conference on Thermoelectrics*, edited by D.M. Rowe (Babrow Press, Cardiff, 1991), pp. 167–173.

5.52 F.M. d'Heurle, R.D. Frampton, E.A. Irene, H. Jiang, C.S. Petersson: Rate of formation of silicon dioxide; semiconducting ruthenium silicide, *Appl. Phys. Lett.* **47**(11), 1170–1172 (1985).

5.53 T. Ohta, A. Yamamoto, T. Tanaka, Y. Sawade, K. Kamisako, T. Horigome: Progress in thermoelectric properties of undoped ruthenium sesquisilicide by HIP method, in: *Proceedings of the 12th International Conference on Thermoelectrics* (Yokohama, 1993), pp. 393–396.

5.54   C.E. Allevato, C.B. Vining: Thermoelectric properties of semiconducting iridium silicides, in: *Proceedings of the 28th Intersociety Energy Conversion Engineering Conference* 1, 1.239–1.243 (1993).

5.55   C.E. Allevato, C.B. Vining: Phase diagram and electrical behavior of silicon-rich iridium silicide compounds, *J. Alloys Comp.* 200(12), 99–105 (1993).

5.56   C.E. Allevato, C.B. Vining: Phase diagram and electrical behavior of silicon-rich iridium silicide compounds, in: *Proceedings of the 27th Intersociety Energy Conversion Engineering Conference* 3, 3.493–3.497 (1992).

5.57   J. Evers: Semiconductor-metal transition in BaSi$_2$, *J. Less-Common Met.* 58(1), 75–83 (1978).

5.58   M. Imai, T. Hirano: Electrical resistivity of three polymorphs of BaSi$_2$ and P-T phase diagram, in: *Silicide Thin Films - Fabrication, Properties, and Applications*, edited by R.T. Tung, K. Maex, P.W. Pellegrini, L.H. Allen (MRS, Pittsburgh, Pennsylvania, 1996), pp. 567–572.

5.59   M. Imai, K. Hirata, T. Hirano: Superconductivity of trigonal BaSi$_2$, *Physica C* 245(1), 12–14 (1995).

5.60   S. Fahy, D.R. Hamann: Electronic and structural properties of CaSi$_2$. *Phys. Rev. B* 41(11), 7587–7592 (1990).

5.61   I.J. Ohsugi, T. Kojima, I.A. Nishida: Temperature dependence of the magnetic susceptibility of a CrSi$_2$ single crystals, *Phys. Rev. B* 42(16), 10761–10764 (1990).

5.62   I. Nishida, T. Sakata: Semiconducting properties of pure and Mn-doped chromium disilicides, *J. Phys. Chem. Solids* 39(5), 499–505 (1978).

5.63   V.E. Borisenko, D.I. Zarovskii, G.V. Litvinovich, V.A. Samuilov: Optical spectroscopy of chromium silicide formed by a flash heat treatment of chromium films on silicon, *Zurn. Prikl. Spectrosk.* 44(2), 314–317 (1986) (in Russian).

5.64   J.C. Lasjaunias, U. Gottlieb, O. Laborde, O. Thomas, R. Madar: Transport and low temperature specific heat measurements of CrSi$_2$ single crystals, in: *Silicide Thin Films - Fabrication, Properties, and Applications*, edited by R.T. Tung, K. Maex, P.W. Pellegrini, L.H. Allen (MRS, Pittsburgh, Pennsylvania, 1996), pp. 343–348.

5.65   V.V. Kleshkovskaya, T.S. Kamilov, S.I. Adasheva, S.S. Khudaiberdiev, V.I. Muratova: Crystal structure of the films of highest manganese silicide on silicon, *Crystallography Reports* 39(5), 815–819 (1994).

5.66   U. Gottlieb, M. Affronte, F. Nava, O. Laborde, A. Sulpice, R. Madar: Some physical properties of ReSi$_{1.75}$ single crystals, *Appl. Surf. Sci.* 91(1), 82–86 (1995).

5.67   A.T. Burkov, A. Heinrich, G. Gladun, W. Pitschke, J. Schumann: Structure and thermoelectric properties of nanocrystalline Re$_x$-Si$_{1-x}$ thin film composites, in: *Proceedings of the 15th International Conference on Thermoelectrics* (IEEE, Inc., Piscataway, NJ, 1996), p. 484.

5.68   V.K. Zaitsev: Thermoelectric properties of anisotropic MnSi$_{1.75}$, in: *CRC Handbook on Thermoelectrics*, edited by D.M. Rowe (CRC Press, New York, 1994), pp. 299–309.

5.69   A.B. Filonov, D.B. Migas, V.L. Shaposhnikov, N.N. Dorozhkin, V.E. Borisenko, H. Lange, A. Heinrich: Electronic properties of semiconducting rhenium silicide, *Europhys. Lett.* 46(3), 376–381 (1999).

5.70   J.-I. Tani, H. Kido: Electrical properties of Co-doped and Ni-doped β-FeSi$_2$, *J. Appl. Phys.* 84(3), 1408–1411 (1998).

5.71   M. Komabayashi, K. Hijikata, S. Ido: Effects of some additives on thermoelectric properties of FeSi$_2$ thin films, *Jpn. J. Appl. Phys.* 30(2), 331–234 (1991).

5.72   T. Kojima: Semiconducting and thermoelectric properties of sintered iron disilicide, *Phys. Stat. Sol. (a)* 111(1), 233–242 (1989).

5.73 U. Birkholz, J.Schelm: Mechanism of electrical conduction in β-FeSi$_2$, *Phys. Stat. Sol.* **27**, 413–425 (1968).

5.74 Y. Isoda, M.A. Okamoto, T. Ohkoshi, I.A. Nishida: Semiconducting properties and thermal shock resistance of boron doped iron disilicide, in: *Proceedings of the 12th International Conference on Thermoelectrics* (Yokohama, 1993), pp. 192–196.

5.75 A. Heinrich, A. Burkov, C. Gladun, G. Behr, K. Herz, J. Schumann, H. Powalla: Thermopower and electrical resistivity of undoped and Co-doped β-FeSi$_2$ single crystals and β-FeSi$_{2+x}$ thin films, in: *Proceedings of the 15th International Conference on Thermoelectrics* (IEEE, Inc., Piscataway, NJ, 1996), pp. 57–65.

5.76 M.I. Fedorov, M.A. Khazan, A.E. Kaliazin, V.K. Zaitsev, N.F. Kartenko, A.E. Engalychev: Properties of iron disilicide doped with Ru, Rh and Pd, in: *Proceedings of the 15th International Conference on Thermoelectrics* (IEEE, Inc., Piscataway, NJ, 1996), pp. 75–78.

5.77 H. von Känel, U. Kafader, P. Sutter, N. Onda, H. Sirringhaus, E. Müller, U. Kroll, C. Schwarz, S. Goncalves-Conto: Epitaxial semiconducting and metallic iron silicides, in: *Silicides, Germanides, and Their Interfaces*, edited by R.W. Fathauer, S. Mantl, L.J. Schowalter, K.N. Tu (MRS, Pittsburgh, Pennsylvania, 1994), pp. 73–84.

5.78 P. Muret, I.Ali: Transport properties of unintentionally doped iron silicide thin films on silicon (111), *J. Vac. Sci. Technol. B* **16**(3), 1663–1666 (1998).

5.79 G. Behr, L. Ivanenko, H. Vinzelberg, A. Heinrich: Single crystal growth of non-stoichiometric β-FeSi$_2$ by chemical transport reaction, (1999) (to be published).

5.80 K. Irmscher, W. Gehlhoff, Y. Tomm, H. Lange, V. Alex: Iron group impurities in β-FeSi$_2$ studied by EPR, *Phys. Rev. B* **55**(7), 4417–4425 (1997).

5.81 M.C. Bost, J.E. Mahan: Optical properties of semiconducting iron disilicide thin films, *J. Appl. Phys.* **58**(7), 2696–2703 (1985).

5.82 S. Brehme, P. Lengsfeld, P. Stauss, H. Lange, W. Fuhs: Hall effect and resistivity of β-FeSi$_2$ thin films and single crystals, *J. Appl. Phys.* **84**(6), 3187–3196 (1998).

5.83 Y. Tomm, L. Ivanenko, K. Irmscher, St. Brehme, W. Henrion, I. Sieber, H. Lange: Effects of doping on the electronic properties of semiconducting iron disilicide, *Mat. Sci. Eng. B* **37**, 215–218 (1996).

5.84 K.G. Lisunov, E. K. Arushanov, Ch. Kloc, U. Malang, E. Bucher: Hopping conductivity in p-type β-FeSi$_2$, *Phys. Stat. Sol. (b)* **195**(1), 227–236 (1996).

5.85 N.F. Mott, E.A. Davies: *Electron Processes in Non-Crystalline Materials* (Clarendon Press, Oxford, 1979).

5.86 B.I. Shklovskii, A.L. Efros: *Electronic Properties of Doped Semiconductors* (Springer, Berlin, 1984).

5.87 D.H. Tassis, C.A. Dimitriadis, O. Valassiades: The Meyer-Neldel rule in the conductivity of polycrystalline semiconducting FeSi$_2$ films, *J. Appl. Phys.* **84**(5), 2960–2962 (1998).

5.88 A. Heinrich, C. Gladun, A. Burkov, Y. Tomm, S. Brehme, H. Lange: Thermopower and electrical resistivity of β-FeSi$_2$ single crystals doped with Cr, Co and Mn, in: *Proceedings of the 14th International Conference on Thermoelectrics,* (Ioffe-Institute, St. Petersburg, 1995), pp. 259–263.

5.89 U. Gottlieb, R. Madar, O. Laborde: Low temperature transport properties of Ru$_2$Si$_3$ single crystals, in: *Silicide Thin Films - Fabrication, Properties, and Applications*, edited by R.T. Tang, K. Maex, P.W. Pellegrini, L.H. Allen (MRS, Pittsburgh, Pennsylvania, 1996), pp. 581–586.

5.90 J. Schumann, D. Elefant, C. Gladun, A. Heinrich, W. Pitschke, H. Lange, W. Henrion, R. Grötzschel: Polycrystalline iridium silicide films. Phase formation, electrical and optical properties, *Phys. Stat. Sol. (a)* **145**(2), 429–439 (1994).

5.91 W. Pitschke, R. Kurt, A. Heinrich, J. Schumann, J. Thomas, M. Mäder: Structure and phase formation in amorphous Ir_xSi_{1-x} thin films at high temperatures, in: *Proceedings of the 15th International Conference on Thermoelectrics* (IEEE, Inc., Piscataway, NJ, 1996), pp. 499–503.

5.92 B.R. Nag: *Electron Transport in Compound Semiconductors* (Springer, Berlin, 1980).

5.93 J.T. Nelson: Electrical and optical properties of Mg_2Sn and Mg_2Si, *Am. J. Phys.* **23**, 390, (1955).

5.94 G. Beddies: Electrical properties of flash vapor-deposited chromium disilicide thin films, *Wiss. Z. - Tech. Hochsch. Karl-Marx-Stadt* **22**(7), 695–707 (1980).

5.95 G.V. Samsonov, L.A. Dvorina, B.M. Rud: *Silicides* (Metallurgiya, Moscow, 1979) (in Russian); G.V. Samsonov, I.M. Vinitskii: *Handbook of Refractory Compounds* (IFI/Plenum, New York, 1980).

5.96 D.J. Oostra, C.W.T. Bulle-Lieuwma, D.E.W. Vandenhoudt, F. Felten, J.C. Jans: β-$FeSi_2$ in (111) Si and in (001) Si formed by ion-beam synthesis, *J. Appl. Phys.* **74**(7), 4347–4353 (1993).

5.97 M. Pauli, M. Dücker, M. Döscher, J. Müller, W. Henrion, H. Lange: Characterization of the heterostructure between heteroepitaxially grown β-$FeSi_2$ and (111) silicon, *Mater. Sci. Eng.* **B21**, 270–273 (1993).

5.98 M. Döscher, B. Selle, M. Pauli, F. Kothe, J. Szymanski, J. Müller, H. Lange: Influence of stoichiometric variations and rapid thermal processing of β-$FeSi_2$ thin films on their electrical and microstructural properties, in: *Silicide Thin Films - Fabrication, Properties, and Applications,* edited by R.T. Tung, K. Maex, P.W. Pellegrini, L.H. Allen (MRS, Pittsburgh, Pennsylvania, 1996), pp. 325–330.

5.99 K. Herz, M. Powalla: Electrical and optical properties of thin β-$FeSi_2$ films on Al_2O_3 substrates, *Appl. Surf. Sci.* **91**(1), 87–92 (1995).

5.100 D.H. Tassis, C.A. Dimitriadis, S. Boultadakis, J. Arvanitidis, S. Ves, S. Kokkou, S. Logothetidis, O. Valassiades, P. Poulopoulos, N.K. Flevaris: Influence of conventional furnace and rapid thermal annealing on the quality of polycrystalline β-$FeSi_2$ thin films grown from vapor-deposited Fe/Si multilayers, *Thin Solid Films* **310**, 115–122 (1997).

5.101 W. Shockley, in: *Electrons and Holes in Semiconductors* (D. Van Nostrand Company, Inc., Princeton, New Jersey, 1950), Chap.8.

5.102 A.B. Filonov, I.E. Tralle, N.N. Dorozhkin, D.B. Migas, V.L. Shaposhnikov, G.V. Petrov, A.M. Anishchik, V.E. Borisenko: Semiconducting properties of hexagonal chromium, molybdenum, tungsten disilicides, *Phys. Stat. Sol. (b)* **186**(1), 209–215 (1994).

5.103 A.E. White, K.T. Short, D.J. Eaglesham: Electrical and structural properties of $Si/CrSi_2/Si$ heterostructures fabricated using ion implantation, *Appl. Phys. Lett.* **56**(13), 1260–1262 (1990).

5.104 N.E. Christensen: Electronic structure of β-$FeSi_2$, *Phys. Rev. B* **42**(11), 7148–7153 (1990).

5.105 E. Arushanov, Y. Tomm, L. Ivanenko, H. Lange: Hole mobility in Cr-doped p-type β-$FeSi_2$ single crystals, *Phys. Stat. Sol. (b)* **210**(1), 187–194 (1998).

5.106 W. Henrion, St. Brehme, I. Sieber, H. von Känel, Y. Tomm, H. Lange: Optical and electrical properties of iron disilicide with different degree of structural perfection, *Solid State Phenom.* **51/52**, 341–346 (1996).

5.107 E. Arushanov, Ch. Kloc, H. Hohl, E. Bucher: The Hall effect in β-$FeSi_2$ single crystals, *J. Appl. Phys.* **75**(10), 5106–5109 (1994).

5.108 S. Teichert, G. Beddies, Y. Tomm, H.-J. Hinneberg, H. Lange: A pronounced hysteresis effect observed in Hall measurements on β-$FeSi_2$ single crystals at 4.2 K, *Phys. Stat. Sol. (a)* **152**(1), K15–K18 (1995).

5.109 H. Lange: Electronic properties of semiconducting silicides, *Phys. Stat. Sol. (b)* **201**(1), 3–65 (1997).

5.110 A.B. Filonov, D.B. Migas, V.L. Shaposhnikov, N.N. Dorozhkin, G.V. Petrov, V.E. Borisenko, W. Henrion, H. Lange: Electronic and related properties of crystalline semiconducting iron disilicide, *J. Appl. Phys.* **79**(10), 7708–7712 (1996).

5.111 P. Lengsfeld, S. Brehme, G. Ehlers, H. Lange, N. Stüsser, Y. Tomm, W. Fuhs: Anomalous Hall effect in β-FeSi$_2$, *Phys. Rev. B* **58** (24), 16154–16159 (1998).

5.112 J. Smith: The spontaneous Hall effect in ferromagnetics II, *Physica* **24**(1), 39–51 (1958).

5.113 L. Berger: Side-jump mechanism for the Hall effect of ferromagnets, *Phys. Rev. B* **2**(11), 4559–4566 (1970).

5.114 J.L. Regolini, F. Trincat, I. Sagnes, Y. Shapira, G. Brémond, D. Bensahel: Characterization of semiconducting iron disilicide obtained by LRP/CVD, *IEEE Trans. Electron Devices* **39**(1), 200–201 (1992).

5.115 A.B. Filonov, I.E. Tralle, D.B. Migas, V.L. Shaposhnikov, V.E. Borisenko: Transport properties simulation of p-type β-FeSi$_2$, *Phys. Stat. Sol. (b)* **203** (1), 183–187 (1997).

5.116 N.P. Barradas, D. Panknin, E. Wieser, B. Schmidt, M. Betzl, A. Mücklich, W. Skorupa: Influence of the ion irradiation on the properties of β-FeSi$_2$ layers prepared by ion beam assisted deposition, *Nucl. Instrum. Methods Phys. Res. B* **127/128**, 316–320 (1997).

5.117 D.H. Tassis, C.L. Mitsas, T.T. Zorba, M. Angelakeris, C.A. Dimitriadis, O. Valassiades, D.I. Siapkas, G. Kiriakidis: Optical and electrical characterization of high quality β-FeSi$_2$ thin films grown by solid phase epitaxy, *Appl. Surf. Sci.* **102**, 178–183 (1996).

5.118 C.B. Vining, C.E. Allevato: Progress in doping of ruthenium silicide (Ru$_2$Si$_3$), in: *Proceedings of the 27th Intersociety Energy Conversion Engineering Conference*, vol. **3** (San Diego, CA, 1992), pp. 3.489–3.492.

5.119 Y. Arita, T. Miyagawa, T. Matsui: Thermoelectric properties of Ru$_2$Si$_3$ prepared by FZ and arc-melting methods, in: *Proceedings of the 17th International Conference on Thermoelectrics* (IEEE, Inc., 1998), pp. 394–397.

5.120 A. Yamamoto, T. Ohta, Y. Sawade, T. Tanaka, K. Kamisako: Investigation of dopants for ruthenium silicide, in: *Proceedings of the 14th International Conference on Thermoelectrics* (Ioffe-Institute, St. Petersburg, 1995), pp. 264–268.

5.121 I. Nishida: Stabilitiy of silicide compounds, *J. Mater. Sci. Soc. Jpn.* **15** (1978) 72–86 (in Japanese).

5.122 C.B. Vining: Silicides as promising thermoelectric materials, in: *Proceedings of the 9th International Conference on Thermoelectrics* (JPL, CalTech, Pasadena, 1990), pp. 249–259; Thermoelectric properties of silicides, in: *CRC Handbook on Thermoelectrics*, edited by D.M. Rowe (CRC Press, 1994), pp. 277–285.

5.123 U. Birkholz, E. Gross, U. Stoehrer: Polycrystalline iron disilicide as a thermoelectric generator material, in: *CRC Handbook on Thermoelectrics*, edited by D.M. Rowe (CRC Press, 1994), pp. 287–298.

5.124 C.B. Vining: The thermoelectric limit ZT ≈ 1: Fact or artifact, in: *Proceedings of the 11th International Conference on Thermoelectrics* (University of Texas at Arlington, Arlington, 1992), pp. 223–231.

5.125 Y. Noda, H. Kon, Y. Furukawa, N. Otsuka, I.A. Nishida, K. Masumoto: Preparation and thermoelectric properties of Mg$_2$Si$_{1-x}$Ge$_x$ (x=0.0–0.4) solid solution semiconductors, *Mater. Trans., JIM* **33**(9), 845–850 (1992).

5.126 C.B. Vining: Extrapolated thermoelectric figure of merit of ruthenium silicide, *Proc. Symp. on Space Nuclear Power Systems, AJP Conf. Proc.* **246**, 338–342 (1992).

5.127 G.A. Slack: New materials and performance limits for thermoelectric cooling, in: *CRC Handbook on Thermoelectrics*, edited by D.M. Rowe (CRC Press, 1994), pp. 407–440.

5.128 M.S. Dresselhaus, T. Koga, X. Sun, S.B. Cronin, K.L. Wang, G. Chen: Low dimensional thermoelectrics, in: *Proceedings of the 16th International Conference on Thermoelectrics* (IEEE, Inc., Piscataway, NJ, 1997), pp. 12–20.

5.129 R. Venkatasubramanian, T. Colpitts: Enhancement in figure of merit with super-lattice structures for thin film thermoelectric devices, in: *Thermoelectric Materials – New Directions and Approaches*, edited by T.M. Tritt, M.G. Kanatzidis, H.B. Lyon, G.D. Mahan (MRS, Pittsburgh, Pennsylvania, 1997), pp. 73–84.

5.130 M.C. Nicolaou: The magnesium silicide germanide stannide alloy: a new concept in thermoelectric energy conversion, in: *Proceedings of the 4th International Conference on Thermoelectrics* (University of Texas at Arlington, Arlington, 1982), pp. 83–88.

5.131 H.T. Kaibe, Y. Noda, Y. Isoda, I.A. Nishida: Temperature dependence of thermal conductivity for $Mg_2Si_{1-x}Ge_x$ solid solution, in: *Proceedings of the 16th International Conference on Thermoelectrics* (IEEE, Inc., Piscataway, NJ, 1997), pp. 279–282.

5.132 N.D. Marchuk, V.K. Zaitsev, M.I. Fedorov, A.E. Kalizin: Thermoelectric properties of some cheap n-type materials, in: *Proceedings of the 8th International Conference on Thermoelectrics* (Inst. Nat. Polytechnique Lorraine, Nancy, 1989), p. 210.

5.133 T. Kajikawa, K. Shida, K. Shiraishi, T. Ito, M. Omori, T. Hirai: Thermoelectric figure of merit of impurity doped and hot-pressed magnesium silicide elements, in: *Proceedings of the 16th International Conference on Thermoelectrics* (IEEE, Inc., Piscataway, NJ, 1997) pp. 362–369.

5.134 M. Riffel, J. Schilz: Influence of production parameters on the thermoelectric properties of Mg_2Si, in: *Proceedings of the 16th International Conference on Thermoelectrics* (IEEE, Inc., Piscataway, NJ, 1997), pp. 283–286.

5.135 I. Ohsugi, T. Kojima, M. Sakata, I. Nishida: Anisotropic thermoelectricity of a $CrSi_2$ single crystal, *J. Jpn. Inst. Metals* **58**(9), 985–988 (1994).

5.136 F.A. Sidorenko, I.Z. Radovsky, I.P. Zelenin, P.V. Geld: Electrical and magnetic properties of solid solutions of vanadium and titanium disilicides in chromium disilicide, *Proshkovaya Metallurgia* **9**, 67–74 (1966) (in Russian).

5.137 H. Hohl, A.P. Ramirez, T.T.M. Palstra, E. Bucher: Thermoelectric and magnetic properties of $Cr_{1-x}V_xSi_2$ solid solutions, *J. Alloys Comp.* **248**(1–2), 70–76 (1997).

5.138 A.O. Avetisian, Yu.M. Gorachev, B.A. Kovenskaja, T.M. Armola: Electrophysical properties of alloys of chromium disilicide and highest manganese silicide, *Dokl. Akad. Nauk Ukrain. SSR, Seria Fiziko-Matemat. Techn. Nauki* (10), 64–67 (1980) (in Russian).

5.139 T. Ohkoshi, I. Nishida, K. Masumoto, H. Kaibe, Y. Isoda, S. Ishida: Slip casting and thermoelectric property of $CrSi_2$, *Trans. Jpn. Inst. Met.* **29**(9), 756–766 (1988).

5.140 J. Schumann, C. Gladun, J.-I. Moench, A. Heinrich, J. Thomas, W. Pitschke: Nanodispersed Cr_xSi_{1-x} thin films: Transport properties and thermoelectric applications, *Thin Solid Films* **246**(1–2), 24–29 (1994).

5.141 B.K. Voronov, L.D. Dudkin, N.N. Trusovo: Anisotropy of thermoelectric properties in chromium and higher manganese silicide crystals, *Kristallografia* **12**(3), 519–521 (1967).

5.142   A.T. Burkov, A. Heinrich, M.V. Verdernikov: Anisotropic thermoelectric materials: Properties and applications, in: *Proceedings of the 13th International Conference on Thermoelectrics* (AIP Press, 1995), vol. 316, pp. 76–80.

5.143   E. Gross: Powder metallurgy of HMS and FeSi$_2$ for the thermoelectric energy conversion, *PhD Thesis* (University of Karlsruhe, Germany, 1993) (in German).

5.144   M.V. Vedernikov, A.E. Engalychev, V.K. Zaitsev: Thermoelectric properties of materials based on the higher manganese silicide and cobalt monosilicide, in: *Proceedings of the 7th International Conference on Thermoelectrics* (University of Texas at Arlington, Arlington,1988), pp. 150–155.

5.145   E. Gross, M. Riffel, U. Stoehrer: Thermoelectric generators made of FeSi$_2$ and HMS: Fabrication and measurement, *J. Mater. Res.* **10**(1), 34–40 (1995).

5.146   D.J. Bergman, O. Levy: Thermoelectric properties of a composite medium, *J. Appl. Phys.* **70**(11), 6821–6833 (1991).

5.147   A.T. Burkov, A. Heinrich, C. Gladun, W. Pitschke, J. Schumann: Effect of interphase boundaries on resistivity and thermopower of nanocrystalline Re-Si thin film composites, *Phys. Rev. B* **58**, 9644–9647 (1998).

5.148   C. Kleint, A. Heinrich, H. Griessmann, D. Hofmann, H. Vinzelberg, J. Schumann, D. Schlaefer, G. Behr, L. Ivanenko: Thermoelectric transport properties of ReSi$_{1.75}$ thin films, in: *Thermoelectric Materials 1998 - The Next Generation Materials for Small-Scale Refrigeration and Power Generation Applications*, edited by T.M. Tritt, H.B. Lyon, Jr., G. Mahan, M.G. Kanatzidis (MRS, Pittsburgh, Pennsylvania, 1999).

5.149   R.M. Ware, D.J. McNeill: Iron disilicide as a thermoelectric generator material, *Proc. IEE* **111**(1), 178–182 (1964).

5.150   I. Nishida: Study of semiconductor-to-metal transition in Mn-doped FeSi$_2$, *Phys. Rev. B* **7**(6), 2710–2713 (1973).

5.151   Y. Isoda, T. Ohkoshi, I. Nishida, H. Kaibe: Si-composition and thermoelectric properties of Mn and Co doped FeSi$_2$, *J. Mater. Sci. Soc. Jpn.* **25**, 311–319 (1989).

5.152   U. Stöhrer, R. Voggesberger, G. Wagner, U. Birkholz: Sintered FeSi$_2$ for thermoelectric power generation, *Energy Convers. Mgmt.* **30**, 143–147 (1990).

5.153   T. Tokiai, T. Uesugi, K. Koumoto: Thermoelectric properties of p-type iron disilicide ceramics fabricated from the composite powder prepared by the precipitation method, *J. Ceram. Soc. Jpn.* **103**(7), 660–675 (1995).

5.154   H. Takizawa, P.F. Mo, T. Endo, M. Shimada: Preparation and thermoelectric properties of β-Fe$_{1-x}$Ru$_x$Si$_2$, *J. Mater. Sci.* **30**(16), 4199–4203 (1995).

5.155   A. Heinrich, G. Behr, H. Griessmann, S. Teichert, H. Lange: Thermopower, electrical and Hall conductivity of undoped and doped iron disilicide single crystals, in: *Thermoelectric Materials – New Directions and Approaches*, edited by T.M. Tritt, M.G. Kanatzidis, H.B. Lyon, G.D. Mahan (MRS, Pittsburgh, Pennsylvania, 1997), pp. 255–266.

5.156   H.P. Geserich, S.K. Sharma, W.A. Theiner: Some structural, electrical and optical investigations on a new amorphous material: FeSi$_2$, *Philos. Mag.* **27**, 1001 (1973).

5.157   J. Schumann, H. Griessmann, A. Heinrich: Doped β-FeSi$_2$ thin film thermoelement sensor material, in: *Proceedings of the 17th International Conference on Thermoelectrics* (IEEE, Inc., 1998), pp. 221–225.

5.158   K. Matsubara, H. Kuno, Y. Okuno, H. Takaoka, T. Takagi: Thin film type thermoelectric devices using amorphous iron disilicide (FeSi$_2$) prepared by ionized-cluster beam technique, *Proc. Int. Ion Eng. Congr.* **2**, 1221–1226 (1983).

5.159   K. Matsubara, K. Kishimoto, K. Nagao, O. Ueda, T. Miki, T. Koyanagi, I. Fujii: Iron disilicides: The possibility of improving thermoelectric figure of merit values by RF-plasma processing, in: *Proceedings of the 12th International Conference on Thermoelectrics* (Yokohama, 1993), pp. 223–230.

5.160   R. Kurt, W. Pitschke, A. Heinrich, H. Griessmann, J. Schumann, K. Wetzig: Effect of doping on the thermoelectric properties of iridium silicide thin films, in: *Proceedings of the 17th International Conference on Thermoelectrics* (IEEE, Inc., Nagoya, 1998), pp. 249–252; R. Kurt: Structure, phase formation and thermoelectric transport properties of Ir-Si thin films, *PhD Thesis* (Technical University Dresden, Germany, 1998) (in German).

5.161   R. Kurt, W. Pitschke, A. Heinrich, J. Schumann, J. Thomas, K. Wetzig, A. Burkov: Phase formation process of Ir_xSi_{1-x} thin films – structure and electrical properties, *Thin Solid Films* **310**(1), 8–18 (1997).

Subject Index

A

absorption coefficient 185, 186, 218, 221, 222, 223, 225, 226, 235, 236, 238

atomic positions 8, 9, 10, 13, 14, 18, 19, 91, 213

B

band structure 182, 183, 185, 186, 187, 190, 191, 194, 195, 196, 197, 198, 200, 205, 209, 210, 211, 212, 216, 226, 227, 246

barium silicides 3, 8, 29, 88, 128, 138, 167, 190, 252
   atomic positions 8
   band structure 190
   crystal structure 2, 8
   density 3
   electrical resistivity 252
   interatomic distances 8
   lattice constants 3
   single crystal growth 138, 167
   space group 3, 8
   thermodynamic parameters 29
   thin film formation 88, 128

C

chromium silicides 3, 9, 31, 32, 55, 89, 95, 112, 115, 118, 128, 132, 139, 152, 160, 161, 167, 168, 169, 191, 215, 217, 226, 235, 236, 237, 254, 257, 272, 288
   absorption coefficient 226, 235, 236
   atomic positions 9
   band structure 191, 226
   crystal structure 9
   density 3
   density of states 193
   dielectric function 215, 216, 217, 236
   effective mass 194, 273
   electrical resistivity 254, 257, 288
   figure of merit 289
   interatomic distances 9

lattice constants 3, 161, 168, 169
mobility 272
Raman spectra 237
reflectance 235, 236
silicidation kinetics 55
single crystal growth 139, 152, 160, 167
space group 3, 9, 235
thermodynamic parameters 31, 32
thermopower 288
thin film formation 89, 95, 112, 115, 118, 128, 132

crystal structure 2, 8, 9, 10, 13, 14, 15, 18, 19, 196, 197, 200, 205, 209, 211

D

density of states 183, 190, 193, 199, 203, 207, 212, 216, 226, 234, 247

dielectric function 185, 213, 214, 215, 216, 217, 218, 219, 220, 221, 236

E

effective mass 183, 187, 190, 194, 195, 197, 200, 204, 208, 213, 245, 247, 273, 279, 281, 285, 289, 296

electrical resistivity 186, 249, 250, 252, 254, 257, 258, 259, 262, 264, 266, 267, 268, 269, 283, 288

extinction index 185

F

figure of merit 244, 284, 285, 287, 289, 290, 291, 293, 294, 295, 297

H

Hall coefficient 269, 276, 278, 279, 280, 283, 284, 297

I

interatomic distances 2, 8, 9, 10, 13, 14, 18, 20

Springer Series in
MATERIALS SCIENCE

Editors: U. Gonser · R. M. Osgood, Jr. · M. B. Panish · H. Sakaki
Founding Editor: H. K. V. Lotsch

* The 2nd edition is available as a textbook with the title: *Laser Processing and Chemistry*

Druck: Strauss Offsetdruck, Mörlenbach
Verarbeitung: Schäffer, Grünstadt